Macromolecules in the Functioning Cell

Macromolecules in the Functioning Cell

Edited by

Francesco Salvatore and Gennaro Marino

University of Naples
Naples, Italy

and

Pietro Volpe

Italian National Research Council
Naples, Italy

PLENUM PRESS · NEW YORK AND LONDON

Library of Congress Cataloging in Publication Data

Soviet-Italian Symposium on Macromolecules in the Functioning Cell, 1st, Capri, 1978.

Macromolecules in the functioning cell.

"Proceedings of the First Soviet-Italian Symposium on Macromolecules in the Functioning Cell held in Capri, Italy, May 24–27, 1978."

Includes index.

1. Molecular biology – Congresses. 2. Macromolecules – Congresses. I. Salvatore, Francesco, 1934- II. Marino, Gennaro. III. Volpe, Pietro. IV. Title. [DNLM: 1. Macromolecular systems – Congresses. W3 S0781J 1st 1978m/QU55 S729 1978m]

QH506.S65 1978 574.8'8 78-27547

ISBN 0-306-40146-0

Proceedings of the First Soviet–Italian Symposium on Macromolecules in the Functioning Cell held in Capri, Italy, May 24–27, 1978

A Division of Plenum Publishing Corporation
227 West 17th Street, New York, N.Y. 10011

Printed in the United States of America

PREFACE

This volume contains nineteen contributions on some of the most relevant topics in modern molecular biology and biochemistry presented by leading scientists of the USSR and Italy.

One group of papers are mainly concerned with the structure and functions of the genetic elements in eukaryote cells; among the topics are the following: Nucleosome structure, characterization of the nuclear matrix, ribosomal gene organization, gene expression during the cell cycle, and mapping of the mitochondrial transcripts. Several other aspects of macromolecule structure and function have been discussed: tRNA modification, translation factors, RNA interaction with RNA-polymerase, DNA-dependent ATPases, proteins involved in active transport, enzyme induction, iron and sulfur proteins, etc. Furthermore, some studies on macromolecule changes in embryonic development and cell differentiation have been presented, including DNA methylation and macromolecular synthesis in sea urchins, polyribosomes in loach, and histone modifications in spermiogenesis. Finally, a stimulating and brilliant presentation on protein—nucleic acid interaction by Professor Engelhardt closes the scientific contributions.

The papers collected in this volume have been presented at the First Soviet-Italian Symposium on "Macromolecules in the Functioning Cell" sponsored by the Italian Society of Biochemistry, the USSR Academy of Sciences, and the Capri Center for Cell Biology and Natural Sciences, and supported mainly by the Italian National Research Council (CNR). The efforts of Professor A. Ruffo, as Chairman of the Scientific Committee, have to be particularly acknowledged as being instrumental in the success of the symposium.

The inclusion of this scientific activity within the frame of the bilateral agreement between the USSR Academy of Sciences and CNR in Italy adds a significant value to the scientific success of the meeting. Therefore, the first Symposium will presumably be followed by others at a two-year interval, alternately in the Soviet Union and Italy.

The Editors wish to express their thanks to the authors of the papers and to Plenum Publishing Company for having made the prompt publication of this volume possible.

Francesco Salvatore
Gennaro Marino
Pietro Volpe

Naples, September 1978

CONTENTS

PART I: STRUCTURE AND FUNCTIONS OF THE GENETIC ELEMENTS

PART II: MACROMOLECULE STRUCTURE AND FUNCTION

PART III: MACROMOLECULE CHANGES IN EMBRYONIC DEVELOPMENT AND CELL DIFFERENTIATION

PART I

STRUCTURE AND FUNCTIONS OF THE GENETIC ELEMENTS

YEAST RIBOSOMAL GENES

A.A. Bayev, K.G. Skryabin, V.M. Zaharyev,
A.S. Krayev, and P.M. Rubtsov

Institute of Molecular Biology of the
USSR Academy of Sciences, Moscow, USSR
Moscow, 117312, Vavilov str., 32

The aim of this paper is to consider the structure of ribosomal operon of baker's yeasts Saccharomyces cerevisiae. Yeast as a primitive eukaryotic microorganism is a very convenient object for genetic and biochemical studies. One can grow it without difficulties and its genetics have been greatly promoted.

The general goal of our studies is the investigation of eukaryotic operons. We began to deal with ribosomal operons because of their convenient identification and cloning. They are a very popular subject and many laboratories are involved in their study, first of all J. Cramer's laboratory at the University at Wisconsin and W. Rutter's laboratory at the University of California.

The organization of yeast ribosomal DNA (rDNA) is rather well known at least in its general outlines. Ribosomal repeating unit (ribosomal operon) consists of four genes coding for 5S, 5.8S, 16S, 25S rRNA, internal TSi, and external TSe spacers. It is highly reiterated - about 140 copies for each haploid yeast genome (see Fig. 1).

The polarity of rRNA genes is 5'-5S, 18S, 5.8S, 25S-3'. This order was established by the first studies, mainly with the help of recombinant DNA techniques (1-8). Repeating units are tandemly organized in clusters, 30-40 units in each (9), and about 70% of them are located in chromosome 1 (10, 11).

The molecular weight of the yeast repeating unit is 5.61×10^6 daltons, ~8650 bp. This is almost two times greater than E. coli, but less than other primitive eukaryotes (Tetrahymena, Physarum,

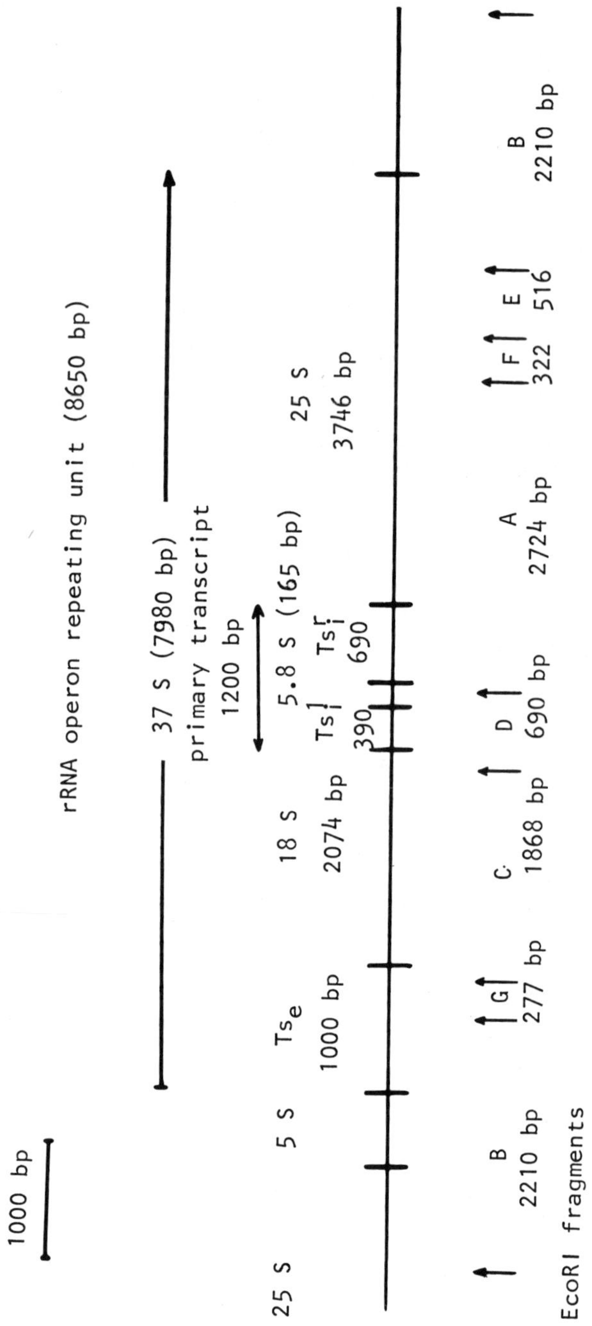

Fig. 1: Map of rRNA operon repeating unit.

Dictiostelium) with repeating units of 6.2, 19.5, and 25 x 10^6 daltons, respectively, or higher eukaryotes such as *Drosophila melanogaster* (7-11 x 10^6), *Xenopus laevis* (6.8-10.5 x 10^6), and *Bos taurus* (21 x 10^6) (8).

18S, 25S, and 5.8S rRNA genes are transcribed by RNA polymerase 1 (13) as a single 37S precursor with a molecular weight of 2.77 x 10^6 daltons (7980 bp) (A.A. Khadjiolov, personal communication). 5S rRNA transcription proceeds in the opposite direction (12) and gives a separate transcript; RNA polymerase III takes part in this process (14).

Processing of the 37S precursor proceeds in several steps, similar in its general traits to all eukaryotes.

We are interested in the regulatory parts of the yeast genome, mainly the promotor and terminator of 5S rRNA, 37S precursor, and the spacers.

Fragments of rRNA operon were isolated with recombinant DNA technology by T.D. Petes, L.M. Herdford and K.G. Skryabin (16-18). DNA of *Saccharomyces cerevisiae* diploid strain A364a+D4 was sheared and used for preparing recombinant DNA with the help of dAdT connectors and vector plasmid pMB9. 75 recombinant plasmids which contained the rDNA, as revealed by hybridization with ^{32}P RNA probes, were isolated and denoted by acronym pC, or pYIr. Repeating rDNA unit has 6 recognition sites for the restriction endonuclease EcoR1 and after digestion it gives seven fragments: A, B, C, D, E, F, G. Each recombinant plasmid of pYIr series contains 1 or more fragments of rDNA.

For our work we used a collection of recombinant plasmids pYIrA-3 (C, D, G), pYIrB-9 (D, A), pYIrC-4 (D), and pYIrC-1 (A, C, D, F, G). Plasmid DNA was isolated according to (19).

Several endonucleases (EcoR1, Hind III, Hae, Sma, Taq and others) were used for the isolation of rDNA fragments.

Products of rDNA digestion with restriction endonucleases were separated by sucrose gradient centrifugation and/or with electrophoresis in 8% polyacrylamide gel.

The ^{32}P-5'-terminal labeling, chain and endonuclease digestion product separation, and nucleotide sequencing were done as described by Maxam and Gilbert (20). Acid apurinisation was used for purine position determination. The reaction was performed in its traditional variant with diphenylamine (21-24) or with piperidine treatment (60 min, 90°C) of DNA fragments, obtained by 66% formic acid cleavage.

Endonuclease EcoRI was isolated according to (25), endonuclease hind III as in (26) and polynucleotideligase as described in (27).

EXTERNAL SPACER TSe AND THE 5S rRNA GENE

Maxam *et al*. (6) determined the sequence of 5S RNA gene (121 bp), 103 bp stretch of the preceding promotor region and 59 bp of the left region. This structure, with some minor exceptions, is in accordance with the described primary structures of 5S RNA of *S. cerevisiae* (28), S. *carlsbergensis* (29), and *Torulopsis utilis* (30). Valenzuela *et al*. (32) also carried out the analysis of this part of rDNA and adjacent regions.

We sequenced a part of the external spacer near the 18S gene (31), i.e. on the right of the fragment, sequenced by Valenzuela *et al*. (32) (see Fig. 2). This stretch consists of 472 bp and its beginning GAATTC overlaps with the end of Valenzuela's fragment.

According to Planta *et al*. (33) the 18S gene of *S. carlsbergensis* starts with the pUAUCUG sequence. Complementary sequence

5'TATCTG
ATACAC

was found in $H_{111}C_1$ fragment derived from R_1G at a distance of 115 bp on the right of the EcoR1 site. The orientation of this sequence coincides with the direction of rDNA transcription in this region. One can suppose that the beginning of the structural 18S RNA gene is established with some probability. It means that fragment, consisting of nucleotides from -1 to -400, is a part of the external spacer.

The sequence of the external spacer has no specific feature, and it differs greatly from the internal spacer.

18S rRNA GENE

The beginning of 18S gene was sequenced during the course of the external spacer analysis. The sequencing of the end of this gene was carried out independently. The sequence of the 335 bp fragment was determined (see Fig. 3). Only the stretch 35 bp separates the fragment from the end of 18S gene.

Fig. 4 presents the possible secondary structure of the 18S RNA fragment deduced from the DNA sequencing.

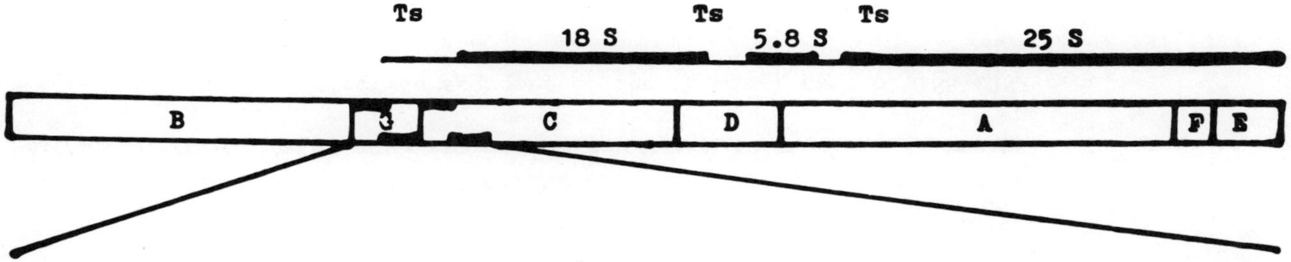

GAATTCGAGTAGGCTTGTCGTTCGTTATGTTTTTGTAAATGGCCTCGTCAAACGGTGGAGAGAGTCGCTAGGTGATCGTCAGATCTGCCTAGTCTCTATA
CTTAAGCTCATCCGAACAGCAAGCAATACAAAAACATTTACCGGAGCAGTTTGCCACCTCTCTCAGCGATCCACTAGCAGTCTAGACGGATCAGAGATAT

CAGCGTGTTTAATTGACATGGGTTGATGCGTATTGAGAGATACAATTTGGGAAGAAATTCCCAGAGTGTGTTTCTTTTGCGTTTAACCTGAACAGTCTCA
GTCGCACAAATTAACTGTACCCAACTACGCATAACTCTCTATGTTAAACCCTTCTTTAAGGGTCTCACACAAAGAAAACGCAAATTGGACTTGTCAGAGT

TCGTGGGCATCTTGCGATTCCATTGGTGAGCAGCGAAGGATTTGGTGGATTACTAGCTAATAGCAATCTATTTCAAAGAATTCAAACTTGGGGGAATGCC
AGCACCCGTAGAACGCTAAGGTAACCACTCGACGCTTCCTAAACCACCTAATGATCGATTATCGTTAGATAAAGTTTCTTAAGTTTGAACCCCCTTACGG

TTGTTGAATAGCCGGTCGCAAGACGTGATTCTTCAAGTGTAACCTCCTCTCAAATCAGCGATATCAAACGTACCATTCCGTGAACACCGGGGTATCTGTT
AACAACTTATCGGCCAGCGTTCTGCACTAAGAAGTTCACATTGGAGGAGAGTTTAGTCGCTATAGTTTGCATGGTAAGGCACTTGTGGCCCCATAGACAA

TGGTGGAACCTGATTAGAGGAAACTCAAAGAGTGCTATGGTATGGTGACGGAGTGCGCTGGTCAAGAGCGTAAGCTT
ACCACCTTGGACTAATCTCCTTTGAGTTTCTCACGATACCATACCACTGCCTCACGCGACCAGTTCTCGCATTCGAA

Fig. 2: Nucleotide sequence of part of the external spacer near the 18S gene.

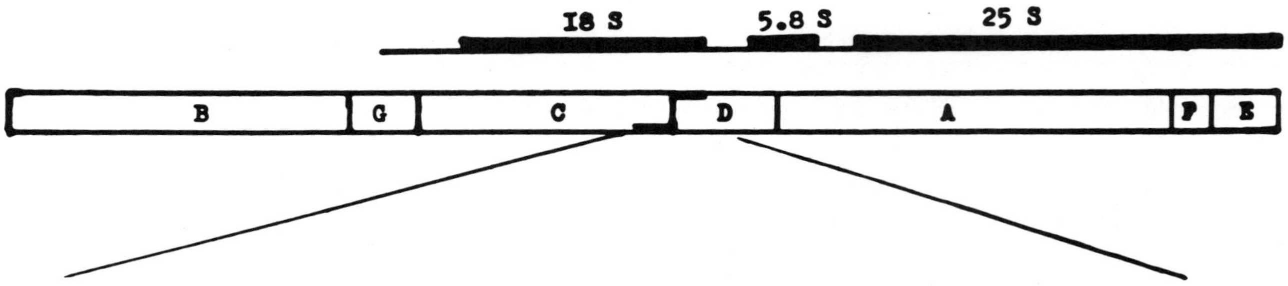

ACACAAACAATTTTATCTATTCATTAAATTTTTGTCAAAAACAAGAATTTTCGTAACTGGAAATTTTAAAATATTAAAAACTTTCAACAACGGATCTCTT
TGTGTTTGTTAAAATAGATAAGTAATTTAAAAACAGTTTTTGTTCTTAAAAGCATTGACCTTTAAAATTTTATAATTTTTGAAAGTTGTTGCCTAGAGAA

GGTTCTCGCATCGATGAAGAACGCAGCGAAATGCGATACGTAATGTGAATTGCAGAATTCCGTGAATCATCGAATCTTTGAACGCACATTGCGCCCCTTG
CCAAGAGCGTAGCTACTTCTTGCGTCGCTTTACGCTATGCATTACACTTAACGTCTTAAGGCACTTAGTAGCTTAGAAACTTGCGTGTAACGCGGGGAAC

GTATTCCAGGGGGCATGCCTGTTTGAGCGTCATTTCCTTCTCAAACATTCCGTTTGGTAGTGAGTGTACTCTTTGGAGTTAA
CATAAGGTCCCCCGTACGGACAAACTCGCAGTAAAGGAAGAGTTTGTAAGGCAAACCATCACTCACATGAGAAACCTCAATT

Fig. 3: Nucleotide sequence of the terminal part of 18S rRNA gene.

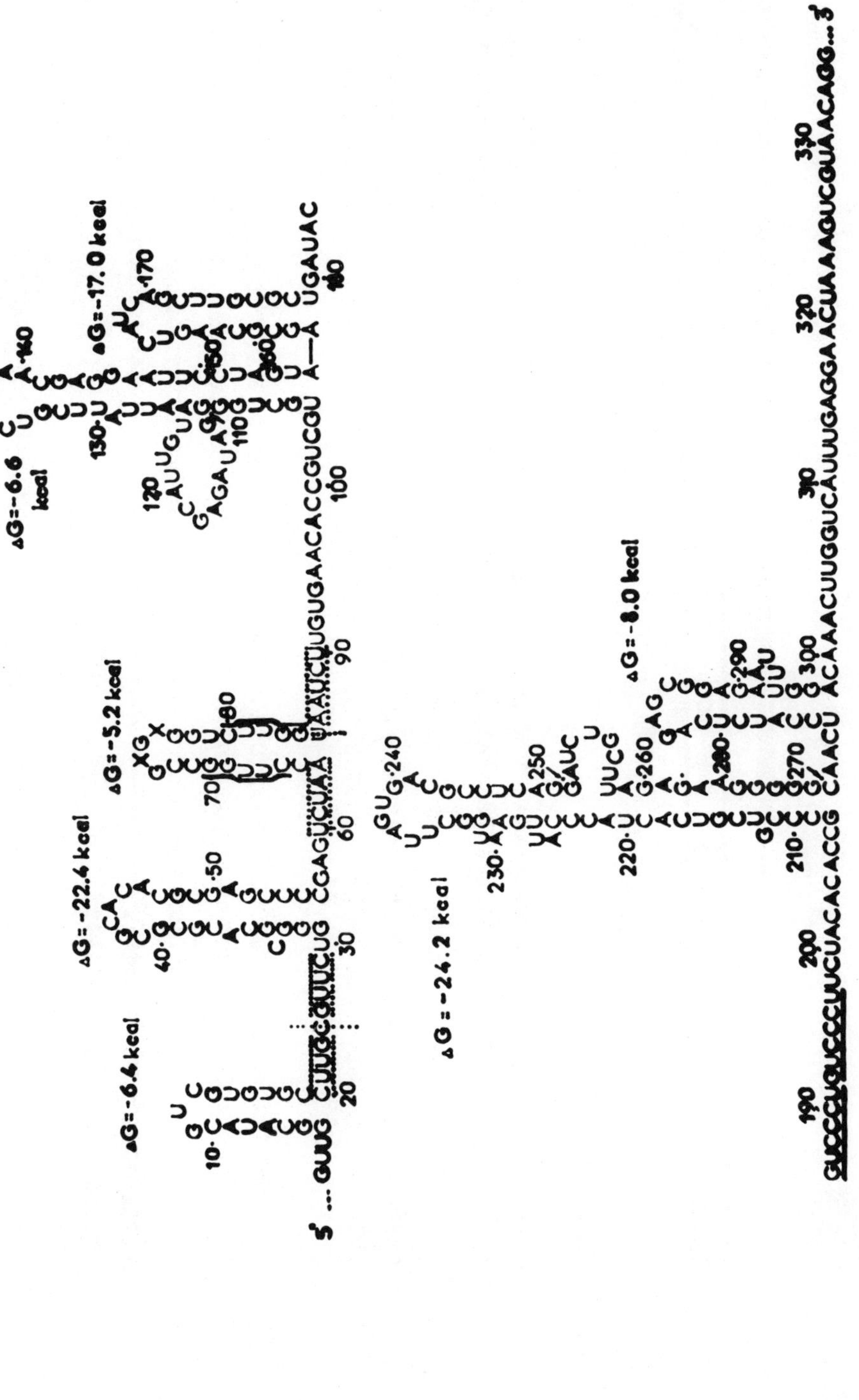

Fig. 4: Possible secondary structure of 18S RNA fragment, deduced from DNA sequencing.

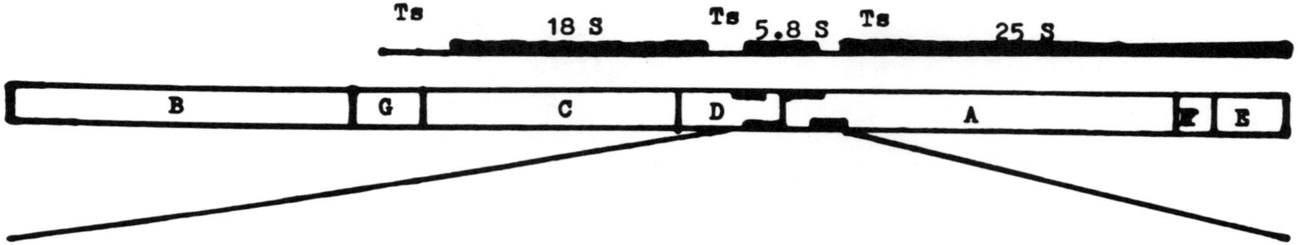

...ACACAAACAATTTTATCTATTCATTAAATTTTTGTCAAAAACAAGAATTTTCGTAACTGGAAATTTTAAAATATTAAAAACTTT
...TGTGTTTGTTAAAATAGATAAGTAATTTAAAAACAGTTTTTGTTCTTAAAAGCATTGACCTTTAAAATTTTATAATTTTTGAAA

CAACAACGGATCTCTTGGTTCTCGCATCGATGAAGAACGCAGCGAAATGCGATACGTAATGTGAATTGCAGAATTCCGTGAATCATC
GTTGTTGCCTAGAGAACCAAGAGCGTAGCTACTTCTTGCGTCGCTTTACGCTATGCATTACACTTAACGTCTTAAGGCACTTAGTAG

GAATCTTTGAACGCACATTGCGCCCCTTGGTATTCCAGGGGGCATGCCTGTTTGAGCGTCATTTCCTTCTCAAACATTCCGTTTGGT
CTTAGAAACTTGCGTGTAACGCGGGGAACCATAAGGTCCCCCGTACGGACAAACTCGCAGTAAAGGAAGAGTTTGTAAGGCAAACCA

AGTGAGTGTACTCTTTGGAGTTAA...
TCACTCACATGAGAAACCTCAATT...

Fig. 5: Nucleotide sequence of 5.8S rRNA gene.

5.8S GENE AND THE INTERNAL SPACER TS_i

5.8 rRNA gene is localized in the ribosomal operon between the 18S and 25S genes (6). The sequence is in accordance with the primary structure of yeast 5.8S rRNA described by Rubin (34). Aside from these, 165 nucleotide parts of the TS_i^R (47 bp) and TS_i^L (70 bp) were sequenced (see Fig. 5).

The sequence of TS_i^L has some special features, first of all the high content of AT pairs, the AT clusters. This spacer has palindromes in common with those of 5.8S RNA gene.

Rubin found that *S. cerevisiae* ribosomes have three forms of rRNA with the following sequences at 5'-ends:

AUAUUAAAAAC..........IIB (5%)

UAUUAAAAAAC..........IIA (5%)

AAAC..........I (90%)

There are two possible explanations of this fact. According to the first, three kinds of 5.8S genes exist in yeast genome; the second explanation assumes three forms of processing of the rRNA precursor. In the first case the most prominent form with a shortened initial sequence will be detected in the analysis; in the second case, the longest one will be more probable.

We studied 3 clones of recombinant plasmid containing the fragment R_ID, where 5.8S RNA gene is localized, and in all three cases found the same electrophoretically determined length and the same sequence in the initial part of 5.8S gene, namely

5' ATATTAAAACC ....

3' TATAATTTTG ....

corresponding to the 5'-end of the 5.8S rRNA

5' AUAUUAAAAC ....

One can conclude that 5.8S RNA gene codes in yeasts form only one kind of rRNA and the heterogeneity of 5.8S RNA depends on the processing.

CONCLUDING REMARKS

The rRNA operon of S. cerevisiae has a molecular weight of 5.61 x 10^6 daltons or $\sim$8500 bp. The sum of all sequenced parts is 2100 bp or 25%. It excludes the possibility of broad conclusions, and we shall confine ourselves to some short remarks.

The organization of the ribosomal operon and its transcription are rather clear - it is a positive fact. But at the same time, established sequences are functionally dumb in spite of their characteristic features. For instance, a cluster of A.T. pairs is located left of the 5S gene and some repeating sequences; small A.T. clusters are located to the right. There is not any similarity to bacterial promotors with their "Pribnow boxes".

An impression arises that nature includes some structural ideas in genome construction but that we are not able to understand them.

There are two more considerations. The first one concerns the length of external spacer TSe. Its length is estimated to be about 1000 bp, after rather complex calculations. It is a transcribed sequence but its processing is unknown. The mapping of the left part of rRNA operon is not clear.

The question of promotors and terminators of 5S gene and 37S precursor is intersting as well. What is their mutual location? Are they brought together (maybe overlapped) or are they separated by a spacer? We are not able to answer these questions.

SUMMARY

The study of ribosomal genes was carried out using the recombinant plasmids which contained the fragments of the operon. By sequencing the 500 nucleotide fragments, the position of 18S RNA structural gene and its outer transcribed spacer was determined. The structure was compared with the inner transcribed spacers. The possible processing mechanisms of 35S rRNA precursor were discussed. The sequence of fragments of 18S RNA structural gene was also presented.

REFERENCES

1. Retel, J., and Planta, R., Biochim. Biophys. Acta, 169, 416 (1968).
2. Schweizer, E., MacKechnie, C., and Halvorson, H.O., J. Mol. Biol., 40, 261 (1969).
3. Kaback, D.B., Bhargava, M.M., and Halvorson, H.O., J. Mol. Biol., 79, 735 (1973).
4. Udem, S.A., and Warner, J.R., J. Mol. Biol., 65, 227 (1972).
5. Rubin, G.M., and Sulston, J.E., J. Mol. Biol., 79, 521 (1973).
6. Maxam, A.M., Tizard, R., Skryabin, K.G., and Gilbert, W., Nature, 267, 643 (1977).
7. Cramer, J.H., Farrelly, F.W., Barnitz, J.T., and Rownd, R.H., Mol. Gen. Genet., 151, 229 (1977).
8. Bell, G.I., De Gennaro, L.J., Gelfand, D.H., Bishop, R.J., Valenzuela, P., and Rutter, W.J., J. Biol. Chem., 252, 8118 (1977).
9. Cramer, J.H., Bhargava, M.M., and Halvorson, H.O., J. Mol. Biol., 71, 11 (1972).
10. Goldberg, S., Dyen, T., Idriss, I.M., and Halvorson, H.O., Mol. Gen. Genet., 116, 139 (1972).
11. Finkelstein, D.B., Blamire, J., and Marmur, I., Nature New Biology, 240, 279 (1972).
12. Aarstad, K., and Øyen, T.B., FEBS Letters, 51, 227 (1975).
13. McLaughlin, C.S., in: Ribosome, M. Nomura, A. Tissieres, P. Lengyel (Eds.), N.Y. Cold Spring Harbor Press, 815.
14. Weinmann, R., and Roeder, R.G., Proc. Nat. Acad. Sci., U.S.A. 71, 1790 (1974).
15. Hadjiolov, A.A., and Nikolaev, N., Progr. Biophys. Molec. Biol., 31, 95 (1976).
16. Petes, T.D., Hereford, L.M., and Skryabin, K.G., J. Bacteriol., 134, 295 (1978).
17. Skryabin, K.G., Maxam, A.M., Petes, T.D., and Hereford, L., J. Bacteriol., 134, 306 (1978).
18. Oetes, T.D. et al., in preparation.
19. Tanaka, T., and Weisblum, B., J. Bacteriol., 121, 354 (1975).
20. Maxam, A.M., and Gilbert, W., Proc. Nat. Acad. Sci., U.S.A. 74, 560 (1977).
21. Burton, K., in: Methods in Enzymology, v. XII, part A, p. 222, L. Grossman, K. Moldave (Eds.), Acad. Press, New York and London (1967).
22. Sverdlov, E.D., Monastyrskaya, G.S., Chestukhin, A.V., and Budowsky, E.I., FEBS Letters, 33, 15 (1973).
23. Sverdlov, E.D., and Levitan, T.L., Bioorgan. Chem., (USSR), 3, 206 (1977).
24. Korobko, V.G., and Grachev, S.A., Bioorgan. Chem., (USSR), 3, 1420 (1977).
25. Freene, P.J., Betlach, M.C., and Boyer, H.W., in: "Methods in Molecular Biology", v. 7, DNA replication, p. 87 (1974).

26. Smith, H.O., in: "Methods in Molecular Biology", v. 7, DNA replication, p. 72 (1974).
27. Van de Sande, J.H., Kleppe, K., and Khorana, H.G., Biochemistry, 12, 5050 (1973).
28. Miyazaki, M., J. Biochem., (Tokyo), 75, 1407 (1974).
29. Hindley, J., and Page, S.M., FEBS Letters, 26, 157 (1972).
30. Nishikawa, K., and Takemura, S., FEBS Letters, 40, 106 (1974).
31. Skryabin, K.G., Zaharyev, V.M., and Bayev, A.A., Dokl. Akad. Nauk. (USSR), 241, 488 (1978).
32. Valenzuela, P., Bell, G.I., Venegas, A., Sewell, E.T., Masiarz, F.R., De Gennaro, L.J., Weinberg, F., and Rutter, W.J., J. Biol. Chem., 252, 8126 (1977).
33. De Jonge, P., Klootwijk, J., and Planta, R.J., Eur. J. Biochem., 72, 361 (1977).
34. Rubin, G.M., J. Biol. Chem., 248, 3860 (1973).

CHARACTERIZATION OF THE NUCLEAR MATRIX OF RAT LIVER AND HEPATOMA 27

I.B. Zbarsky, S.N. Kuzmina and T.V. Buldyaeva

N.K. Koltzov Institute of Developmental Biology, USSR Academy of Sciences
117808 Moscow 117334, Vavilov Street, 26. USSR

ABSTRACT

High salt extraction of isolated cell nuclei combined with nucleases treatment results in a preparation of acidic nuclear proteins which represent a nuclear matrix consisting of the nuclear envelope, nucleolus, and intranuclear fibrils and granules. After detergent treatment a nuclear skeleton remains which consists mostly of protein. Polyacrylamide gel electrophoresis with SDS reveals three protein bands of molecular weight 65 to 70000 D to be predominant components of the matrix.

The electrophoretic pattern of hepatoma 27 matrix differs largely from that of liver. Three distinct bands with molecular weight higher than 100000 D are here very prominent as well as relatively low molecular weight bands in region of 20000 D and lower.

Dilute alkali (0.01 - 0.05 N) extracts the typical matrix protein triplet of molecular weight 65 - 70 000 D. Resulting nuclear residual protein preparations from liver contain no tryptophan and consist mostly of proteins of molecular weight lower than 60000 D. However, in tumor residual protein preparations the high molecular weight bands peculiar of the hepatoma are still more prominent than in matrix preparations.

The features characteristic of the tumor are not revealed in corresponding preparations from regenerating liver.

INTRODUCTION

Our earlier work showed that isolated cell nuclei, after high salt extraction of chromatin deoxyribonucleoprotein, preserve their shape and retain acidic proteins of non-histone character. Exhaustive extraction of this material with dilute alkali (0.01 - 0.05 N NaOH) leaves about 5% of the nuclear substance as the "residual protein" fraction containing no tryptophan (1,2). Later on the fractions subsequently extractable with 0.14 N NaCl, 1-2 M NaCl, and 0.02 N NaOH were identified as the material of the nuclear sap ("globulin fraction"), the chromatin ("deoxyribonucleoprotein fraction"), the nucleoli and residual chromosomes ("acidic protein fraction"), and of the nuclear envelope ("residual protein fraction") respectively (3, 4, 5).

Electron microscopy of those fractions demonstrated that a material containing remnants of the nuclear envelope, nucleoli, and intranuclear ribonucleoprotein network, preserving the shape of the nucleus, remained after salt extraction of isolated nuclei, i.e. after elimination of the chromatin and the nuclear sap. This structure, unaffected by DNase treatment, was regarded as a nuclear skeleton to which the chromatin is attached (6, 7). Similar results were later on published by other workers (8, 9, 10, 11).

Recently, various preparations corresponding to the above, have been obtained by different workers. In addition to the solvents used earlier, a treatment with a non-ionic detergent, Triton X-100 has been employed. This treatment, as well as extractions with buffered physiological saline, high salt (1 - 2 M) solutions, and treatment with DNase and RNase, have been used with various tissues in different sucsession. Nevertheless, the preparations obtained were very similar in their structure and composition and essentially corresponded to the above (6 - 11).

Such preparations have been isolated from rat liver (12 - 14) and described as "nuclear matrix" or "nuclear protein matrix", from Krebs ascites tumor cells (15), - termed as "nuclear ribonucleoprotein network", from Chinese hamster ovary (CHO) cell culture (16), from HeLa cell culture (17, 18) - as "nuclear ghosts", from HeLa cells infected with adenovirus 2 (19), *Tetrahymena pyriformis* macronuclei (20, 21), mouse liver (22), and HeLa and kangaroo rat cell cultures (23). Among all these denominations the "nuclear matrix" has become most generally used.

The nuclear protein matrix consists essentially of protein and contains some 2% RNA, 1% DNA, and 1 - 5% phospholipid. Omiting RNase treatment the RNA content can exceed 10%, while the amount of phospholipid largely depends on the condition of detergent treatment and on the concentration of the latter.

Polyacrylamide gel electrophoresis in the presence of sodium dodecyl sulfate (SDS) reveals 20 - 30 protein components, however three protein bands with a molecular weight between 65000 and 70000 D are very prominent and are considered as the main substance of filamentous matrix network (12 - 14, 19 - 22). The matrix is ascribed to be important in the initiation of DNA replication (13, 24), transcription (15), and other cellular life processes (25).

The nuclei of malignant tumor cells were shown to be rich in acidic and especially residual protein fractions; the latter, contrary to that of normal tissues, was found to contain tryptophan (1, 2, 5, 7, 26, 27). Then, the residual protein fraction was demonstrated to increase along with the growth of a transplanted tumor (28). Treatment with organic solvents resolved the tumor nuclear residual protein to an extractable amorphous lamellar fraction containing the tryptophan, and absent from cell nuclei of normal tissues, and a non-extractable fraction devoid of tryptophan and similar to the residual nuclear protein of normal cells (29, 30, 7).

In conformity with the above data on peculiarities of the tumor cell nuclei, we studied the nuclear matrix and residual protein of rat hepatoma 27 and quiescent and regenerating rat liver.

MATERIALS AND METHODS

Male Wistar rats weighing 100 - 150 g were used. Subcutaneously growing rat hepatoma 27 was transplanted by implantation of a chopped tumor tissue suspension. The tumor was studied in 4, 6, and 8 weeks after transplantation. The regenerating liver was taken from rats sacrificed 24 hours after partial hepatectomy according to Higgins and Anderson (31).

For isolation of cell nuclei a modified Blobel and

Potter method (32) was used. 10% tissue homogenate in 0.25 M sucrose containing TKM buffer (0.025 M KCl, 0.05 M $MgCl_2$, 0.01 M Tris-HCl, pH 7.5) was overlayered on 0.32 M sucrose containing the same buffer with a cushion of 2.2 M sucrose, and centrifuged at 40000 g for 40 minutes. The nuclear pellet was washed with 0.25 M sucrose containing the TKM buffer. To prepare the nuclear matrix, the pellet was suspended in 0.2 M sucrose containing the buffer A (2 mM $MgCl_2$, 3mM $CaCl_2$, 20 mM Tris-HCl, pH 7.4), supplemented with Triton X-100 to a final concentration of 0.05%, incubated at 0°C for 10 minutes, and centrifuged at 1000 g for 10 minutes. The pellet was washed in buffer A, and resuspended in 5 ml of the same buffer containing 50 mM NaCl. 650 µg electrophoretically pure DNase I (Worthington) and 650 µm RNase I (Reanal) were added to the suspension and the latter was incubated at room temperature for 1 hour. Then 45 ml of 2.2 M sucrose in buffer A was added, and deoxyribonucleoprotein (DNP) material was extracted at room temperature for 10 minutes with shaking. The pellet was suspended in 5 ml of buffer A, containing 100 µg each of the same pancreatic DNase and RNase preparations, and washed with buffer A (20).

For polyacrylamide gel electrophoresis by molecular weight, samples were incubated in 0.05 M Tris-HCl buffer, pH 6.8 containing 2% SDS, 12% glycerol, and 5% mercaptoethanol at 100°C till solubilization. Then a sample containing 100 - 200 µg protein was applyed on concentrating gel, consisting of 2.5% acrylamide, 0.6% methylbisacrylamide, 20% sucrose and 2% SDS. Resolving gel contained 12% acrylamide, 0.3% methylbisacrylamide, dissolved in 0.43 M Tris-HCl buffer, pH 8.9.

The electrophoresis was run at 4°C for 3 - 3.5 hours in a Tris-HCl buffer containing 0.6 M glycine and 0.1% SDS at pH 8.3 and a current of 1 ma during the first 20 minutes, and 3.5 ma during remaining time, per a gel. The gels were fixed in a mixture of ethanol: acetic acid: water (4.5: 4.5: 1) for 18 hours and stained with 0.125% Coumassie blue (Ferak, Berlin) dissolved in the same mixture. The superfluous dye was washed off with 7% acetic acid (33).

For electron microscopy the preparations were fixed in 2.5% glutaraldehyde, postfixed in 2% OsO_4, dehydrated in graded ethanol series, and embedded in Epon 812-Araldyte. Ultrathin sections cut with an LKB ultratom were stained with uranyl acetate according to Reynolds and examined by use of an JEM 7A electron microscope.

RESULTS AND DISCUSSION

Morphologically, the matrices conserve the shape of the nuclei, but their diameter is about 30 - 50% smaller. On electron micrographs (figures 1 and 2) the outlines of the nuclear envelope, pore complexes and both nuclear membranes are seen (higher detergent concentrations remove the remnants of the membranes); the remnants of the nucleoli (their fibrillar parts and fibrillar centers) are also evident, as well as the intranuclear fibrillar network and ribonucleoprotein granules, the quantity of which may be reduced in parallel with the RNase treatment. The electron microscopic appearence of regenerating liver and hepatoma 27 matrices does not differ essentially from that of normal quiescent liver (figs. 1 and 2).

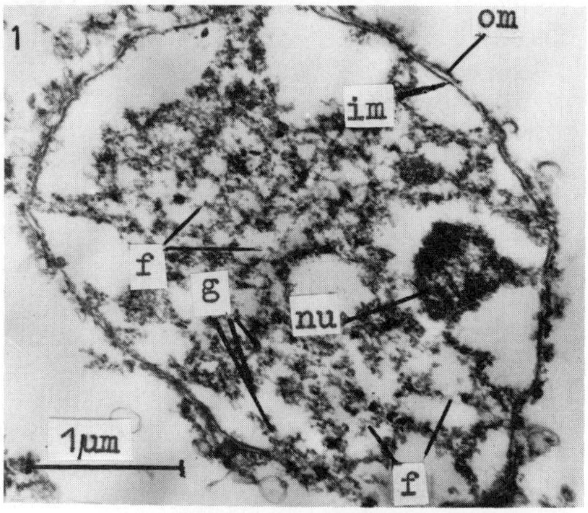

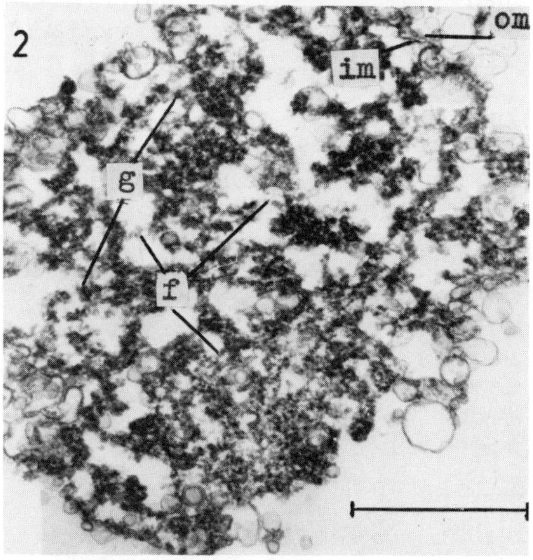

Figures 1 and 2. Electron microscopy of rat liver (fig. 1, 20000x) and hepatoma 27 (fig. 2, 25000x) nuclear matrix.

om-outer nuclear membrane
im-inner nuclear membrane
nu-nucleolus
f -fibrils
g -granules

The matrix preparations consist on 96-98% of protein and contain about 2% RNA and less than 1% of DNA and of phospholipid. If RNase treatment be omited the RNA content may amount to or even exceed 10%, while phospholipid content essentially depends on the manner of detergent treatment and on its concentration.

The gross composition (i.e. the ratio of protein, nucleic acid and phospholipid) in matrix preparations from resting and regenerating liver and hepatoma 27 at different stages of its growth has been essentially similar.

The nuclear matrices of quiescent and regenerating liver and of hepatoma 27, at early stages of its growth, contain about 10 - 12% of the total nuclear protein. However, along with the growth of the tumor the percentage of matrix protein, as well as that of the resudual protein fraction, increases and reaches 27% in 8 weeks after the tumor implantation.

Electrophoretic studies of matrix proteins demonstrate many bands. On an electrophoregramm of nuclear protein matrix from normal liver at least 20 - 25 bands can be detected (fig. 3). The pattern of these protein bands is essentially different from those of deoxyribonucleoprotein or globulin fractions of the same nuclei (fig. 3). While in DNP fraction histones are predominant and among non-histone proteins a band of molecular weight about 35 - 40000 D is well defined; in the globulin fraction the bulk of proteins are revealed in clusters of bands between 20000 and 60000 D, among matrix proteins in accordance with other data (12, 14, 16, 19, 20, 21, 22, 25) three bands with molecular weight between 65000 and 70000 D are pronounced (fig. 3). Similar bands were also reported in isolated nuclear pore complexes-lamina preparations (34, 35), and were regarded by the authors as the principal component of the fibrillar matter of the dense lamina of the nuclear envelope and of the nuclear matrix (22, 35). On this assumption a term "matrixin" for these

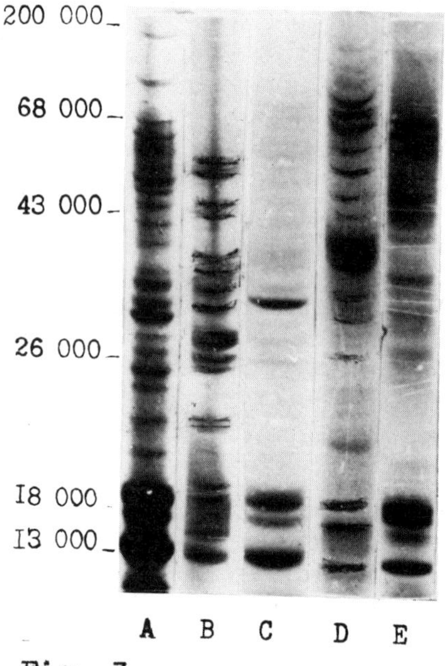

Fig. 3.
PAAG electrophoresis of rat liver nuclear protein fractions

A-Nuclei, B-Globulin fraction, C-Deoxyribonucleoprotein fraction, D-Matrix, E-Residual protein fraction

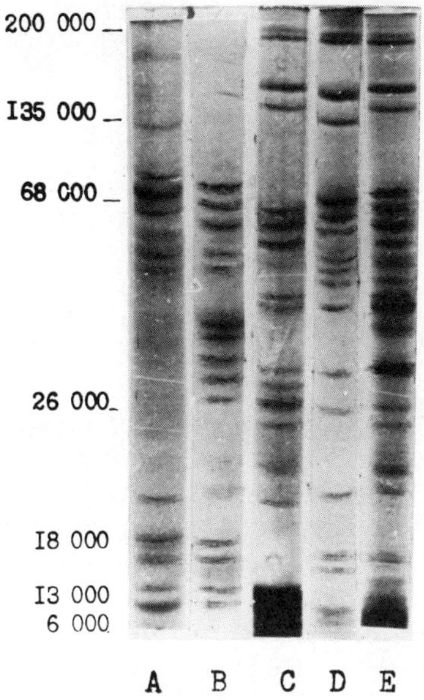

Fig. 4. PAAG electrophoresis of nuclear protein matrix
A-Rat liver, B-Regenerating rat liver (24 h after hepatectomy), C-Hepatoma 27 (4 weeks after implantation), D-Hepatoma 27 (6 weeks after implantation), E-Hepatoma 27 (8 weeks after implantation)

Molecular weight markers: Lactate dehydrogenase-135000, beef serum albumin-68000, Egg albumin-43000, Chymotrypsin-26000, cytochrome c-13000, insulin-6000.

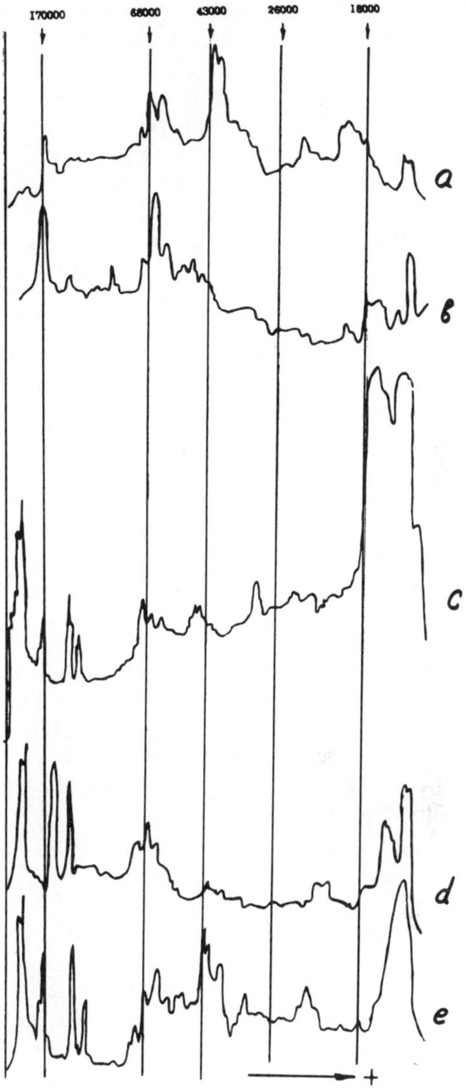

Fig. 5. Electrophoretic profiles of the nuclear protein matrix
(a)-rat liver, (b)-regenerating rat liver, (c)-hepatoma 27 (4 weeks after implantation), (d)-hepatoma 27 (6 weeks after implantation), (e)-hepatoma 27 (8 weeks after implantation)

proteins was proposed (22). Apart of this protein cluster, a group of proteins with molecular weight between 40000 and 60000 D and several less pronounced bands in low molecular weight (lower than 20000 D) and high molecular weight (100000 - 200000 D) regions are seen.

The pattern of regenerating liver nuclear matrix does not differ largely from that of normal resting liver. However, in hepatoma 27 the differences are clearly pronounced (fig. 4). These differences are still better displayed at densitometric electrophoretic profiles (fig. 5) quite consistent with the visible appearence of the gels.

At early stage of growth (4 weeks after transplantation) a protein cluster with relatively low molecular weight (around 10000 D) is predominant. This cluster hardly belongs to histones for the molecular weight of its bands are lower and those bands are not pronounced in the DNP fraction (fig. 3). These low molecular weight bands are also rather slightly detected in matrices from quiescent or regenerating liver (figs. 3 and 5).

A question naturally arises, whether these low molecular weight polypeptides might be the degradation products due to proteolysis during the preparation of the matrix. However, this explanation is hardly probable for the nuclear matrices isolated at 37°C, i.e. in a condition at which the proteolytic degradation should be more pronounced, without RNase or detergent treatment when the degradation should be hampered (fig. 6), as well as with the use of an inhibitor of nuclear proteases, phenylmethylsulfonyl fluoride (PMSF, 36), did not differ in their electrophoretic patterns.

Three high molecular weight bands (about 135000, 150000, and 200000 D, the latter band being double) are pronounced in hepatoma 27 matrices at each stage of its growth studied (4, 6, and 8 weeks after implantation, figs. 4 and 5). Only one of those bands is definitely detected in quiescent or regeneratiog liver nuclear matrices (fig. 5). These bands are not reported by other workers studied the nuclear matrix. However, in nuclear matrix of HeLa cells (19), a group of proteins with molecular weight between 90000 and 135000 D is described and shown to be probably glycoproteins. It is not clear whether these proteins correspond to the proteins revealed by us in tumor matrices. In conformity with predominance of protein bands characteristic of the tumor matrices, the cluster of three bands with molecular weight between 65000 and 70000 D prevailing in nuclear matrix from normal liver, is cosiderably less defined in hepatoma matrix preparations.

In electrophoregramms of the residual protein fraction (fig. 7) the protein triplet in region of 65000 - 70000 D is hardly detected. The corresponding bands can be revealed in alkaline extracts of isolated nuclei or matrices indicating in conformity with the observations of other workers (22, 37) that these proteins are extractable with alkaline solutions. In residual protein fractions of normal liver, proteins with molecular weight between 25000 and 70000 D, especially between 45 000 and 65000 D, and a protein cluster with a mobility close to that of histones are most prominent. It is not clear whether this cluster corresponds to histones; however, for histones as proteins with pronounced cationic properties are poorly soluble in alkali it is possible

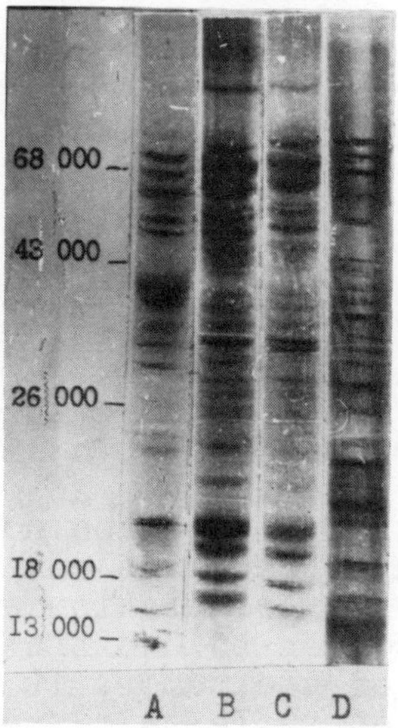

Fig. 6. PAAG electrophoresis of nuclear protein matrix isolated in different conditions. (A)-according to Wunderlich (21), (B)- the same but without RNase, (C)-at 37°C, (D)-with 1% Triton X-100

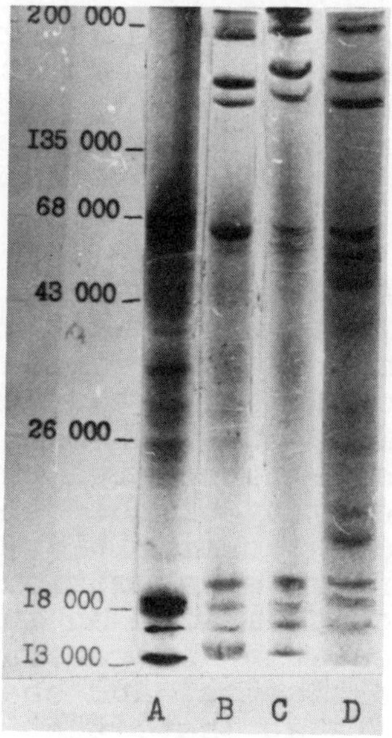

Fig. 7. PAAG electrophoresis of residual protein fraction (A)-rat liver, (B)-hepatoma 27 (4 weeks after implantation), (C)-hepatoma 27 (6 weeks after implantation), (D)-hepatoma 27 (8 weeks after implantation)

that they could remain, rather in some altered state, in the residual protein preparations.

Electrophoretic patterns of the residual protein fractions from hepatoma are still more different from those of normal liver than the patterns of nuclear matrix preparations. The high molecular weight group of proteins pronounced in hepatoma residual proteins is practically absent from those of normal liver. However, a cluster of polypeptides with molecular weight around 10000 D, is here much less defined than in matrices; probably these proteins are rather soluble in dilute alkali. Protein bands with molecular weight at 35000-65000 D, especially with molecular weight of 40000 D and 60000 D are also detected in the hepatoma residual protein fraction electrophoretic pattern.

Thus we have found essential peculiarities in hepatoma 27 nuclear protein matrix and residual protein fraction protein patterns which are markedly different from those of quiescent or regenerating rat liver. These peculiarities are reproduced in many experiments and may be regarded as characteristic of the tumor. The patterns are not identical at different stages of the tumor growth.

A question naturally arises as to the biological significance of these prominent differences of the nuclear matrix and residual protein of the tumor from normal condition, not observed in whole tissue or other cell fractions? It may be supposed that these features were associated with rapid proliferation and high mitotic activity of the tumor tissue. However, the absence of these peculiarities in regenerating liver does not support this assumption. Another interpretation could connect them with an inhibition of differentiation and manifestation of general features common to undifferentiated tissues, and characteristic of the tumor. It is possible at last, that the alteration reflects a specific rearrangement of the genetic or epigenetic apparatus characteristic of the tumor growth.

This possibility is supported by our preliminary results showing similar peculiarities in the nuclear matrix of another transplanted tumor of non-hepatic origin and by our earlier data showing the resemblance of homologous preparations among different tumors of man and animals (1, 2, 5, 7, 26, 27, 28). At last, in a sole communication on an electrophoretic study of Zajdela hepatoma nuclear protein matrix (38), a finding is also reported of high molecular weight polypeptide, absent from matrix preparations of normal liver. It is worth to mention also that the appearence of characteristic features, associated with uncontrolled tumor growth, in the nuclear matrix consists well with an observation that after infection of HeLa cells with adenovirus 2, the virus-coded proteins appear at first namely in the nuclear matrix (19).

It is possible that some peculiarities of protein composition of the nuclear matrix might be due to the condition of tumor growth, its stage, or some secondary degenerative or other alteration processes in the tumor tissue. From this point of view it is important to differentiate between essential and non-essential features for the tumor growth among the peculiarities observed at different its stages.

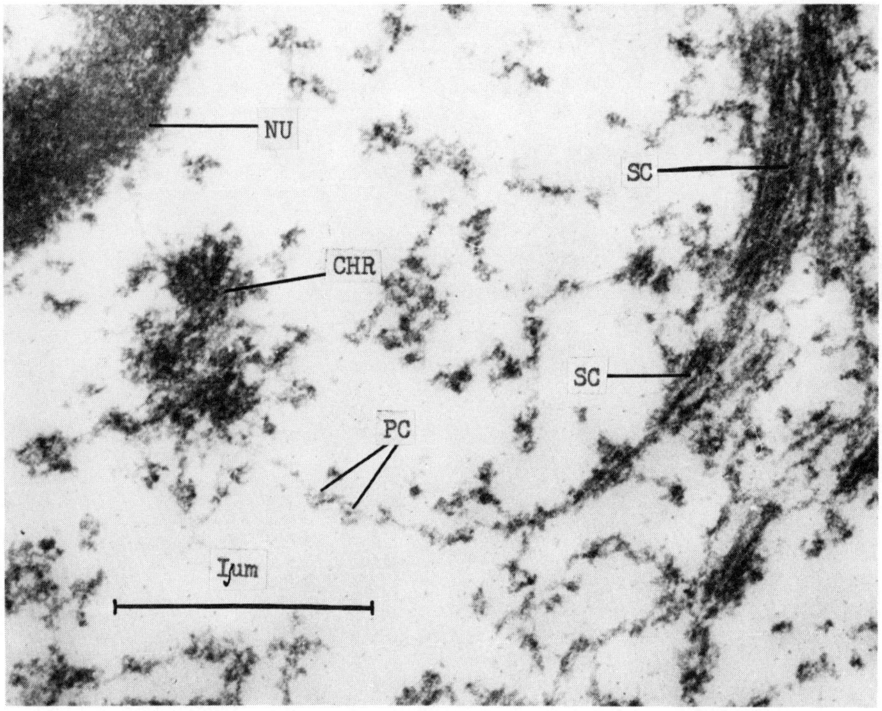

Fig. 8. Electron microscopy of Rana temporaria late oocyte karyosphere. 39000x.
NU-Nucleolus, CHR-Chromatin, SC-Synaptinemal complexes, PC-Pore complexes

An indication that some features of the tumor nuclear matrix, namely the presence of high molecular weight polypeptide bands, could be characteristic of non-differentiated cell, may be the finding of resembling bands in late oocyte karyosphere of a frog (Rana temporaria).

The karyosphere consists essentially of fibrillar nucleoli, synaptinemal complexes, and fibrillar-granular structures, the "pseudomembranes", containing the nuclear pore complexes (39, 40, 41).

In collaboration with Dr. L.S. Filatova we have isolated those karyospheres from late oocyte nuclei of

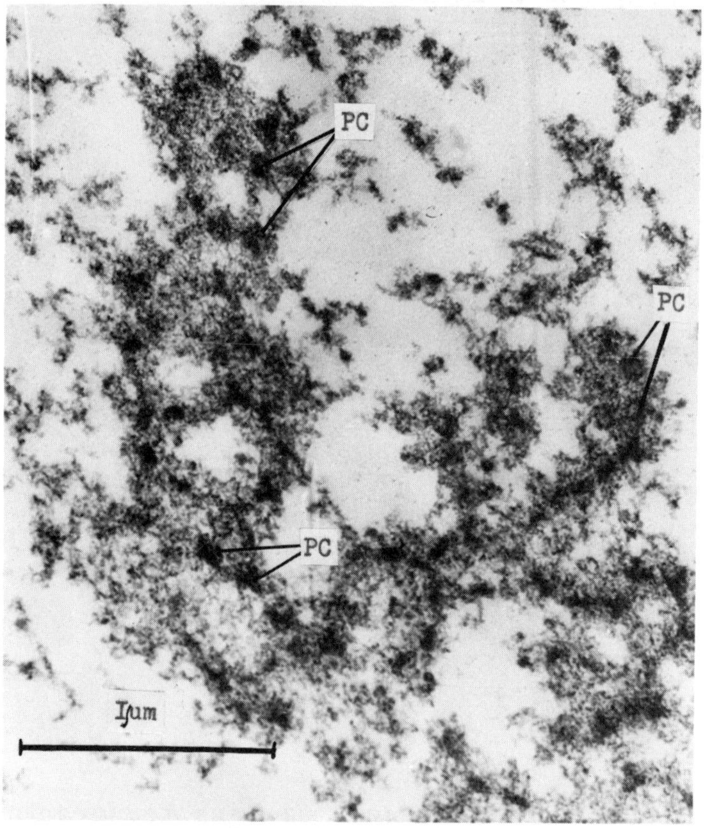

Fig. 9. Electron microscopy Rana temporaria late oocyte karyosphere, 39000x.
Pseudomembranes with the pore complexes (PC) are seen

Rana temporaria and found that their electron microscopic appearence was very similar to that of the nuclear matrix (fig. 8). Practically all the components seen on the electron micrograph including the synaptinemal complexes (22) have the properties of integral parts of the nuclear protein matrix. At fig. 9 it is clearly seen that the fibrillar structures described as "pseudomembranes" do contain the nuclear pore complexes.

Polyacrylamide gel electrophoresis of isolated karyospheres in the same condition as that of the nuclear matrix, reveales high molecular weight components, the major part of which are similar to those found in the nuclear matrix of rat hepatoma 27 (fig. 10).

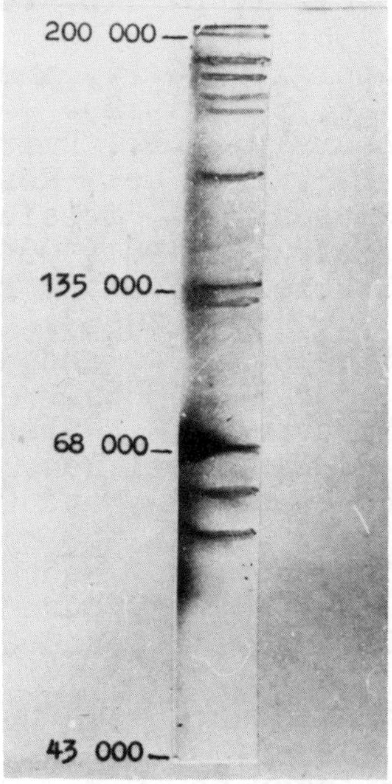

Fig. 10. PAAG electrophoresis of Rana temporaria late oocyte karyosphere

As the nuclear matrix is ascribed to be important for genome structure, initiation of DNA replication, transcription and transport of RNA and ribonucleoproteins, and for other processes essential for the cell life (13, 15, 22, 25) further studies of peculiarities of the tumor nuclear protein matrix and its structural and chemical components, may be of prominent interest for the understanding the molecular mechanisms of the tumor growth, and possibly of normal and pathological development in general.

REFERENCES

1. Zbarsky, I.B., and Debov, S.S. Dokl. Akad. Nauk SSSR 62, 795 (1948).
2. Zbarsky, I.B., and Debov, S.S. Biokhimiya 16, 390 (1951).
3. Zbarsky, I.B., and Georgiev, G.P. Biochim. Biophys. Acta 32, 301 (1959).
4. Soudek, D., and Beneš, L. Cesk. Biol. 4, 416 (1955).

5. Zbarsky, I.B. Proc. Int. Congr. Biochem. 5-th, 1961. Vol. 2, p. 116. Pergamon Press and PWN, Warsaw, 1963.
6. Georgiev, G.P., and Chentsov, Y.S. Exp. Cell Res. 27, 570 (1962).
7. Zbarsky, I.B., Dmitrieva, N.P., and Yermolayeva, L.P. Exp. Cell Res. 27, 573 (1962).
8. Smetana, K., Steele, W.J., and Busch, H. Exp. Cell Res. 31, 198 (1963).
9. Steele, W.J., and Busch, H. Biochim. Biophys. Acta 119, 501 (1966).
10. Steele, W.J., and Busch, H. Biochim. Biophys. Acta 129, 54 (1966).
11. Narayan, K.S., Steele, W.J., Smetana, K., and Busch, H. Exp. Cell Res. 46, 65 (1967).
12. Berezney, R., and Coffey, D.S. Biochem. Biophys. Res. Commun. 60, 1410 (1974).
13. Berezney, R., and Coffey, D.S. Science (Wash. D.C.) 189, 291 (1975).
14. Berezney, R., and Coffey, D.S. J. Cell Biol. 73, 616 (1977).
15. Faiferman, I., and Pogo, A.O. Biochemistry 14, 3808 (1975).
16. Hildebrand, C.E., Okinaka, R.T., and Gurley, L.R. J. Cell Biol. 67, 169a (1975).
17. Riley, D.E., Keller. J.M., and Byers, B. Biochemistry 14, 3005 (1975).
18. Keller, J.M., and Riley, D.E. Science (Wash. D.C.) 193, 399 (1976).
19. Hodge, L.D., Mancini, P., Davis, F.M., and Heywood, P. J. Cell Biol. 72, 194 (1977).
20. Herlan, G., and Wunderlich, F. Cytobiologie 13, 291 (1976).
21. Wunderlich, F., and Herlan, G. J. Cell Biol. 73, 271 (1977).
22. Comings, D.E., and Okada, T.A. Exp. Cell Res. 103, 341 (1976).
23. Ghosh, S., Paweletz, N., and Ghosh, I. Exp. Cell Res. 111, 363 (1978).
24. Fedorov, N.A., Ovcharuk, I.N., Borisov, B.N., and Vanyushin, B.F. Dokl. Akad. Nauk SSSR 236, 1256 (1977).
25. Berezney, R., and Coffey, D.S. Adv. Enzyme Regul. 14, 63 (1976).
26. Debov, S.S. Biokhimiya 16, 314 (1951).
27. Zbarsky, I.B. Usp. Sovr. Biol. 52, 164 (1961).
28. Saidov, S.M. Vopr. Onkol. 1, 86 (1955).
29. Zbarsky, I.B., and Debov, S.S. Vopr. Med. Khimii 1, 198 (1955).

30. Zbarsky, I.B., Yermolayeva, L.P., and Dmitrieva, N.P. Vopr. Med. Khimii 8, 218 (1962).
31. Higgins, G.M., and Anderson, R.M. Arch. Pathol. 12, 186 (1931).
32. Blobel, G., and Potter, V.R. Science (Wash. D.C.) 154, 1662 (1966).
33. Davis, B.J. Ann. New York Acad. Sci. 121, 404 (1964).
34. Aaronson, R.P, and Blobel, G. Proc. Natl. Acad. Sci. USA 72, 1007 (1975).
35. Dwyer, N., and Blobel, G. J. Cell Biol. 70, 581 (1976).
36. Carter, D.B., and Chi-Bom Chae Biochemistry 15, 180 (1976).
37. Jackson, R.C. Biochemistry 15, 180 (1976).
38. Berezney, R., and Hughes, B.B. J. Cell Biol. 75, 406a (1977).
39. Chentsov, Y.S., and Polyakov, V.Y. Ultrastruktura Kletochnogo Yadra (Ultrastructure of the Cell Nucleus), Nauka, Moscow, 1974.
40. Gruzova, M.N., and Parfenov, V.N. Tsitologiya 18, 261 (1976).
41. Gruzova, M.N., and Parfenov, V.N. J. Cell Sci. 28, 1 (1977).

THE PHYSICAL MAP OF THE VARIOUS TRANSCRIPTS OF RAT LIVER MITOCHONDRIAL DNA

C. Saccone[+], G. Pepe[+], H. Bakker[x], M. Greco[+], C. De Giorgi[+] and A.M. Kroon[x]

[+]Istituto di Chimica Biologica, Università di Bari Bari, Italy and [x]Laboratory of Physiological Chemistry State University, Groningen, The Netherlands

It is well known that mitochondria and chloroplasts possess their own DNA. Eukaryotic cell contains therefore at least two or, in the case of plants and plastid-containing micro-organisms, even three different genetic systems (1-4). The organelle genomes show a pattern of cytoplasmic and maternal inheritance. For a number of features in lower organisms this has been known already for a long time. Recently it has been shown that the same holds for the mitochondrial genome of animal cells (5-8). Although the genetic function of the organelle genomes is not yet known in all details, it is evident they are indispensable and vital for eukaryotic organisms. For the expression of their genomes, the organelles depend on the nucleus. They constitute heteronomous genetic entities, their expression and continuity being ensured by the existence within the organelles of elements coded for and synthesized by the main genetic system of the cell, the nuclear-cytoplasmic genetic system. This implies that a much higher level of genetic complexity is reached in the eukaryotic cell as compared to the prokaryotic cell. Although also in prokaryotes extrachromosomal genetic elements such as plasmids may exist, the obvious difference is that the products of transcription and translation of these elements are not obligatory for the cells to survive as is the case for the mitochondrial and chloroplast genomes. It is, therefore, necessary to envisage a very delicate interplay of control and regulation processes taking place inside the eukaryotic cell which are probably accomplished in a two-way direction, from the nuclear-cytoplasmic system to the organelle system and viceversa. Since 1965, after the isolation for the first time of intact mitochondrial DNA molecules from animal cells the study of Mitochondrial Biogenesis has been mainly concentrated on elucidation of the properties and meaning of the mitochondrial genetic system.

In about ten years most of the problems of this apparently complicated system have been unravelled taking also advantage of the use of organisms like yeast and Neurospora in which genetic approaches can be devised. The main characteristics of the mitochondrial genetic system and various properties of its constituents, were already known more than five years ago (1-4). Other problems, like the isolation of mitochondrial messenger RNAs and of proteins coded for by mitochondrial DNA still remain, largely, open questions. Now however the use of modern techniques for the study of nucleic acids, particularly the availability of restriction endonuclease enzymes and methods for DNA sequencing, represent a powerful tool to clarify the structure of the mitochondrial genome at molecular level and to reach a complete understanding of its functional significance. Physical maps of mitochondrial genome have been constructed by using different organisms both lower and higher eukaryotes. It is well known that in animal cells the size of mitochondrial genome, the smallest that can be found in nature, is roughly constant, about 10×10^6 daltons or 5 μ contour length. From animal cells furthermore mitochondrial DNA molecules can easily be extracted undegraded in their original closed duplex circular form. Physical maps of mitochondrial DNA from a variety of animals including man have been constructed (2-4). In this report we wish to summarize the studies already published on the physical map of rat liver mitochondrial DNA (9) and to add some new data which may contain useful information about the mapping of mitochondrial gene products.

Endonuclease restriction map of rat liver mitochondrial DNA. The number of fragments obtained after digestion of mitochondrial DNA with various restriction endonuclease enzymes is shown in Table I. Capital letters designate the terminal fragments in order of increasing electrophoretic mobility. The fragments were usually detected in composite slabgels consisting of a small sealing layer of 10% polyacrylamide, a layer of about 8-10 cm of 3% polyacrylamide and a 20 cm layer of 0.7% agarose. However in order to detect very short fragments longer 10% and 3% polyacrylamide layers and only a small agarose layer was used in some experiments. The order of the fragments obtained with a single restriction endonuclease and the overlapping of various fragments after digestion with different enzymes was established by using different approaches such as: a) the analysis of the length of restriction fragments of partially and completely digested mtDNA; b) analysis of double digests of total mtDNA and of the fragment patterns obtained after digestion of isolated restriction fragments with a second endonuclease; c) electron-microscopical length measurements of the fragments; d) identification of the fragments of complete, single and double digestions and of partially digested fragments containing the base sequences complementary to the mitochondrial rRNAs, using the stripfilter hybridization technique. By using all these techniques we arrive at the physical map of rat liver mitochondrial DNA shown in Fig. 1 in which 37 cleavage sites have been quite well localized.

TABLE I

Restriction Enzyme Fragments of Rat Liver mtDNA

Enzyme		Eco RI	Hind III	Bam HI	Hpa I	Hha I	Hind II	Hap II
Recognition sequence		GAATTC	AAGCTT	GGATCC	GTTAAC	GCGC	GTPyPuAC	CCGG
Nr of fragments		8	6	2	2	4	5	11
Size	A	5500	5950	10000	12500	8300	6900	3550
(bp's)	B	3550	3750	4450	2300	3900	4900	2850
	C	2650	2300	--	--	1900	2150	1750
	D	1800	1900	--	--	780	630	1630
	E	650	800	--	--	--	250	1550
	F	400	150	--	--	--	--	1120
	G	125	--	--	--	--	--	970
	H	100	--	--	--	--	--	680 (2x)
	I	--	--	--	--	--	--	180
	J	--	--	--	--	--	--	110
	Σ	14775	14850	14450	14800	14880	14830	15020

For experimental details see reference 9, 10.

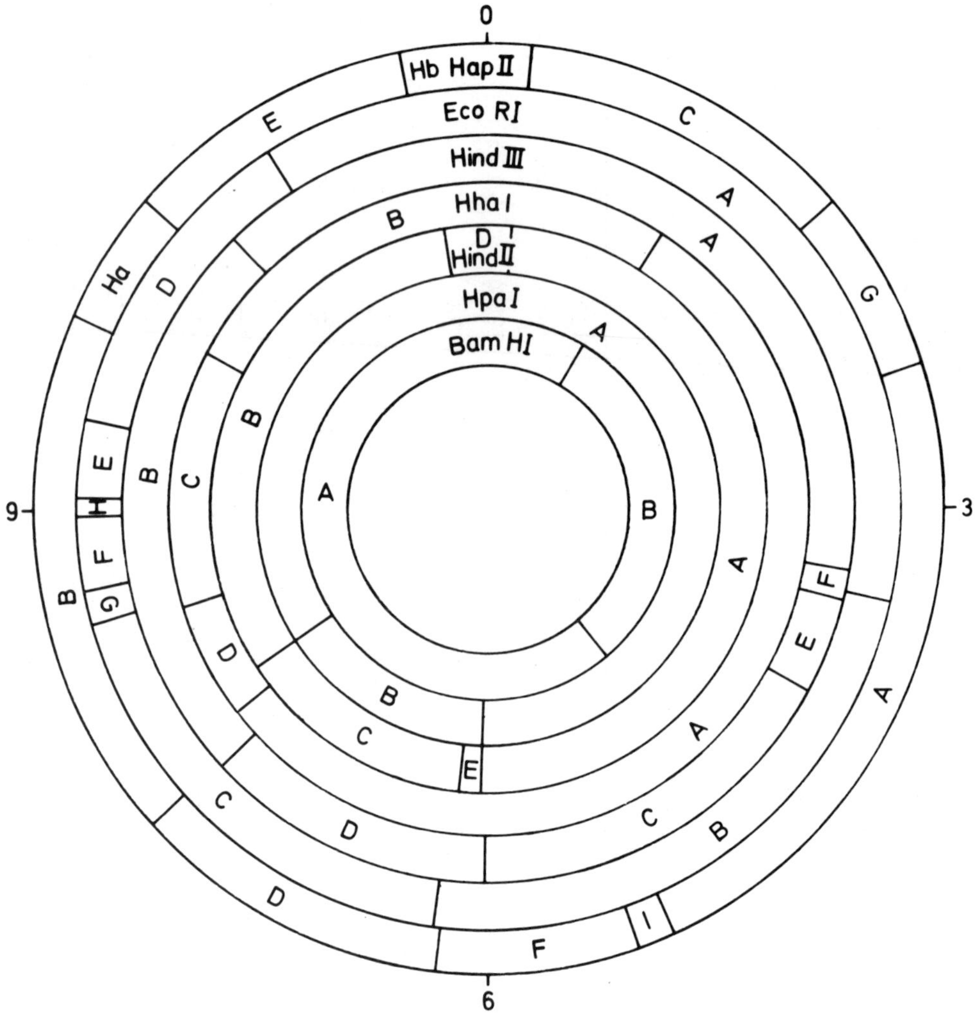

Figure 1. Map of rat-liver mitochondrial DNA obtained by using various restriction enzymes.

Mapping of ribosomal and transfer RNAs. The regions of mitochondrial DNA containing sequences complementary to mitochondrial rRNAs and tRNAs were identified by using the stripfilter hybridization technique. Mt rRNAs and tRNAs were isolated and iodinated with ^{125}I. After hybridization the mitochondrial DNA fragments containing radioactivity were visualized by autoradiography and subsequently counted in a scintillator counter in order to obtain quantitative hybridization data. These latter measurements enabled us to calculate the space occupied by single genes, their relative position and, in the case of tRNA, tentatively, also the number of genes coded for by mitochondrial DNA (9-10). Table II shows the quantitation of the strip-filter hybridization data with 16S and 12S RNAs extracted from mitochondrial ribosomes. Since it was found that the small rRNA species, 12S rRNA, was contaminated with fragments of 16S, in most hybridization experiments 16S cold rRNA was added as competitor. Therefore in Table II the radioactivity found on 12S rRNA strip-filters was corrected for 16S rRNA contamination (for further details see ref. 9). The results of the calculations led to the conclusion that about 5/6 of the 12S rRNA gene was lying on Hap E. Since Hap E also contains about half of the 16S rRNA gene it follows that the two ribosomal RNA genes are closely linked to each other, leaving a gap between them of about 200 base pairs. Furthermore taking the number of base pairs of HapHa as 44% of the length of the 16S rRNA we can calculate a molecular weight of 16S rRNA of about 0.5×10^6

TABLE II

Quantitation of the Stripfilter Hybridization with 16S and 12S mt rRNA

	Fragments Hybridizing with mt rRNA					
	HindA	HindB	HapB	HapE	HapHa	HapHb
Fragment length (bp's)	5940	3750	2850	1550	680	680
16S rRNA % hybridization	24	76	5	51	44	0
12S rRNA % hybridization corrected for contaminating 16S rRNA	100	0	0	84	0	16

Experimental details as in reference 9.

TABLE III

Localization of Mitochondrial Transfer RNA Genes on Various Restriction Fragments of Rat-Liver Mitochondrial DNA

Fragment	^{125}I-mt-tRNA bound % of total	number of tRNA genes	fragment	^{125}I-mt-tRNA bound % of total	number of tRNA genes
Eco A	36	7-8	HhaA	61	12-13
Eco B	28	5-6	HhaB	18	3-4
Eco C	24	4-5	HhaC	16	3-4
Eco D	4	0-1	HhaD	5	1
Eco E	3	0-1			
Eco F	4	0-1	HIIIA	41	8-9
			HIIIB	34	6-7
Hap A	27	5-6	HIIIC	19	3-4
Hap B	26	5-6	HIIID	5	1
Hap C	14	2-3			
Hap D	8	1-2	HIIIB	37	7-8
Hap E	4	0-1	HB1	28	5-6
Hap F	10	2	HB2	18	3-4
Hap G	5	1	HIIID	14	2-3
Hap H	6	1-2	HB3	4	1

For experimental details see reference 10

daltons which is in good agreement with the molecular weight of larger ribosomal RNA species extracted from other animal cells (1). It should be stressed that the space occupied by 16S rRNA gene on mtDNA obtained with this calculation is also in excellent agreement with electron microscopical measurements of DNA-RNA hybrids (2). In the case of tRNA,by expressing the results of hybridization experiments as percentage of the total radioactivity bound to the fragments on the strip-filters, we found that the minimum percentage bound to any fragment varied around 5%. This value was therefore considered as corresponding to one tRNA gene and the number of tRNA genes on each fragment was provisionally calculated by dividing the % of total radioactivity bound by 5. In table III the number of tRNA genes and their localization on various restriction fragments obtained by using this approach are shown. A total number of tRNA genes between 16 and 23 was obtained in this way and this value is in good agreement with previous findings with mtDNA from other animal cells (10).

Mapping of high molecular weight-non ribosomal RNA species. In order to localize on the physical map of rat liver mtDNA the regions coding for RNA species which are neither ribosomal not transfer RNA, we have extracted from mitochondria the high molecular

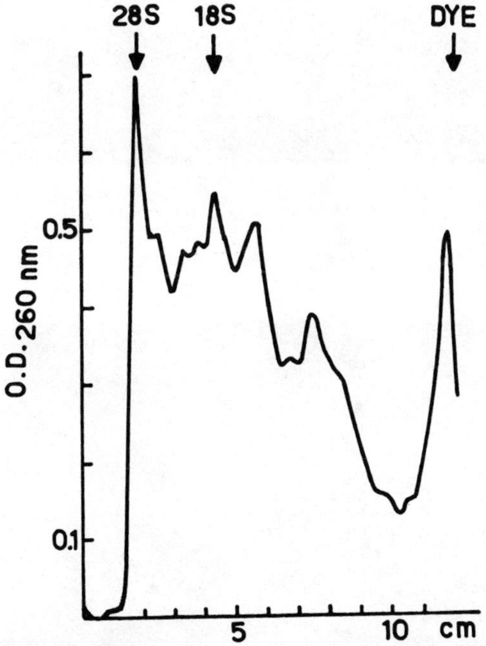

Figure 2. Polyacrylamide-gel electrophoresis of high molecular weight mtRNA.The mtRNA extracted as described in the text was layered on a 2.7% polyacrylamide-gel. Electrophoresis was carried out for 2 h at 10°C and 5 mA/gel. The gel was washed for 2 h in distilled water and scanned at 260 nm.

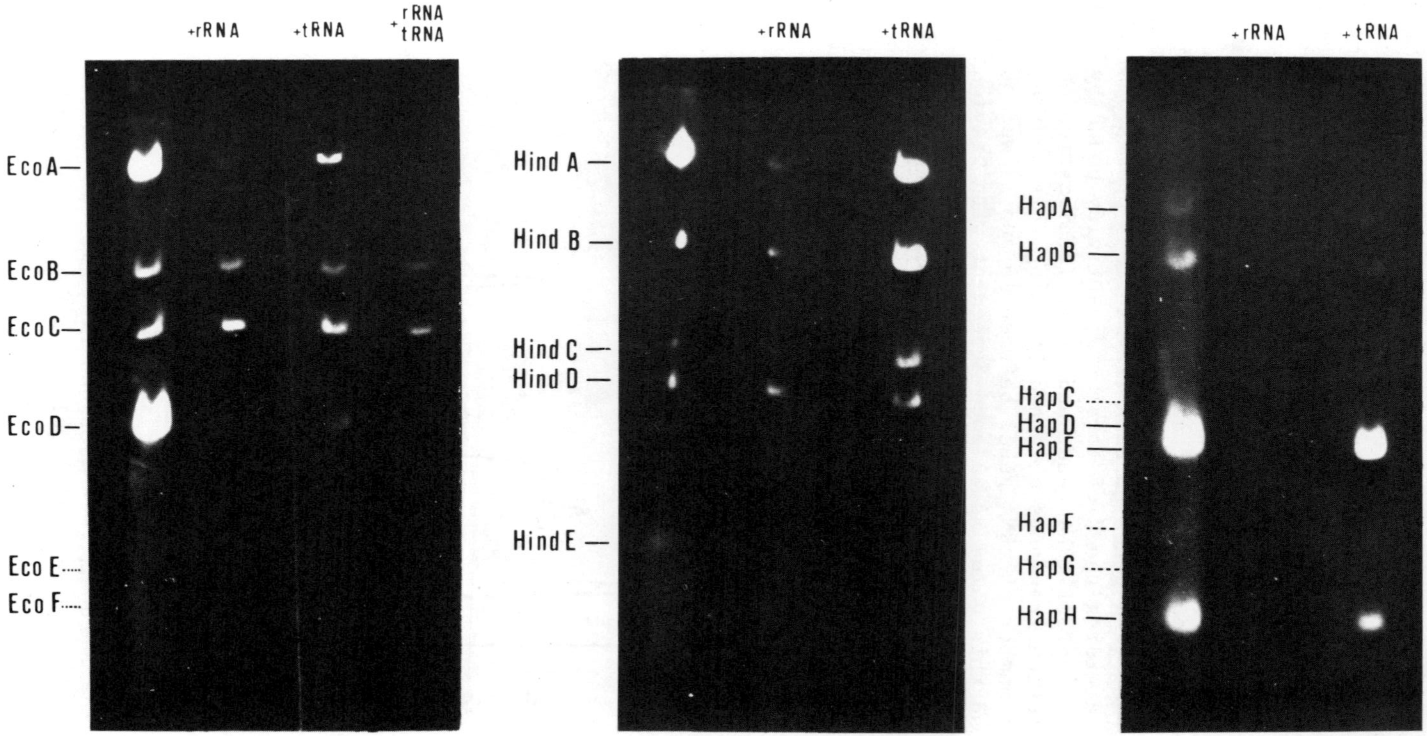

Figure 3. Hybridization of high molecular weight mtRNA with various restriction fragments of mtDNA. Strip-filters containing the denatured restriction fragments of rat liver mtDNA were incubated with ^{125}I-iodinated mtRNA (14×10^6 cpm/µg). The positions of the various fragments visible on the autoradiograms are indicated by the black lines. The experiments were performed either in absence or in the presence of cold mt rRNA (40 µg) and/or tRNA (80 µg) as competitor.

TABLE IV

Hybridization of High Molecular Weight non Ribosomal RNA Species with Various Restriction Fragments of Rat Liver mtDNA.

Fragments	^{125}I mtRNA bound							
	- compet.		+ rRNA		+ tRNA		+ rRNA + tRNA	
	cpm	% total	cpm	% total	cpm	% total	cpm	% total
Eco RI total	36350		9390		14920		5400	
Eco A	11268	31	1408	15	2984	20	540	10
Eco B	3635	10	3474	37	2835	19	1458	27
Eco C	2544	7	3474	37	5371	36	3402	63
Eco D	18902	52	1033	11	3730	25		
Hind III total	25100		6910		29220			
Hind A	15562	62	1935	28	10227	35		
Hind B	3263	13	1658	24	11396	39		
Hind C	1004	4	967	14	3799	13		
Hind D	1757	7	1520	22	2922	10		
Hind E	3514	14	829	12	877	3		

For experimental details see figure 3.

weight RNA fraction. After treatment with urea and LiCl the bulk of tRNA is excluded since it remains in the supernatant. The high-molecular weight RNA, purified through CsCl gradient (11), appeared undegraded. The electrophoretic pattern is shown in Fig. 2. After alcohol precipitation the RNA was iodinated with ^{125}I and used in hybridization experiments. The hybridization experiments were performed either in the absence or in the presence of a large excess of cold mt rRNA and/or tRNA as competitor. The results of hybridization experiments with Eco RI, Hind III and Hap II are shown in Fig. 3. It can be observed that when total mtRNA was used, the competition was mainly afforded by ribosomal RNA but also to a lesser extent by the transfer fraction.This suggests that, although the presence of tRNA species could not be detected from the electrophoresis pattern (Fig. 2), our RNA preparation was still contaminated with traces of transfer RNAs or it contains transfer precursors of higher molecular weight. The quantitative results, reported in Table IV indicate a number of interesting data. 1) The competition afforded by tRNA fraction, which is clearly higher than expected on the basis of tRNA content of our preparation, confirms our previous results (10) that this fraction contains ribosomal contaminant. Of course it is not possible to know whether fragments of messenger RNA are also present as contaminant. 2) Comparing percentages of radioactivity bound to Eco D fragment in the presence of ribosomal RNA alone or ribosomal plus transfer RNA we have to assume that at least one tRNA gene lyes on Eco D fragment probably located between the two ribosomal RNA genes as clearly suggested in a previous paper (10). 3) Comparing the percentage of total counts bound to Eco RI fragments in absence of competitor RNA to that in the presence of rRNA or rRNA plus tRNA, we can tentatively calculate that the steady state RNA population extracted from mitochondria in our experimental conditions seems to consist of 75% ribosomal RNA, 10% transfer and 15% of species which can be only messenger RNA species or may contain parts of precursor molecules which are subsequently processed to give final mitochondrial products. This calculation is obviously related to our hybridization conditions. From the hybridization data obtained with Hind III fragments about the same competition of rRNA can be calculated. The higher value of total radioactivity bound in the presence of tRNA in this case may be due to different amount of mtDNA transferred to strip-filters. 4) From the overlapping of Eco RI and Hind III fragments we can suggest that the region corresponding to Eco C and to part of Hind B plus Hind H is the region where mitochondrial messenger RNAs are mostly concentrated. In order to separate the poly(A) containing fraction, in other experiments, the iodinated RNA was passed through an Oligo(dT) cellulose column at 2°C. The bound and non bound fractions were collected and used in strip-filter hybridization experiments. The hybridization of poly(A)-RNA with Eco RI fragments (Fig. 4) clearly shows that the fraction bound to Oligo(dT) cellulose column contains ribosomal RNA sequnces which could be either simply present as

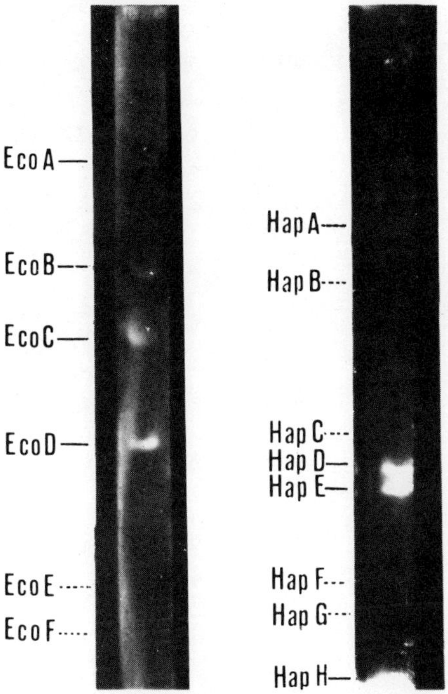

Figure 4. Hybridization of mt poly(A)-RNA with mt Eco RI and Hap II fragments. The iodinated mtRNA was passed through an Oligo(dT)-cellulose column. The bound fraction ($1.2x10^5$ cpm) was collected and used in strip-filters hybridization experiments. The hybridization with Eco RI fragment was carried out in presence of tRNA (80 μg) as competitor.

contaminant or as polyadenylated ribosomal precursors. The quantitative data concerning poly(A)-RNA hybridization shown in Table V indicate that this fraction does not preferentially bind to any fragment and seems to contain the same species, although in different proportions, present in total mtRNA fraction or in the fraction which does not bind to the column (results not shown). This poses the problem of the presence of polyadenylated RNA species in mitochondria. Data in literature suggest that mitochondria from lower eukaryotes, do not contain poly(A)-RNA or contain RNA species with very short poly(A) sequences (12-13). On the other hand animal mitochondria seem to contain and to be able to synthesize poly(A)-RNA. For HeLa cells the presence of multiple poly(A)-RNA species in mitochondria has

TABLE V

Hybridization of Mitochondrial Poly(A)-RNA with Restriction Fragments of mtDNA

Fragments	^{125}I poly(A)-RNA bound cpm	% total
Eco RI total	2534	
Eco A	329	13
Eco B	633	25
Eco C	786	31
Eco D	786	31
Hap II total	792	
Hap A	79	10
Hap B	24	3
Hap C	--	--
Hap D + E	636	80
Hap F	55	7

For experimental details see figure 3.

been reported (14-15). From our results, however, we are forced to conclude that the percentage of mtRNA linked to Oligo(dT)-column is very low and therefore that the major part of high molecular weight, probably messenger RNA species, are either free from adenylate sequences or contain a very short poly(A) tail. Further investigations could probably clarify this point. Taking into account all the above mentioned results we can provisionally conclude that genes for messenger RNA species are mainly localized between 10 min and 45 min of the physical map of rat liver mitochondrial DNA where also the D-loop and the direction of the replication are indicated (Fig. 5).

The translation of mtRNA species. It is well known that RNA preparation from rat liver mitochondria is highly contaminated by extramitochondrial ribosomal RNA and it probably contains cytoplasmic mRNA as well. These contaminating messenger chains are expected to contain longer poly(A) sequences than those of mitochondrial mRNA molecules (14). It has been shown (16) that using Oligo(dT) cellulose column chromatography at two different

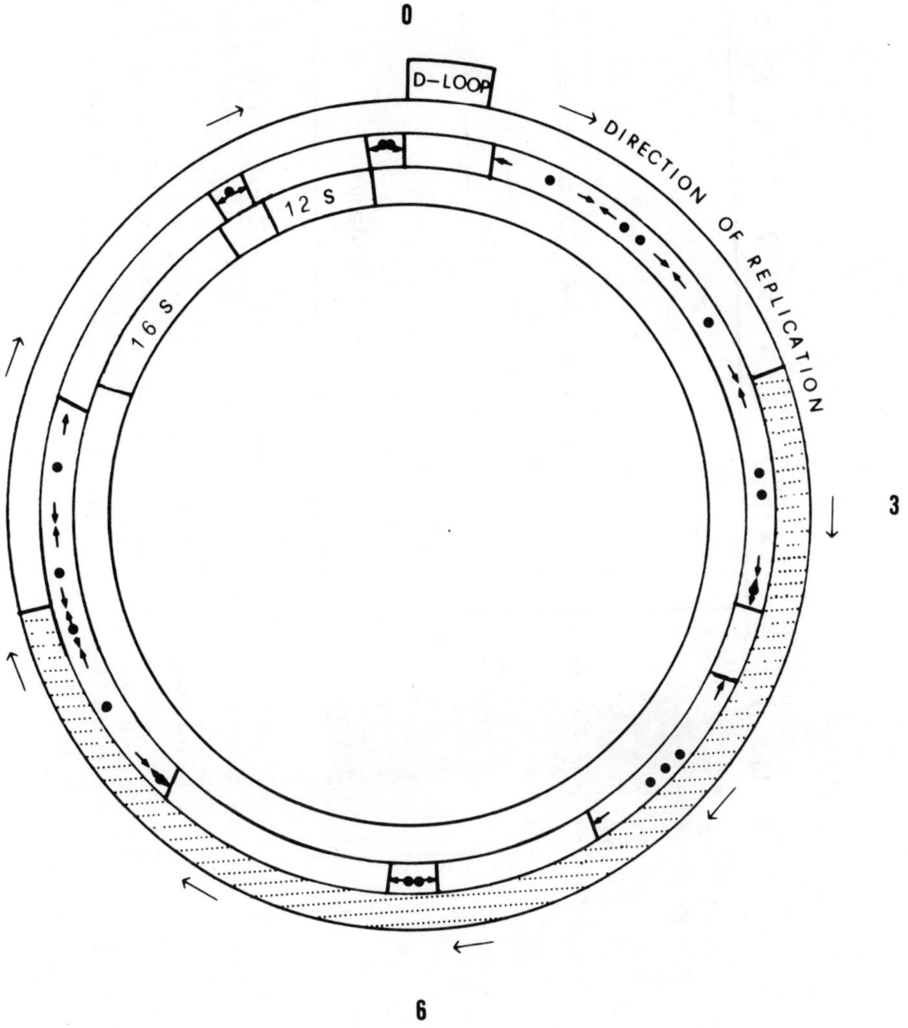

Figure 5. Map of rat-liver mitochondrial DNA with the position of a number of genetic markers. The inner circle shows the position of the rRNA genes. On the middle circle the arrows indicate areas in which tRNAs are located: the provisional number of these genes corresponds with the number of black circles in each area. The outer circle shows the part of the genome which contains mRNAs genes (between about 10 min and 45 min). Outside the D-loop and the direction of replication are indicated. For experimental details see reference 9,10.

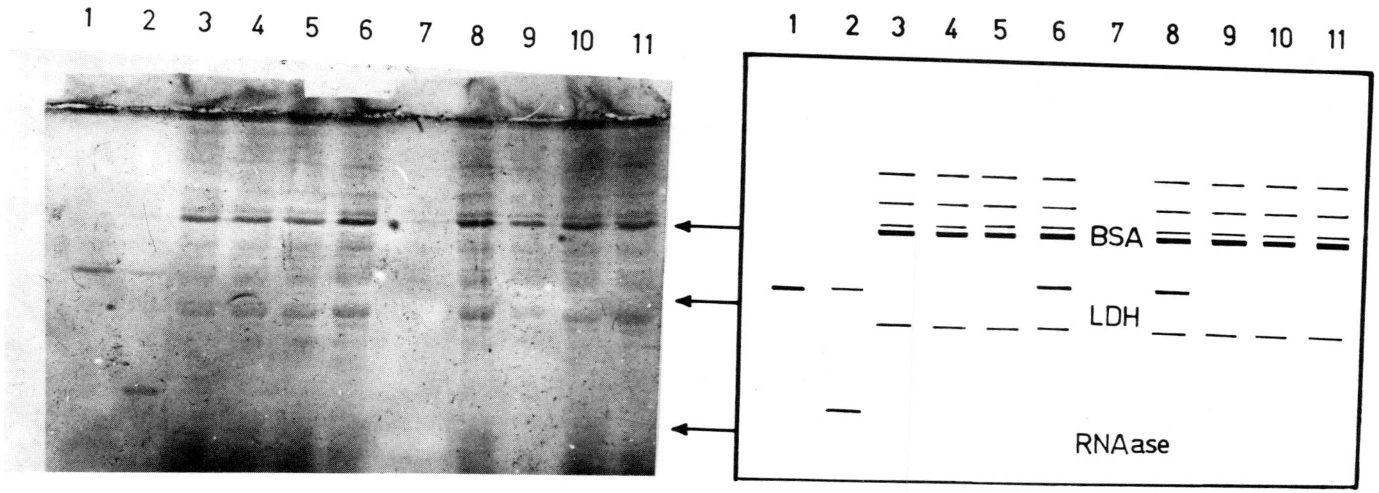

Figure 6. SDS gel electrophoresis of polypeptides synthesized by mtRNA in a cell-free system and by isolated mitochondria. The mtRNA was chromatographed on Oligo(dT) column at room temperature, than the unbound peak was rechromatographed at 4°C. The protein synthesis was measured as reported by Moorman (17) using ^{35}S -methionine as labelled amino acid. After incubation samples were heated at 100°C in 0.23 M 2-mercaptoethanol and 1% SDS and layered on 10-20% polyacrylamide gel gradient. The gel was stained with Coomassie Brillant blue, dried and autoradiographed. Unlabelled proteins: Bovine serum albumin 68,000 M.W.; lactate dehydrogenase 36,000 M.W.; ribonuclease 13,700 M.W..

temperatures it is possible to separate the RNA containing long poly(A) sequences from an RNA population with shorter adenylate residues: in fact at room temperature Oligo(dT) cellulose binds only RNA chains with relatively long poly(A) sequences; however rechromatography of non bound material at 2°C enables RNA molecules with poly(A) tails as short as 16 adenylate residues to bind to the column. In order to test the biological activity of RNA containing different lengths of poly(A), the RNA retained on the Oligo(dT) column at room temperature, the RNA non bound and bound in the cold room to the Oligo(dT) column, were examined for their ability to stimulate the translation system of an E. coli extract. In order to compare the electrophoretic mobility of the labelled polypeptides synthesized in a cell-free system with the authentic products of mitochondrial protein synthesis, isolated mitochondria were incubated in the presence of ^{35}S-methionine in the same conditions and run on the same SDS gel electrophoresis. The results are reported in Fig. 6 which shows the autoradiography of electropherogram of the SDS polyacrylamide gel. The most pronounced bands are also reported in the scheme. In slot 1 the main product synthesized in vitro by isolated mitochondria is a 45,000 daltons M.W. polypeptide. If isolated mitochondria are incubated in the presence of E. coli extract (slot 2) it is possible to see the 45,000 M.W. band and an additional band of 18,000 daltons. In both cases (slot 1 and 2) the endogenous labelled proteins from E. coli, well detectable when E. coli extract was incubated without any further addition (slot 11), appeared as faint bands probably because the presence of proteolitic enzymes within the mitochondrial preparation caused their partial degradation. Total mtRNA was tested with E. coli extract and no synthesis of new polypeptide was observed (slot 4). Likewise the RNA bound to the Oligo(dT) cellulose column at room temperature and thus containing long poly(A) tails, did not show messenger activity in our conditions (slot 5) whereas the RNA eluted from the Oligo(dT) in the cold room directed the synthesis of a polypeptide of 45,000 M.W. visible in slots 6 and 8 and probably the same polypeptide was synthesized also after addition of the cold room-bound RNA (slots 9 and 10). These results suggest that the messenger activity we find in our conditions is present in the mtRNA preparation and this mRNA is without the poly(A) tail or has a poly(A) sequence so short that it is at the limit of binding to Oligo(dT) cellulose even at 2°C. This mRNA apparently directs the synthesis of a protein having the same electrophoretic mobility as one of the products synthesized in vitro by isolated mitochondria. The synthesis of two polypeptides having a molecular weight of 45,000 and 18,000 in addition to other products, has been reported recently also in yeast by Groot et al. (18).

Concluding remarks. From the experiments presented above, it follows that much information has still to be obtained. Combination of the three approaches available may hopefully lead to a rapid further unravelment of the complex functions of the rat liver mitochondrial DNA. In the first place the mtDNA is sufficiently characterized to enable us to start the further characterization at the level of base sequence analysis. This should provide the theoretical genetic potential of this relatively small DNA. In the second place further and defined studies of the mtRNA along the pathways outlined may offer a more precised insight in the mtDNA sequences that are actually transcribed and in how they are processed. Finally the use of differential inhibitors of cytoplasmic and mitochondrial protein synthesis, in vivo as well as in vitro may shed light on how these mtRNAs are actually expressed.

Acknowledgement. The cooperation between the two laboratories of the authors has been highly facilitated by NATO Research Grant No. 1484.

REFERENCES

1. Saccone, C. and A.M. Kroon. The biogenesis of mitochondria. (1974) Academic Press, New York.
2. Saccone, C. and A.M. Kroon. The genetic function of mitochondrial DNA. (1976) North-Holland, Amsterdam.
3. Bücher, Th., W. Neupert, W. Sebald and S. Warner. Genetics and biogenesis of chloroplasts and mitochondria. (1976) North-Holland, Amsterdam.
4. Bandlow, W., R.J. Schweyen, K. Wolf and F. Kaudewitz. Mitochondria 1977. Genetics and biogenesis of mitochondria. (1977) Walter de Gruyter, Berlin.
5. Hutchison, C.A. III, J.E. Newbold, S.S. Potter and M.H. Edgell. Maternal inheritance of mammalian mitochondrial DNA. Nature (1974) 251, 536-538.
6. Buzzo, K., D.L. Fouts and D.R. Wolstenholme. Eco RI cleavage site variants of mitochondrial DNA molecules from rats. Proc. Natl. Acad. Sci. USA (1978) 75, 909-913.
7. Kroon, A.M., W.M. de Vos and H. Bakker. The heterogeneity of rat liver mitochondrial DNA. Biochim. Biophys. Acta. In press.
8. Hayashi, J., H. Yonekawa, O. Gotoh, J. Motohashi and Y. Tagashira. Two different molecular types of rat mitochondrial DNAs. Biochim. Biophys. Res. Commun. (1978) 81, 871-877.
9. Kroon, A.M., G. Pepe, H. Bakker, M. Holtrop, J.E. Bollen, E.F.J. van Bruggen, P. Cantatore, P. Terpstra and C. Saccone. The restriction fragment map of rat-liver mitochondrial DNA. Biochim. Biophys. Acta (1977) 478, 128-145.
10. Saccone, C., G. Pepe, H. Bakker and A.M. Kroon. The genetic organization of rat liver mitochondrial DNA. Mitochondria 1977. Genetics and biogenesis of mitochondria. Walter de Gruyter, Berlin (1977) pp. 303-315.
11. Glišin, V., R. Crkvenjakov and C. Byus. Ribonucleic acid isolated by cesium chloride centrifugation. Biochem. (1974) 13, 2633-2637.

12. Groot, G.S., R.A. Flavell, G.J. van Ommen and L.A. Grivell. Yeast mitochondrial RNA does not contain poly(A). Nature, Lond. (1974) 252, 167-169.
13. Hendler, F.J., G. Padmanaban, J. Patzer, R. Ryan and M. Rabinowitz. Yeast mitochondrial RNA contains a short polyadenylic acid segment. Nature (1975) 258, 357-359.
14. Hirsch, M. and S. Penman. Mitochondrial polyadenylic acid-containing RNA: localization and characterization. J. Mol. Biol. (1973) 80, 379-391.
15. Ojala, D. and G. Attardi. Expression of the mitochondrial genome in HeLa cells XXII. Identification and partial characterization of multiple discrete poly(A)-containing RNA components coded for by mitochondrial DNA. J. Mol. Biol. (1974) 88, 205-219.
16. Nudel, U., H. Soreq and U.Z. Littauer. Globin mRNA species containing poly(A) segments of different lengths. Eur. J. Biochem. (1976) 64, 115-121.
17. Moorman, A.F.M., F. Lamie and L.A. Grivell. A coupled transcription-translation system derived from Escherichia coli: the use of immobilized deoxyribonuclease to eliminate endogenous DNA. FEBS Lett. (1976) 71, 67-72.
18. Groot, G.S.P., N. van Harten-Loosbroek and J. Kreike. Electrophoretic behaviour of yeast mitochondrial translation products. Biochim. Biophys. Acta (1978) 517, 457-463.

ORGANIZATION OF LAC REPRESSOR, RNA POLYMERASE AND HISTONES OF DNA

A. Mirzabekov, R. Beabealashvilli,
A. Kolchinsky, A. Melnikova, V. Schick, and
A. Belyavsky

Institute of Molecular Biology
Açacemy of Sciences of USSR
Moscow, USSR

INTRODUCTION

Most of DNA in the eukaryotic genome appears to be complexed with histones. Little is known about the state of DNA in the course of its replication, transcription, and recognition by specific proteins. It may be asked whether DNA remains bound to histones during these processes or histones are displaced from DNA and, in particular, whether DNA covered with histones can be recognized by proteins within its major and minor grooves.

In this paper, we shall analyze the arrangement in the DNA grooves, as well as the unwinding of DNA, with bacterial RNA polymerase, lac repressor, and eukaryotic histones. The comparison of prokaryotic and eukaryotic proteins suffers some obvious limitations and is justified only by the fact that bacterial proteins are much more available. In addition, the sequence of histone arrangement on DNA in the nucleosome core particles will be presented. For studying the organization of proteins on DNA, a number of new techniques have been developed in our laboratory. Methylation of DNA with dimethyl sulfate has been introduced to localize ligands in the DNA grooves as well as to measure DNA unwinding (I-3). Sequencing of histones along DNA was carried out by cross-linking histones to partly depurinated DNA (4), scission of one DNA strand at the point of cross-linking, and measuring the size of DNA fragments cross-linked to each histone fraction (5,6).

METHODS

Localization of Ligands in the DNA Grooves and Measurement of DNA Unwinding

Dimethyl sulfate (DMS) is well known to methylate mainly the N-7 atom of guanine in the major groove, the N-3 of adenine in the minor groove, and the N-I of adenine in the single-stranded DNA. Location of a ligand in the major and minor DNA grooves sterically blocks and therefore inhibits the formation of 7-methyl guanine and 3-methyl adenine, respectively. The presence of single-stranded regions in DNA significantly increases the formation of I-methyl adenine. Thus, the quantitative comparison of methylated bases formed upon methylation of DNA and DNA complexes indicates the localization of a ligand in the DNA grooves and enables to estimate the DNA unwinding. Methylation was carried out with ^{3}H DMS at its low concentration under trace labeling conditions, and did not appear to appreciably affect the structure of DNA complexes (I-3).

Sequencing Histones along DNA

DNA in the nucleosome core particles was methylated with DMS (Fig.I). Methylated purine bases labilize glycosyl bonds and can therefore be partly removed under mild conditions at neutral pH and 45°C. The aldehyde groups formed in depurinated sites of DNA react with the ε-amino groups of the adjacent lysine residues of histones and form the Schiff bases. The Schiff bases catalyze the β-elimination reactions of quantitative splitting of the phosphodiester bond at the 3'-OH group of the depurinated nucleotides. As a result, histones become cross-linked only to the 5'-terminal fragments of DNA. The covalent bonds between histones and DNA through the Schiff base can be stabilized by reduction with $NaBH_4$. This procedure (4,5) avoids drastic treatment and hardly influences the nucleosome structure. The size of the DNA fragment cross-linked to histone molecule determines the position of the histone on one DNA strand relatively to its 5'-end. Sizing of single-stranded ^{32}P labeled DNA fragments cross-linked to each histone

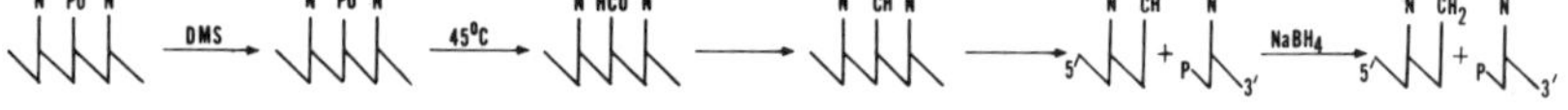

Fig. I. The scheme of reactions resulting in cross-linking histones to DNA and scission of one DNA strand at the point of cross-linking

fraction and identification of these ^{125}I labeled histones were carried out using electrophoresis in two two-dimensional polyacrylamide slab gel systems as described (5,6).

RESULTS

Arrangement of Proteins in the DNA Grooves

Table I summarizes the shielding of the DNA grooves with proteins against methylation with dimethyl sulfate. Lac repressor binds rather well to many different DNAs with a preference for the A-T rich sequences (7). In non-specific complexes with DNAs of Tetrahymena (70% of the A-T base pairs), calf thymus (57% of A-T), and non--glucosylated T4 DNA, lac repressor preferentially protects the minor groove against methylation by II-I8%. The major groove does not appear to be involved in the non-specific interaction with the repressor since glucosylated within the major groove and non-glucosylated T4 DNA are shielded similarly against methylation (8). Thus, lac repressor is likely to bind to non-specific DNA preferentially within the minor groove. This conclusion is also supported by a number of other experiments: effective binding of the repressor to DNA filled, within the major groove, with glucosyl residues (7) or HgX residues (X-mercaptans) (9), and the failure of UV-induced cross-linking of the repressor to poly d(-BrdUrd) from the side of the major groove (IO). On the other hand lac repressor protects against methylation and appears to make contacts with three adenines in the minor groove and four guanines in the major groove (II), moreover, it can be cross-linked to several thymines from the side of the major groove (IO) of the lactose operator. The two-step model for binding of the repressor to DNA has recently been proposed on the basis of these data (8). Initially, looking for specific sequences, lac repressor may interact with the sugar--phosphate backbone and DNA bases preferentially along the minor DNA groove. When the proper DNA segment is roughly recognized in such a manner, the repressor begins to interact with both DNA grooves.

RNA polymerase. The specific binding of the RNA polymerase holoenzyme of E.coli to T7 phage DNA does not affect methylation of DNA in the major groove but decreases methylation in the minor groove by IO%. With poly d(A-T) the polymerase shields the minor groove by 8%, and the shielding is enhanced about twice as much when initiation of RNA synthesis or RNA synthesis occurs. Thus, RNA polymerase in specific and non-specific complexes with DNA seems to interact slightly with the minor groove and to leave the major groove predominantly exposed (3). It is also possible that the shielding of the N-3 atom of adenine by the polymerase may occur in the unwound region of DNA (see below). These data represent the overall feature of interaction between RNA polymerase and DNA and, of course do not rule out

TABLE I Shielding of the Major and Minor DNA Grooves and Unwinding of DNA with Proteins as Measured by Methylation of DNA with Dimethyl Sulfate

Protein-DNA complexes	Shielding (%)		Unwinding****		References
	major gr	minor gr	%	base pairs	
Lac repressor + non-specific DNA	-	II-I8*	no	no	(8)
Lac-repressor + lactose operator	4 Gua	3 Ade			(II)
RNA polymerase holoenzyme + poly d(A-T) or T7 DNA	no	8-I0**	35	I7	(3)
poly d(A-T) + nascent RNA	no	I7**	25	I2	(3)
poly d(A-T) (initiated RNA synthesis)	no	I4**	36	I7	(3)
Chromatin	I4***	no	no	no	(2)
Mono-, di- & tri-nucleosomes	I8-20*	-	no	no	(I6)
Histones H2A, H2B, H3 or H4 + DNA	I5-I8*	-	no	no	(I6)
Histone HI+DNA	-*	-	no	no	(I6)

* preferential shielding of one of the DNA grooves measured according to 7-methyl guanine/3-methyl adenine ratio; ** and *** determined according to a decrease in the formation of methylated bases in DNA complexes measured at one time point or in the course of the reaction, respectively; **** was measured for DNA segments of about 50 base pairs long covered with an RNA polymerase molecule.

the possibility of interactions of the polymerase with a few bases within the major groove (L. Johnsrud, unpublished results).

Likewise, alterations of the DNA structure that affect the minor grooves of the A-T and G-C base pairs seem to be particularly effective in blocking the utilization of a promoter site by T7 RNA polymerase (I2).

Histones. Histones shield the major groove of DNA from methylation to a rather low degree, by I4-20% in chromatin, mono-, di- and trinucleosomes, but leave the minor groove well exposed. The latter is in a good agreement with a high accessibility of the minor groove of DNA in chromatin to a reporter molecule (I3) and to the antibiotics netropsin and distamycin A (I4,I5). This shielding is hardly related to the organization of histones in nucleosomes since the same shielding of the major groove by I5-I8% has been demonstrated in the complexes of DNA reconstituted with individual histone fractions H2A, H2B, H3, H4. On the contrary, histone HI is unlikely to shield any DNA grooves (I6). It appears that histones are only partly buried in the DNA major groove by interacting with the sugar-phosphate backbone mainly from the side of the major groove, and do not essentially penetrate into the minor groove.

Unwinding of DNA

Table I summarizes also the inwinding of DNA with different proteins as measured by methylation of the N-I atom of adenine in DNA with DMS. Lac repressor in non-specific complexes with DNA as well as histones do not seem to unwind DNA. It has been reported that the repressor does not significantly unwind the lactose operator (7) and binds well to poly d(A-U-HgX) with cross-linked chains (9).

RNA polymerase. It has been shown by measuring the degree of superhelicity of covalently closed DNA that the RNA polymerase holoenzyme unwinds upon its binding about 7 base pairs of DNA; an uncertainty in this assessment was, however, as high as a factor 2 (I8). In our experiments, specific and non-specific binding of RNA polymerase to DNA significantly increases the formation of I-methyl adenine upon methylation of their complexes. This increase corresponds to unwinding of about 35% of DNA or about I5-I7 base pairs in the DNA segment about 50 base pairs long which is covered with RNA polymerase. The presence of the nascent RNA chain in the RNA polymerase-DNA complex reduces the size of the single-stranded DNA regions by about IO nucleotides. This is in a good agreement with an early made suggestion that a nascent RNA chain forms a heteroduplex with one of the unwound DNA strands of about IO base pairs in length (I9).

Arrangement of Histones on DNA

The following arrangement of histones along one strand of DNA in the nucleosome core particles has been determined by the sequencing procedure described above (6): H2B(20-30,30-40)-H4(40-50,50-60,60-70)-H2A & H3(70-80)-H3 & H4(80-90)-H3(90-I00)-H2B(I00-II0,II0--I20)-H2A(I20-I30)-H3(I30-I45).

The figures in parenthesis indicate the measured distance in nucleotides (±3-5 nucleotides) of a DNA segment cross-linked to a histone from the 5'-end of a DNA strand. The model of the linearized structure of the nucleosome core particle based on these data is shown in Fig. 2.

In the model, different regions of histones molecules cover a little less than one turn of the DNA helix and acquire therefore the shape of incomplete helical clamps. All the gaps in the clamps face one side of the DNA helix. As a result of such orientation of the gaps, histones do not appear to form topological locks around DNA and would allow the core particle to dissociate easily into the histone octamer and DNA.

DISCUSSION

Histones bound to DNA are partly buried in the major groove and seem to leave the minor groove of DNA well exposed. On the other hand, RNA polymerase shields

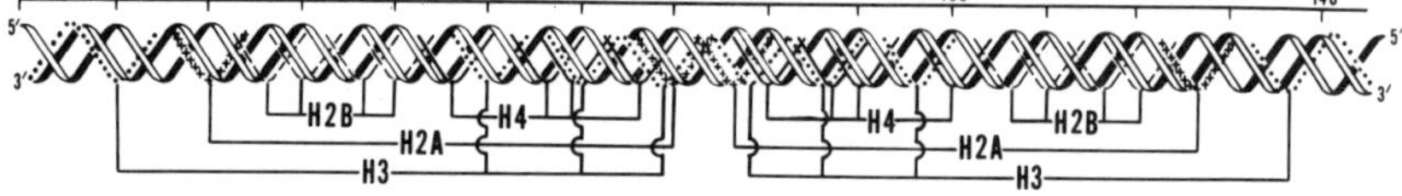

Fig. 2. The linearized model of the nucleosome core particle. Histones are arranged along the sugar-phosphate backbone from the side of the major groove of the DNA double helix. Histones appear to interact with different DNA segments with the N-, C-terminal and central regions of the molecules (6)

the minor groove, and lac repressor in non-specific complexes with DNA appears to be arranged in the minor groove too. The A-T specific antibiotics, netropsin and distamycin A, and G-C specific actinomycin D also shield only the minor groove against methylation (19). These data raise the possibility that the major groove of DNA is preferentially occupied with proteins organizing the chromosomal DNA, e.g. histones, but the minor groove is more essential for recognition of DNA or the primary binding of specific proteins to DNA. This makes it possible that DNA covered with histones might nevertheless be recognized by proteins along the minor groove.

In the structural model of the nucleosome core particle, histones do not form the topological locks around DNA. Therefore, specific proteins bound to the nucleosomal DNA first along the minor groove might then extract some DNA segments from the protein core of the nucleosome (folded or unfolded). These locally released DNA segments could be the sites of replication, transcription, DNA modification and unwinding, etc. Alternatively, one may envisage the simultaneous presence of histones and other proteins in these functioning sites of DNA.

REFERENCES

(I) A.D. Mirzabekov, A.M. Kolchinsky, Localization of some molecules within the grooves of DNA by modification of their complexes with dimethyl sulfate, Mol.Biol.Reports I, 379 (I974).

(2) A. Mirzabekov, D. San'ko, A. Kolchinsky, A.F. Melnikova, Protein arrangement in the DNA grooves in chromatin and nucleoprotamine in vitro and in vivo revealed by methylation, Eur. J. Biochem. 75, 379 (I977).

(3) A.F. Melnikova, R. Beabealashvilli, A.D. Mirzabekov, A study of unwinding of DNA and shielding of the DNA grooves by RNA polymerase by using methylation with dimethyl sulfate, Eur. J. Biochem. 84, 30I (I978).

(4) E.S. Levina, A.D. Mirzabekov, Covalent bonding of proteins to DNA in chromatin, Dokl. Akad. Nauk SSSR 22I, I222 (I975).

(5) A.D. Mirzabekov, V.V. Shick, A.V. Belyavsky, V.L. Karpov, S.G. Bavykin, The arrangement of histones on DNA in nucleosomes, Cold Spring Harbor Symp. Quant. Biol. 42, in press.

(6) A.D. Mirzabekov, V.V. Shick, A.V. Belyavsky, S.G. Bavykin, Primary organization of nucleosome core particles of chromatin. Sequence of histone arrangement along DNA, Proc. Natl. Acad. Sci. USA, in press (I978).
(7) S.-Y. Lin, A.D. Riggs, Lac repressor binding to non-operator DNA: detailed studies and a comparison of equilibrium and rate competition methods, J. Mol. Biol. 72, 67I (I972).
(8) T.J. Richmond, T.A. Steitz, Protein-DNA interaction investigated by binding E.coli lac repressor protein to poly d(A·U-HgX), J. Mol. Biol. I03, 25 (I976).
(9) R. Ogta, W. Gilbert, Contacts between the lac repressor and thymines in the lac operator, Proc. Natl. Acad. Sci. USA 74, 4973 (I977).
(I0) W. Gilbert, A. Maxam, A. Mirzabekov, Contacts between the lac repressor and DNA revealed by methylation in Control of Ribosome Synthesis, eds. Kjeldgaard, N.O. & Maaloe, O. (Munksgaard, Copenhagen), p. I39 (I976).
(II) S.J. Stahl, M.J. Chamberlin, Groups on the outside of the DNA helix affect promoter utilization by T7 RNA polymerase, in RNA Polymerase, eds. Losick, R. & Chamberlin, M., Cold Spring Harbor Laboratory, 429 (I976).
(I2) R.T. Simpson, Interaction of a reporter molecule with chromatin, Biochemistry, 9, 48I4 (I970).
(I3) A.F. Melnikova, A.S. Zasedatelev, A.M. Kolchinsky, G.V. Gursky, A.L. Zhuze, S.L. Grochovsky, A.D. Mirzabekov, Accessibility of the minor groove of DNA in chromatin to the binding of antibiotics netropsin and distamycin A, Mol. Biol. Reports 2, I35 (I975).
(I4) C. Zimmer, G. Luck, E. Sarfert, Approach to the analysis of decondensed DNA structural regions in chromatin using netropsin as a probe, Biologisches Zentralblatt 95, I57 (I976).
(I5) A.M. Kolchinsky, D.F. San'ko, V.V. Shick, A.D. Mirzabekov, A.A. Gineitis, The state of DNA grooves in mono- and oligonucleosomes and complexes of DNA with histone fractions, Molekularnaya biologia I2, 365 (I978).
(I6) J.C. Wang, M.D. Barkley, S. Bourgeois, Measurements of unwinding of lac operator by repressor, Nature 25I, 247 (I974).
(I7) J.C. Wang, The degree of unwinding of the DNA helix by ethidium, J. Mol. Biol. 89, 783 (I974).

(I8) L.P. Savochkina, R.Sh. Beabealashvilli, G.V. Gursky, S.D. Trachanov, A.S. Zasedatelev, Complexes of RNA polymerase with nucleic acids: Actinomycin D binding studies, Studia Biophysica 40, I35 (I973).

(I9) A.M. Kolchinsky, A.D. Mirzabekov, A.S. Zasedatelev, G.V. Gursky, S.L. Grochovsky, A.L. Zhuze, B.P. Gottikh, On the structure of the complexes of distamycin type antibiotics and actinomycin D with DNA: new data on the localization of these antibiotics within the DNA minor groove, Molekularnaya biologia 9, I9 (I975).

ORGANIZATION OF THE RIBOSOMAL GENES CLUSTER OF THE LOACH

M.Ya.Timofeeva, G.I.Eisner, N.S.Kupriyanova, K.G.Skryabin and A.A.Bayev

Institute of Molecular Biology of the USSR Academy of Sciences
Moscow 117312, Vavilov str.,32

ABSTRACT

Ribosomal DNA was prepared from the nucleoli of the loach oocytes. The nucleoli were isolated by the differential centrifugation of homogenates from small growth oocytes and from oocytes at the beginning of vitellogenesis. DNA samples were analysed by density gradient centrifugation and hybridization with labeled rRNA of loach. The samples were enriched 200-300-fold in ribosomal RNA genes. rDNA was digested with EcoRI, Bam H1 and Hind III nucleases and analysis of the restricts by gel electrophoresis was carried out. The Eco R1 and Bam H1 nucleases were shown to have only one restriction site in rDNA of loach amplified nucleoli. The Hind III nuclease has at least five restriction sites. The combination of Eco R1 and Bam H1 nucleases cleaves this rDNA into two fragments of 6.1 and 3.7×10^6 daltons. The data indicate that the ribosomal RNA gene of loach has a repeat unit of $9,8\times10^6$ daltons and is not heterogeneous in the length.

INTRODUCTION

The studies on the organization and the structure of eukaryotic genes, which allow conclusions on their functioning, require isolation of individual genes. This problem is much simpler for repeating genes which are organized in a cluster, for instance, ribosomal RNA genes. The specific GC-content permits to separate ribosomal DNA (rDNA) by using CsCl-density centrifugation (1). This approach was used for isolating rDNA of Amphibia (1,2), Mammals (3), Tetrahymena (4) and Sea Urchin (5). However the relative rDNA content in eukaryotic genome does not exceed 0,2%, thus complicating the isolation of rDNA in amounts required for structural analysis. A modification of this method was proposed by Joseph and Staford (5): they succeeded in isolating DNA preparations from Sea Urchin embryos enriched in rDNA 50-fold, but the molecular weight of these rDNA did not exceed $1{,}9 \times 10^7$ daltons.

We report here the experimental data on the isolation of rDNA from amplified nucleoli of loach oocytes and on the studies of the structural organization of those. The loach (teleost fish) was chosen because of the information available on the general characteristic of genome DNA organization (6). Using Eco R1, Bam H1 and Hind III nucleases, we determined the size of the repeating unit and demonstrated that the length of this one is homogeneous unlike rDNA of many other organisms.

MATERIALS AND METHODS

Nucleoli were isolated from small oocytes. Oocytes were prepared by the method of Ozernyuk (7) and homogenized in a 0,35 M sucrose solution with 0,5% Twin 80, pH 7,5. The homogenate was centrifuged at 4.000 rpm for 5 min in a T23 centrifuge. The pellet was suspended in a 0.25 M sucrose pH 7.5 and purified by centrifugation through a layer of 1.8 M sucrose.

DNA Isolation. Marmur method (8) was used with some modifications.

Density Gradient Centrifugation. Preparative CsCl density gradients were carried out in ultracentrifuge (Model Spinco L5-65, Beckman Instruments) using rotor Ti50 at 36.000 rpm, 20°C for 72 h.

DNA-RNA Hybridization. Aliquots of DNA after denaturation by 0.1 N NaCl treatment were bound to Millipore filters (2B-6, Schleicher, Schuhl) and the hybridization with ^{3}H-labeled rRNA was carried out in 4xSSC and 0.1% SDS, at 62°C, for 18 h (9,10). The filters were washed by 4xSSC and 0,1% SDS at 20°C and 65°C, treated with RNase (50 ug/ml) washed by 4xSSC, dried and radioactivity was measured with dioxan scintillator (Intertechnique).

^{3}H-RNA Preparation. Ribosomal RNA was isolated from loach muscle (11) and labeled by incubation with ^{3}H-dimethylsulfate at -10°C (12). Specific activity of preparations obtained was 30.000 cpm/ug RNA.

Restriction Enzyme Treatment and Agarose Gel Electrophoresis. The nuclease digestion was carried out according to the procedure of Polisky et al.(13). Aliquots of DNA (5 μg) were incubated in appropriate buffer with 25 ml enzyme solution (1 μl enzyme solution digests 1 μg λ fage DNA at 37°C after 1 h incubation). Buffers for EcoR1: 10 mM Tris (pH 7.5), 50 mM NaCl, 5 mM $MgCl_2$; for Bam H1: 10 mM Tris (pH 7,5), 10 mM $MgCl_2$, 6.6 mM βME; for Hind III: 6,6 mM Tris (pH 7,5), 6.6 mM $MgCl_2$, 6.6 mM βME, 60 mM NaCl. For electrophoresis 1% agarose gels were used with λ DNA Hind III fragments as markers for determination of molecular weights. After staining with ethidium bromide, the bands were visualized using 260 nm UV light source, and photographed on Micrad 300 film with orange filter (OS-4).

RESULTS

According to the cytological data (14,15), oocytes of teleost fish contain hundreds of nucleoli. One might expect therefore that even total DNA preparations from oocytes would be enriched several hundred times in ribosomal genes. The content of rDNA should be still higher in preparations obtained from isolated nucleoli. That is why our efforts were concentrated (a) on isolating the nucleoli from oocytes, and (b) on choosing conditions under which rDNA isolated from the nucleoli would possess sufficient size to allow the restriction analysis (at least one repeat unit per molecule).

rDNA was isolated from oocytes of small growth - early vitellogenesis, since by the end of vitellogenesis amplified nucleoli gradually stop functioning and are destroyed (14). The oocytes separated from the stroma of an ovary were divided according to their size, and only those oocytes whose diameter was less than 300 μm were used in experiments (7).

To purify the nucleoli from the yolk oocytes were homogenized in Tris (pH 7.5) and 0.5% Twin 80, then homogenate was passed through a layer of 1.8 M sucrose. The pellet of nucleoli were subjected to lysis in Tris (pH 9.5) containing 1% SDS. DNA was deproteinized using water saturated phenol at pH 7.5 and a mixture of chloroform, cresol and isoamyl alchohol, as was described for isolation of high molecular weight DNA from yolk platels of Xenopus oocytes (16). The DNA preparations thus obtained were analyzed in a CsCl density gradient (fig.1) and by hybridization with ^{3}H-rRNA (table 1).

Fig.1 shows two peaks in nucleolar rDNA a heavy (ρ 1,715) one comprising ~60% of preparation, and a second, lighter peak (ρ 1.705) - a "shoulder", which corresponds to 30-40% of the preparation. The sperm DNA is represented by the main peak that has a buoyant density of 1.699 and only an insignificant portion (0.2%) sediments with buoyant density of 1.728 (a heavy satellite). This heavy satellite was hybridized with ^{3}H-rRNA and hence corresponded to rDNA. The DNA preparations from the nucleoli contained far more of the heavy component and were therefore considerably enriched in rDNA.

rDNA from the amplified nucleoli of oocytes and from sperm of loach were compared in their ability to be hybridized with ^{3}H-rRNA. Table 1 shows

Table 1

The hybridization of sperm DNA and DNA of amplified nucleoli from loach oocytes with ^{3}H-rRNA

Origin of DNA	DNA on a filter (ug)	^{3}H-rRNA add ug	^{3}H-rRNA hybridization, cpm: in a sample	^{3}H-rRNA hybridization, cpm: per 1 ug DNA
Sperm	19.5	2.9	320	16
	19.5	5.8	493	25
	19.5	8.7	266	14
				average 18
Nucleoli	0.25	2.9	137	550
	0.25	2.9	161	640
	0.25	5.8	300	1200
	0.25	5.8	490	1960
	0.25	8.7	510	2010
	0.25	11.8	569	2270

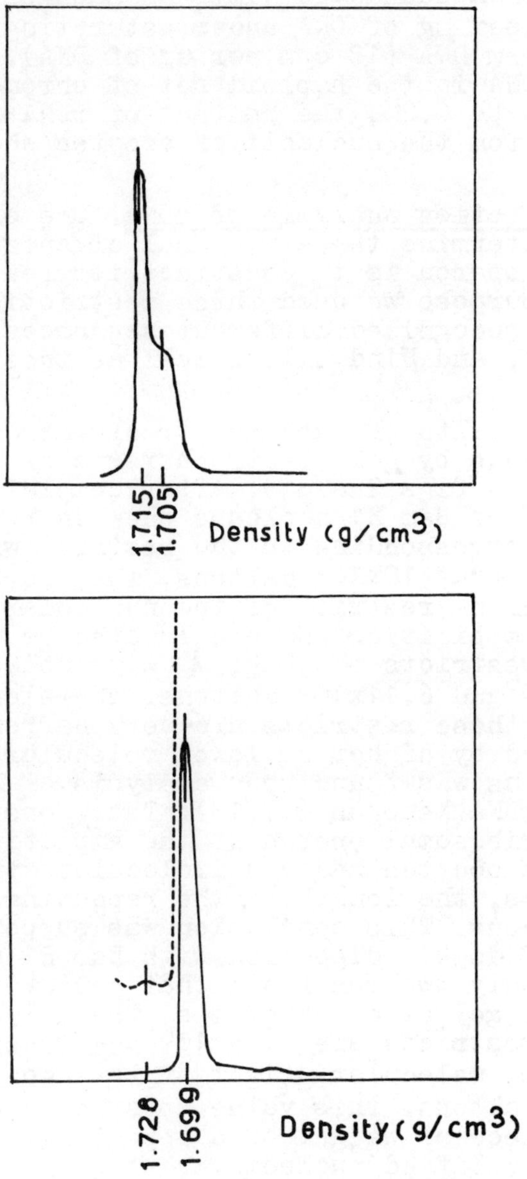

Fig. 1. Preparative centrifugation in CsCl density gradient of loach DNA. Above - DNA from amplified nucleoli of oocytes; Below - sperm DNA. —— OD_{260}, ···· OD_{260} x 10.

the results of such comparison. The preparations of nucleolar DNA bound 120 times as much of ^{3}H-rRNA (2270 cpm per μg of DNA under saturating conditions) as did sperm DNA (18 cpm per μg of DNA). If the content of rDNA in the haploid set of chromosomes is assumed to be 0.2%, the content of rDNA in the DNA prepared from the nucleoli of oocytes should be at least 25%.

Restriction analysis of rDNA. One of the simplest ways to determine the structural organization of the ribosomal operon is to construct its restriction map. For this purpose we used three restriction endonucleases which recognized different sequences in DNA: Eco R1, Bam H1, and Hind III as well as their combinations.

The results of nuclease treatment and analysis of the fragments by gel electrophoresis of nucleolar DNA are shown in figs 2 and 3. After complete digestion with Eco R1 or Bam H1 nuclease one can recognize only one band corresponding to the restrict with molecular weight of about 10x10^6 daltons. The Eco R1 restrict and the Bam H1 restrict of the nucleolar DNA have identical mobilities and are located between the Hind III restricts of phage λ with molecular weights of 14.6x10^6 and 6.11x10^6 daltons. The electrophoretic bands of these restricts are very narrow and are not accompanied by higher or lower molecular weight minor fractions as was found upon analysis of Eco R1 restricts of rDNA Xenopus (17,18). Thus, one can conclude that the ribosomal operon of the amplified nucleoli from loach oocytes had a unimolecular structure. In other words, the length of the repeating unit was not heterogeneous. This conclusion was supported by the results of double digestion with Eco R1 and Bam H1 yielding only two restricts that gave very narrow bands upon gel electrophoresis. The molecular weights of these restricts are 6.15x10^6 and 3.7x10^6 daltons. The overall molecular weight of the two fragments is 9,86x10^6 daltons, this value apparently corresponding to the molecular weight of a repeating unit of rDNA from the amplified nucleoli.

After complete digestion of nucleolar DNA with Hind III nuclease (fig.4) five fragments (3.55, 2.24, 1.47, 1.1 and 0.74x10^6 daltons) were revealed. The total weight of these restricts is 9.1x10^6 daltons, which permits to suggest that there are two fragments

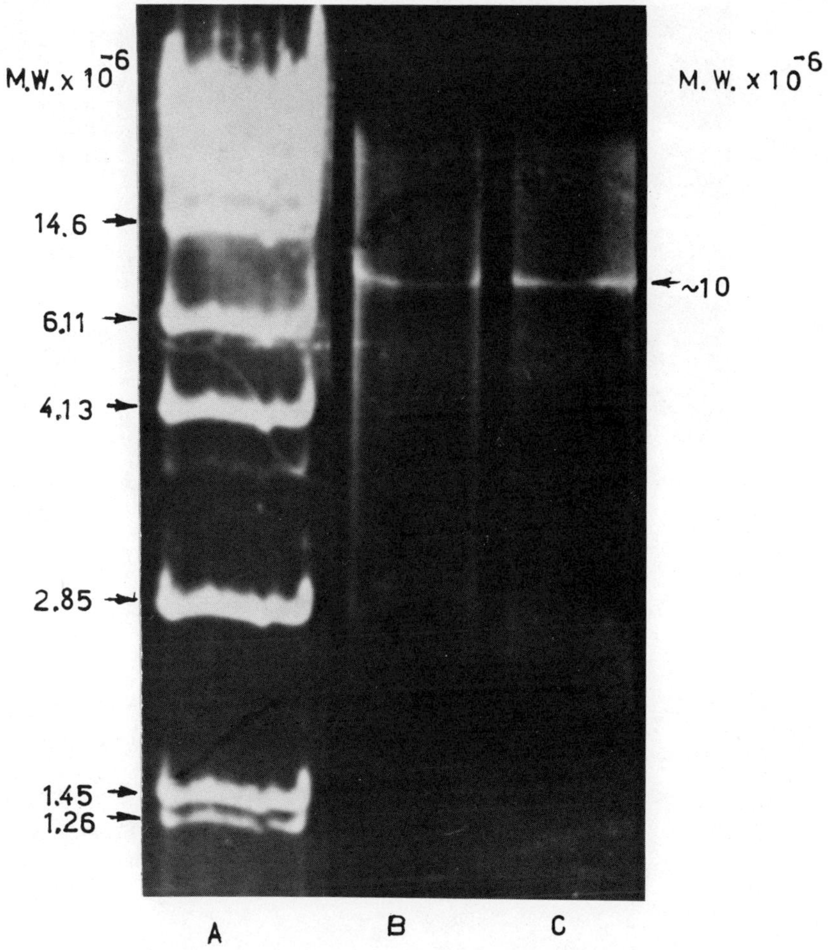

Fig.2. Eco R1 restriction pattern of amplified nucleoli DNA.
a - Hind III restricts of fage DNA.
b,c - Eco R1 restricts of two different DNA preparations from amplified nucleoli.

in the band of 0.74×10^6 daltons (possibly corresponding to subrepeats of a non-transcribed spacer). In this case the sum will be close to the values calculated from the sum of two Eco R1 and Bam H1 restricts.

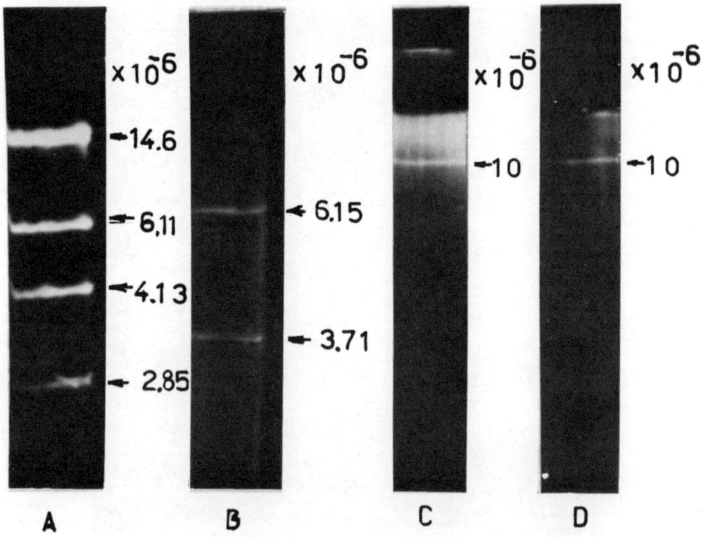

Fig. 3. Eco R1 and Bam H1 restriction pattern of amplified nucleoli DNA.

a - Hind III restricts of λ fage DNA.
b - Double digestion of nucleolar DNA by Eco R1 and Bam H1 nucleases

c - Eco R1 restrict, d-Bam H restrict of of nucleolar DNA.

The fragment which has a molecular weight of 1.47×10^6 dalton disappears after double digestion with Hind III and Eco R1 nucleases whereas the fragment with a molecular weight of 0.95×10^6 dalton can be detected.

DISCUSSION

DNA thus prepared from the isolated amplified nucleoli of loach oocytes is enriched in ribosomal

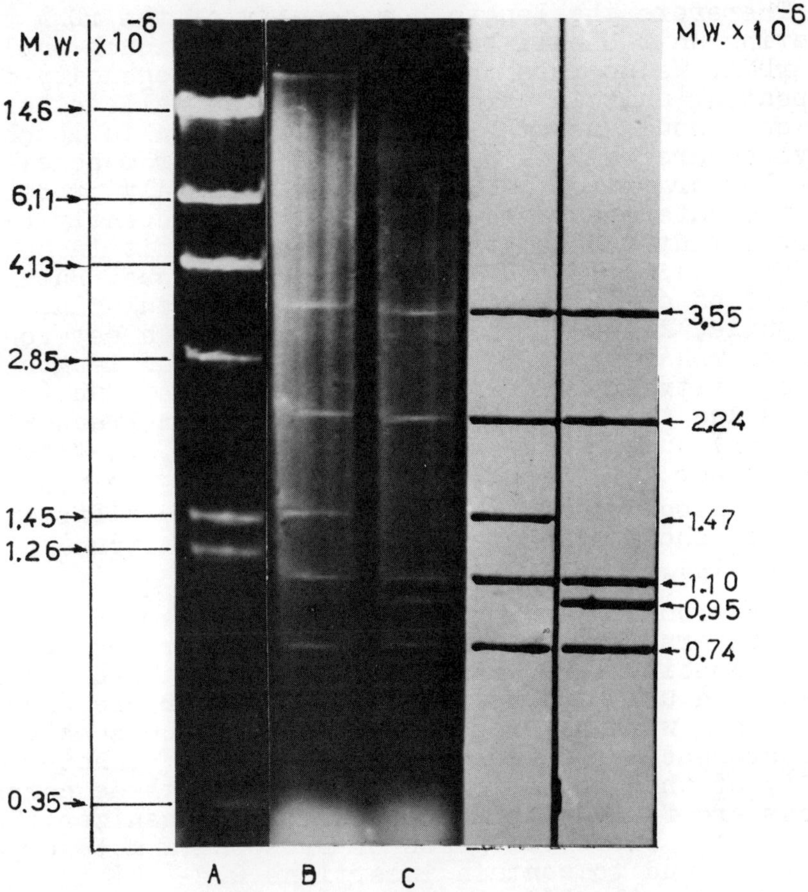

Fig.4. Hind III restriction pattern of rDNA from amplified nucleoli.

a - Hind III λ fage DNA restricts.
b - Hind III nucleolar DNA restricts.
c - Hind III and Eco R1 double digest of the nucleolar DNA.

genes. This DNA can be successfully used both for direct determination of the structural organization of ribosomal genes, and for multiplication by means of gene engineering. The preparations of nucleoli were isolated from the oocytes of different females, and from loaches from different geographical districts of the USSR.

Therefore the length homogeneity of the rDNA repeating unit should be typical for the loach amplified rDNA. We suppose the same length homogeneity for a repeating unit in chromosomal rDNA as well, otherwise one should assume that amplification in loach always occurs only in one locus of the chromosomal nucleolar organizer. This, however is hardly probable from the data available (18,19). It has recently been reported that the length of a repeating unit in yeast (20) and bovine (3) rDNA is also homogeneous. But for the most of organisms investigated (acetabularia, drosophila, Xenopus, mouse, man) the length heterogeneity of rDNA repeating unit was found. The length heterogeneity in Xenopus (18,19), mouse and man (21) rDNA is due to a different size of the non-transcribed spacer. Such a size variability of the non-transcribed spacer in Xenopus rDNA (from 1.8×10^6 to 5.5×10^6 daltons) results from the combination of a different number of short (50 base pairs or 3×10^4 daltons) subrepeats (19).

Electron microscopy revealed an extremely complicated pattern for rDNA organization in various Acetabularia species (22). Homogeneous lengths both of the transcribed unit and nontranscribed spacer are encountered along with heterogeneity of the spacers only, or heterogeneity of the transcribed unit, or heterogeneity of the two regions; and that all these combinations are found within one nucleolar organizer (22). Finally, the ribosomal genes of Drosophila melanogaster were found to contain insertions 0.5-6 kb in length within the 28S gene (23-26), so that the lengths of the repeating units were respectively greater.

These data indicate that different factors may account for heterogeneous length of rDNA repeating units in various organisms; mechanisms responsible for this heterogeneity are not yet clarified. On the other hand, elucidation of the mechanisms that maintain the absolute homogeneity in the length of the repeating unit of chromosomal and amplified rDNAs requires further studies.

REFERENCES

1. D.D.Brown and C.S.Weber. J.Mol.Biol., 34, 1, 661-697 (1968).
2. Birnstiel M., Speirs J., Purdom I., Jones K., Loening U.E. Nature 219, 454-463 (1968).

3. Blin N., Stephenson E.C., Stafford D.W. Chromosoma, 58, 41-50 (1976).
4. Gall J.G. Proc. Nat. Acad. Sci. USA, 71, 3078-3081, (1974).
5. Joseph D.R., Stafford D.W. Biochim. et Biophys. Acta, 418, N 2, 167-174, (1976).
6. Kuprijanova N.S., Timofeeva M.Ja. Europ. J. Biochem. 44, 59-65 (1974).
7. Ozernyuk N.D. Dokl. Akad. Nauk USSR, 207, 974-977 (1972).
8. Marmur J. J.Mol. Biol., 3,208-218 (1961).
9. Gillespie D., Spigelman S. J.Mol.Biol., 12, 829-842 (1965).
10. Barker R.B. Analyt. Biochem., 78, 569-571 (1977).
11. Castels J.J., Wool I.G. Methods in Mol. Biol., 2, 1-28 (1972).
12. Mirzabekov A.A., San'ko D.F., Kolchinsky A.M., Melnikova A.S. Eur. J. of Biochem., 75, 379-382 (1977).
13. Polisky B., Green P., Gartin D.E., McCarthy B.J., Goodman H.M., Boyer H.W. Proc. Natl. Acad. Sci. USA, 72, 3310-3314 (1975).
14. Sakun O.F. Dokl. Akad. Nauk USSR, 137, 749-751 (1961).
15. Chmilevsky D.A. Citologia (Rus.), 13, 1233-1256 (1971).
16. Hanocq F., Kirsch-Volders M., Hanocq-Quertier J., Baltus E., Steinert G. Proc. Nat. Acad. Sci. USA, 59, 1322-1326 (1972).
17. Dawid I.B., Brown D.D., Reeder R.H. J. Mol. Biol., 51, 341-360 (1970).
18. Wellauer P.K., Reeder R.H., Carroll D., Brown D.D., Deutch A., Higashinakagawa T., Dawid I.B. Proc. Nat. Acad. Sci. USA, 71, 2823-2827 (1974).
19. Wellauer P.K., Reeder R.H., Dawid I.B., Brown D.D. J.Mol. Biol., 105, 487-505 (1876).
20. Petes T.D., Hereford L.M., Skryabin K.G. J.Bacteriol., 134, 295-305 (1978).
21. Arnheim N., Southern E.M. Cell, 11, 363-370 (1977).
22. Spring H., Krohne G., Franke W.W., Scheer U., Trendelenburg M.F. J.Microscopic Biol.Cell, 25, 107-116 (1976).
23. Pelligrini M., Manning J., Davidson N. Cell, 10, 213-224 (1977).
24. Wellauer P.K., Dawid I.B. Cell, 10, 193-212 (1977).
25. Glover D.M., Hogness D.S. Cell, 10, 167-176 (1977).
26. White R.L., Hogness D.S. Cell, 10, 177-192 (1977).

A NOVEL TYPE OF GENE ORGANIZATION IN EUKARYOTIC CHROMOSOMES

G.P. Georgiev, Y.V. Ilyin, N.A. Tchurikov,
V.A. Gvozdev and E.V. Ananiev

Institute of Molecular Biology and Institute
of Molecular Genetics, USSR Academy of Sciences
Moscow, USSR

Only two types of organization of the structural genes in the eukaryotes have been known until recently. Most of the genes are represented by unique sequences occurring one per genome. The genes for globins, ovalbumin, fibroin, albumin, crystallins and many other can be mentioned as examples. It is suggested that unique genes are surrounded by DNA sequences which do not encode any protein but may participate in the regulation of expression of the particular gene. It was found with D.melanogaster that one band in a polytene chromosome containing about 20-30 kb per single DNA strand on the average corresponds approximately to one gene or, more precisely, to one complementation group. This indicates the existence of excess DNA in a genome which may be involved in regulation and other service functions. The location of unique genes along the chromosome is strictly fixed.

Besides unique genes, multiple structural genes have been also found. For example, histone genes which are represented in the genome by a number of grouped blocks each containing five structural genes (one for every type of histone) separated by spacers. Again all these blocks are concentrated in one fixed position of the genome. Some other multiple genes were detected but in general they occur more rarely than unique genes.

In the present report we describe the third type of organization of the genetic material in eukaryotes which was discovered while investigating the properties of three *Drosophila* genes isolated from the culture by cloning the recombinant phage λgt (1,2). These recombinant phages contained fragments of *D.melanogaster* DNA prepared by restriction with EcoRI endonuclease. The clones for analysis were selected on the bases of efficient hybridization with messenger RNA from the culture cells of *D.melanogaster*. DNA fragments representing the actively expressed structural genes were recovered by this method. Three clones designated as Dm 225, Dm 234 and Dm 118 were selected. They bound 0.8, 0.2-0.3 and 1.1% of polysomal mRNA, respectively. As the average content of individual mRNA is lower than 0.1%, one may conclude that transcription of the selected genes is much above the average level (3).

Thus, all three genes are actively transcribed. Each of them combines to mRNA of a definite size upon hybridization. Dm 225 DNA hybridizes to 20S mRNA, Dm 234 to 30-35S mRNA, and Dm 118 DNA to 18OS mRNA. These S values correspond to RNAs about 2.3 kb, 4-5 kb and 2 kb, respectively. The cloned DNA fragments comprise 2.9 kb (Dm 225), 1.7 (Dm 234), and 4 kb (Dm 118). Therefore, only Dm 225 and Dm 118 may contain the whole structural gene or a significant part of it while the Dm 234 fragment corresponds to less than a half of the gene.

For gene 225, the data indicating its active translation were also obtained. mRNA was prepared from polysomes of different size and hybridized to Dm 225 DNA (4). The peak of RNA binding was located in polysomes containing about 20-25 ribosomes per mRNA that correlated well with the size of mRNA. In other words, ribosomes are attached along the whole length of mRNA, and it seems very probable that mRNA is in the process of translation. For two other mRNAs such analysis has not yet been done but it seems quite possible that they are also translated because the corresponding mRNAs were prepared from polysomes.

In the next series of experiments, DNA from clones was hybridized to the total DNA of *D.melanogaster* labeled by nick translation (3). Under the saturation conditions, Dm 225 DNA binds about 0.5% of total *Drosophila* DNA. It is a very high figure. If Dm 225 sequence is unique, it should bind only 0.002% of DNA and such binding was really observed when DNA fragments containing unique sequences were studied. Therefore, one may conclude that Dm 225 DNA is represented in the genome by 250 copies. In the same way, we detected the number of copies for gene 234 to be equal to 300, and for gene

118 to about 50. Thus, all the three genes under investigation are multiple genes.

It is important that the structural gene itself is repeated. It was first proved in experiments where the DNA-DNA hybridization was performed in the presence of unlabeled mRNA taken as a competitor. In the case of Dm 225 DNA, the addition of excess mRNA decreased the total DNA binding by about 75%. Thus, at least 75% of the repetitive sequence presented in Dm 225 DNA can be accounted to a structural gene. Even higher competition was obtained in the case of Dm 234 DNA which probably corresponds to a structural gene as a whole. With Dm 118 DNA, the situation is more complex. It consists of two EcoRI subfragments. Both are repeated about 50 times per genome, but only one of them corresponds to a structural gene (3).

A more detailed analysis of the properties of Dm 225 DNA was performed thereafter. The DNA excised with the aid of EcoRI endonuclease was further cleaved to five fragments by Hae III endonuclease. These subfragments were separated electrophoretically and their arrangement in Dm 225 DNA was detected. It is as follows: B(0.76 kb) D(0.32 kb) E(0.11 kb) A(1.1 kb) C(0.5 kb). In hybridization experiments, mRNA combines with all subfragments, Its 3'-end is mapped in the fragment B. Thus, almost all Dm 225 DNA, except a part of the fragment B, corresponds to a structural gene coding mRNA, and probably the whole structural gene is included in the Dm 225 DNA. This result again confirms the repetitiveness of the whole structural gene 225 (4).

The next question solved with the gene Dm 225 was whether all copies of this gene in the genome were identical or not. To answer this question, the total DNA of <u>D.melanogaster</u> was restricted by EcoRI endonuclease, the same enzyme which had been used for original excising Dm 225 DNA. Thereafter, the DNA was fractionated according to the size in polyacrylamide gel, transferred onto a nitrocellulose filter, and hybridized with /^{32}P/ labeled Dm 225 DNA. The distribution of the bound label along the filter was analysed by autoradiography. The distribution of total DNA EcoRI restricts was very heterogeneous along the filter. However, the labeled material was found only in one band. This band contained DNA of 2.9 kb in length. Therefore, all copies of DNA in the genome containing Dm 225 sequences in EcoRI restricts had exactly the same size. If the same experiment was performed using two endonucleases, EcoRI and Hae III, hybridization took place only with fragments equal in size to those obtained from Dm 225 DNA by EcoRI + Hae III treatment. In other words, all 250 copies of Dm 225 gene in the genome of *D.melanogaster* are identical. The same result was obtained in the work with two different lines of cultivated cells (4).

The next question was whether sequences surrounding all copies of the Dm 225 gene in the genome were different. To answer this question, the total DNA of *D.melanogaster* was restricted with Hind III enzyme which did not cut the Dm 225 fragments. Thus, this enzyme excised the investigated gene with its flanking sequences. After blotting to nitrocellulose filter, the hybridization with Dm 225 /^{32}P/ DNA was performed. The label now hybridized with the heterogeneous population of sequences. One may conclude that different copies of Dm 225 gene in the genome have different flanking sequences. Moreover, the distribution of the label in DNA from two different cell sublines of *D.melanogaster* is different. Thus, the sequences surrounding Dm 225 DNA in the two sublines are not the same. It would be relevant to note that these two sublines were originally recovered from one parental line; however, these lines were grown separately for a several hundred generations, one of the lines being adapted to grow in a serum-free medium. Consequently, one can suggest that gene 225 is localized in a number of different sites of the genome and that these sites may vary in different cells (4).

This suggestion has been proved in experiments of the *in situ* hybridization of labeled DNA or labeled cRNA transcribed from cloned DNA to polytene chromosomes of *Drosophila* salivary glands (2,3,5). It was found that the DNAs of each of the three studied genes hybridized to a large number of sites in the chromosomes. All cells obtained from the same animal has the same distribution of hybridization sites. For example, Dm 225 DNA hybridized to 40 sites on polytene chromosomes obtained from heterozygous (gtw^{a} x gt^{13z}) animals. Such heterozygous animals were used in order to obtain a high extent of polyteny and therefore thick chromosomes. In such animals, one can discern sometimes the regions of asynapsis, the regions where the homologous chromosomes are not paired. In these regions, one can follow the distribution of hybridization sites between parental chromosomes. We found that this distribution was quite different. Therefore, the detailed investigation of hybridization with chromosomes from animals of different *Drosophila* stocks was performed. First of all, the gt w^a and gt^{13z} animals were compared. Animals of any of these stocks had 20-25 hybridization sites on their chromosomes but only six of them were common. The localization of all others was different. When individuals from the same stock were compared, these differences were less prominent but still 20-30% of hybridization sites were different.

The genes Dm 225 and Dm 234 have very similar distribution in the genome. The score of chromosomes from 15 animals gave about 70 sites where these two genes may occur. The localization of gene 118 is even less stable. About 150 sites have been found where gene 118 can be located. A simple calculation shows that each site for gene Dm 225 or 234 should contain about 10 copies of these genes on the average while gene 118 should be represented by only a few copies per site. It may explain why the latter is translocated more easily.

A study of the sites where the described genes, in particular Dm 225 and 234, are located demonstrated their non-random distribution in chromosomes, a remarkable coincidence of these sites with the location of the so-called intercalary heterochromatin in D.melanogaster. The intercalary heterochromatin is characterized by the following properties: (1) the occurrence of non-homologous pairing of these regions in polytene chromosomes; (2) late replication; (3) occurrence of underreplication resulting in the so-called "weak spots". The total number of such sites in chromosomes is approximately one hundred. Their functions remain obscure.

We found that about 90% of hybridization sites for genes Dm 225 and 234 coincide with the sites of intercalary heterochromatin known from literature. On the other hand, about half of the latter are covered by hybridization sites for the two above mentioned genes. One may conclude that multiple structural genes, scattered throughout chromosomes, are localized in the regions of intercalary heterochromatin; apparently, these genes are easily transposed from one site of intercalary heterochromatin to another.

The genes with such properties are probably widely spread in Nature. Two families of multiple genes effectively transcribed and scattered throughout the genome were described in Hogness laboratory (6). Recently, we have found a new structural gene coding for about 2% of total mRNA in a close neighbourhood to Dm 225 gene (Tchurikov, Ilyin and Georgiev, in preparation).

It is interesting that grouped multiple genes with stable localization, such as histone or rRNA genes, are also localized in intercalary heterochromatin. Moreover, recent observations have indicated the possibility of their translocation although it is much less prominent than in the case of the three aforementioned genes (7).

A general scheme can be drawn according to which eukaryotic chromosomes contain a number of regions designated as intercalary heterochromatin where different multiple genes are concentrated. Although the nature of these genes remains obscure, one may suggest that they are involved in the synthesis of abundant proteins used for the own needs of the cells. Therefore, the corresponding genes should be actively expressed. These genes may be easily translocated but only among the mentioned regions (nests for multiple genes) which are organized in such a way as to make such translocations possible. It is not excluded that the negative control of gene expression in these sites is less prominent than in other sites of the genome.

Differences in the number and distribution of multiple genes in the genome may account for many individual differences between animals. Further studies of the new type of organization of the genetic material discovered in this work may open several lines of research into the genome of higher organisms.

REFERENCES

(1) Y.V. Ilyin, N.A. Tchurikov, G.P. Georgiev, Selection and some properties of recombinant clones lambda bacteriophage containing genes of *Drosophila* melanogaster, *Nucleic Acids Res*. 3, 2115-2127 (1976).

(2) G.P. Georgiev, Yu.V. Ilyin, A.P. Ryskov, N.A. Tchurikov, G.N. Yenikolopov, V.A. Gvozdev, E.V. Ananiev, Isolation of eukaryotic DNA fragments containing structural genes and the adjacent sequences, *Science* 195, 394-397 (1977).

(3) Y.V. Ilyin, E.V. Ananiev, A.P. Ryskov, G.N. Yenikolopov, S.A. Limborska, N.E. Maleeva, V.A. Gvozdev, G.P. Georgiev, Studies on the DNA fragments of mammals and *Drosophila* containing structural genes and adjacent sequences, *Cold Spring Harbor Symp. Quant. Biol*. 42 (1977) in press.

(4) N.A. Tchurikov, Yu.V. Ilyin, E.V. Ananiev, L.G. Polukarova and G.P. Georgiev, The properties of gene Dm 225, a representative of dispersed repetitive genes in Drosophila melanogaster, Nucleic Acids Res. (1978) in press.

(5) E.V. Ananiev, V.A. Gvozdev, Yu.V.Ilyin, N.A. Tchurikov and G.P.Georgiev, Reiterated genes with variable location in intercalary heterochromatin of D.melanogaster chromosomes, Chromosoma (1978) in press.

(6) D.J. Finnegan, G.M. Rubin, M. Young and D.S. Hogness, Multigene families in the genome of Drosophila melanogaster, Cold Spring Harbor Symp. Quant. Biol. 42 (1977) in press.

(7) R.V. Old, H.G. Kallan, K.W. Gros, Localization of histone gene transcripts in newt lampbrush chromosomes by in situ hybridization, J. Cell Sci. 27, 57-79 (1977).

DIFFERENTIAL GENE EXPRESSION DURING THE CELL LIFE CYCLE

T. Eremenko, T. Menna, and P. Volpe

International Institute of Genetics and Biophysics
CNR, Naples, Italy

Abstract - Karyokinesis appears to be a "dead" point in macromolecular biosynthesis, while interkinesis fully provides the environment for dynamic relations in the working of the DNA⇌RNA→Protein machinery (lytic virus genome and coat protein are also synthesized during interkinesis, and not during mitosis, and virus oncogenes also appear to be integrated and expressed mainly during the S-phase). The model of the cell cycle which emerges is based on the finding of an asymmetric system of temporal "orbits" showing that DNA, RNA and proteins are synthesized, processed and structurally modified at given optimal stages of the cellular cycle. Such a generalization may become a useful analytical basis in genetic engineering of eukaryotes and in studies on cell differentiation or transformation.

I. SPACE AND TIME

The aim of this work is that of drawing attention to the problem of space and time in macromolecular reactions occurring in the functioning cell. The flow of genetic information, from nucleus to cytoplasm, has to do with the spatial organization of given cell compartments, while the life of the cell actually implies a temporal order in the development of the different steps of gene expression.

Knowledge of the space coordinate for the flow of gene information, from cell nDNA via messenger RNA (1) and back to cell nDNA via inverse transcriptase (2), is what molecular biology has successfully defined from 1950 to 1970 (3). However, the way opened in 1961 by the studies on allostery and genetic regulation (4), which appeared to promise some key for solving the mystery of cell differentiation (5), might risk leading to a sort of "staticity" in the absence of a further consideration of the time order in the induction/repression (or feed-back inhibition) mechanisms. Cell cycle studies (6) thus emerged as a necessary strategy in developmental biology (7), as was widely recognized during recent years (8-10). In this spirit, we show here, on the one hand, the cell cycle changes in the pool of precursor molecules and, on the other, we describe how in the living cell - as a function of time - the events which involve biosynthesis, repair, structural modification and processing of macromolecules are subordinated one to another (Fig. 1). These events appear to be genetically encoded and periodic in nature. From one cell division to the next, they follow - we can say - "orbital" pathways in a temporal system which is asymmetric (Fig.3). Such "asymmetry" is interesting not only because of its theoretical analysis which may help remove cell biology from a stage of simple quantitative description to a stage of knowledge of the relativity of interdependent processes, but also because in practice it induces one to suppose that the interaction between the world of a cell and the world of a virus (lytic or oncogenic) should take place at favourable stages of the cell life (11, 12).

II. THE EXPERIMENTAL SYSTEM

In the present research we employed the human cancer *HeLa* cells mainly (13), since in cell biology (14-16) and in studies on cell-virus interactions (17,18), particularly in those involving poliovirus (19), they are analogous to Drosophila in genetics or sea urchins in developmental biology. *HeLa* cells have a rather long cell cycle, lasting 20 hours, with all the four phases, G_1, S, G_2 and M (20,21). They can be grown and synchronized in suspension (22) to yield the large amount of material needed for experimentation with purified molecules (23) and macromolecules (24-30). The method used here for measuring the length of each phase of the cell cycle directly during the course of the experiments in suspension (21,31) is employed elsewhere both in investigations of cell functions (32,33) and virus infection (34). In *HeLa* cells synchro-

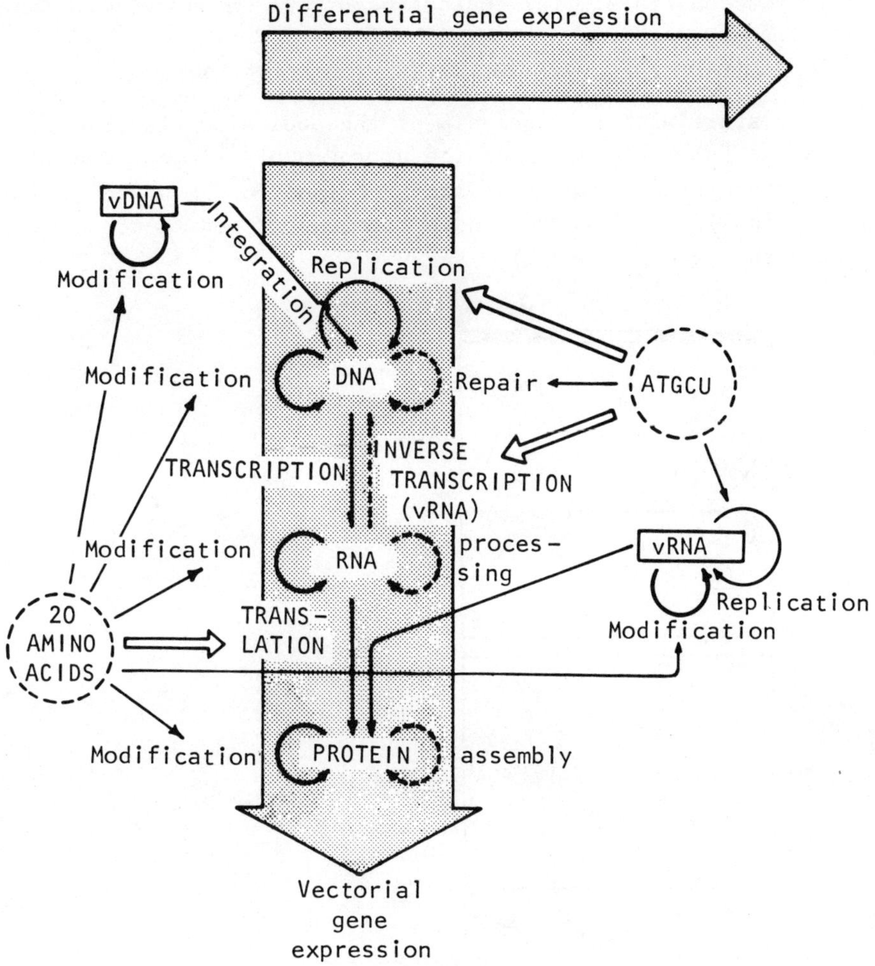

Fig. 1 - SPACE-TIME COORDINATES OF CELL AND VIRUS GENE EXPRESSION. The vertical arrow shows the space polarity of the flow of gene information, from nucleus to cytoplasm (laterally, it is shown how a DNA oncovirus or an RNA lytic virus can interfere in biosynthesis of the cell macromolecules and how the pools of nitrogen bases and amino acids can support the different steps of cell and virus gene expression). The horizontal arrow shows the time polarity of the flow of gene information, from one cell division to the next, i.e. the differential chronology with which the steps of cell and virus gene expression can occur during the mitotic cycle.

nized with thymidine the nDNA replication (which is the basis for gene expression) is similar to that occurring *in vivo* (30).

III. POOLS OF PRECURSOR MOLECULES

Let us start with the analysis of the pool size fluctuations through the *HeLa* cell cycle of the nitrogen bases and amino acids as precursors of nucleic acids and proteins. The amino acid methionine, obviously, is involved in macromolecular modification as donor of methyl groups (Fig.1).

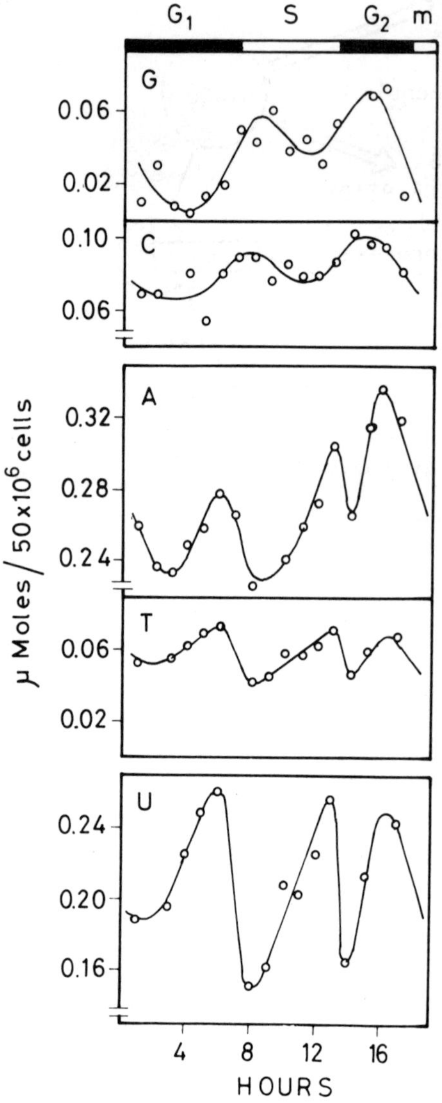

Fig. 2 - CHANGES IN THE INTRACELLULAR POOL CONCENTRATION OF NITROGEN BASES DURING THE HELA CELL CYCLE. G, guanine; C, cytosine; A, adenine; T, thymine; U, uracil.

a - *Nitrogen bases*. Figure 2 shows that at least in one point of the cell cycle (mid-G_2) the five bases are accumulated in the cell in parallel. However, as a whole, the intracellular concentration of guanine and cytosine is shifted in time with respect to the intracellular concentration of adenine, thymine and uracil. Guanine and cytosine show their first oscillation in early S, while adenine, thymine and uracil present a first oscillation in late-G_1 and an additional oscillation in late-S. These patterns are not the mirror picture of the patterns of cell (Fig.3) and virus (Fig.4) nucleic acid biosynthesis illustrated below.

b - *Amino acids*. An analogous situation during the cell cycle was observed with the intracellular amino acid pool fluctuations (23,26,35). Most of the amino acids are accumulated in early-S, at the boundary S/G_2, and during mitosis (23). We shall see later that the bulk of cell translation occurs in G_1 and G_2 (Fig. 3). Hence, decrease of the intracellular amino acid concentration in these two phases is rather reasonable, but decrease of the intracellular amino acid concentration during the phase of nDNA synthesis is unclear (23).

Therefore, the variations of biosynthesis (Fig. 3) and modification (Figs. 6 and 7) of macromolecules during the cell cycle, as shown below, should be attributed only partly to the pool changes of the nitrogen bases and amino acids.

IV. BIOSYNTHESIS OF MACROMOLECULES

We shall analyse the biosynthesis of cell and virus macromolecules, separately.

a - *Cell macromolecules*. nDNA-replication, during the S-phase, and nDNA repair, during the remaining part of the cell cycle (12, 31), were detected at hourly intervals on CsCl-purified polymers (24) following their incorporation both of ^{3}H-thymidine (24) and ^{14}C from the methyl group of methionine which through the C_1-chain enters the purine ring of adenine and guanine and the methyl group of thymine (37,38). Replication of nDNA, during S, and replication of mDNA, during S and G_2 (24,49), were compared instead analyzing their different kinetics of labelling with ^{3}H-thymidine (24). At 1-hr intervals throughout the whole cell cycle, transfer and ribosomal RNAs were fractionated on sucrose gradients (25), while the bulk of pre-mRNA with attached poly-A was purified on oligo-dT cellulose (38,39). During interkinesis and karyokinesis, messenger

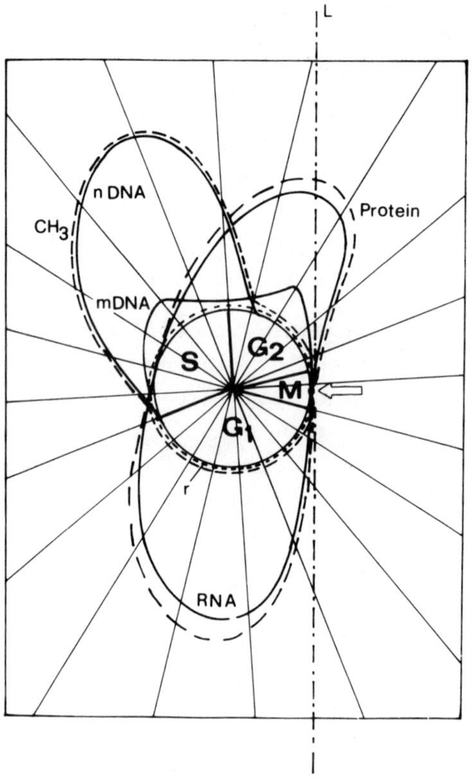

Fig. 3 - MODEL FOR "ORBITAL" GENE EXPRESSION FLOW AROUND THE HELA CELLULAR CYCLE (G_1, S, G_2 and M). At 1-hr intervals during the whole interkinesis and karyokinesis, the cell pulse-labels lasted: 20 min for nDNA and mDNA (with ^{14}C-thymidine); 50 min for RNA (with ^3H-uridine); 3 min for proteins (with ^{14}C -leucine). Macromolecules were purified: nDNA and mDNA, in CsCl; pre-mRNA, on oligo-dT cellulose; nascent peptides, from polysomes. The time, in hr, is marked by radii on the internal circular coordinate functioning as abscissa. The ordinate is represented by any radius whose scale shows the rates of macromolecular biosynthesis. The arrow shows the asymmetric "dead" point in gene expression (M-phase). L, line of asymmetry. "Orbits" from the central circular coordinate: nDNA repair synthesis (r---); nDNA duplicative synthesis (—) and methylation (- - -) in S; mDNA synthesis in S and G_2 (—); transcription (—) and translation (— —) in G_1 and G_2 (Ref.12).

RNA was isolated from previously purified polysomes (26,38). In parallel, the ribosome patterns were used for calculation of the amount of ribosomal subunits, of ribosomes, and of polyribosomes per cell (26). The translational potentiality of the polysome machinery during the cell cycle was judged on the basis of the rate of peptide chain elongation (26), while the incorporation into protein of each of the 20 amino acids was measured (36).

The results of this investigation can be summarized in a model (Fig.3) leading to the following general conclusions (12).

1. Genetic information flows from nDNA to nDNA during the S-phase of the cell cycle (24), in accordance with the previous knowledge (6); however, in *HeLa* cells, most of this information flows from nDNA to RNA in G_1 and G_2 (25-39).

2. Proteins are mainly synthesized in G_1 and G_2 (26,36,40,42) with the exception of histones (41) and some enzymes (35) which are produced in synchrony with nDNA synthesis.

Thus, in *HeLa* cells synchronized with thymidine (30), the bulk of translation follows the bulk of transcription, during the phases G_1 and G_2, always with an 1-hr delay (39), but transcription and translation do not follow or "overlap" nuclear gene duplication which is asymmetrically shifted in time, during the phase S (12).

3. The M-phase is characterized by a sensible block of macromolecular biosynthesis (24-26), confirming the previous information (6,43,44). For this reason, mitosis can be considered as a point of relative "asymmetry" in the cell cycle (12).

The harmony of the temporal "orbital" system of gene expression during the mitotic cycle (Fig.3) can be destroyed by infection with a lytic virus, such as poliovirus (18,19).

b - *Lytic virus macromolecules.* RNA genome of poliovirus can be replicated in the cytoplasm of *HeLa* cells (52). We have shown that minus strand poliovirus RNA (which requires 2 hrs to be formed after infection) is synthesized continuously throughout interkinesis and karyokinesis (53,54). However, plus strand poliovirus RNA (which is the genome of the progeny virus formed 2 to 4 hrs after the beginning of infection) is replicated during the S-phase only (Fig.4). The S-phase dependence both of RNA and DNA lytic virus genomes is now a generally accepted phenomenon (34,55-59). Poliovirus coat proteins appear to be synthesized also during the *HeLa* S-phase (11). This cell cycle stage, as shown in figure 3, is characterized by a relatively low rate of cell translation (26,36,40). Hence, the S-phase free ribosomes (26) might be available for translation of virus mRNA to a higher extent (12). In conclusion:

1. In the *HeLa* cell, which is not infected, the nuclear gene duplication is shifted in time from transcription and translation (Fig.3).

2. On the other hand, in the *HeLa* cell, which is infected with poliovirus, the S-phase provides favourable conditions for synthesis of both virus mRNA$^+$ (Fig.4) and virus proteins (11,53).

The reason for partial inhibition or absolute complete shut-off of the host cell as well as viral gene expression in given cell cycle stages, particularly in mitosis, is not yet elucidated (12, 60). However, the information collected suggests that regulation is of a different nature in the two cases. For instance, the rea-

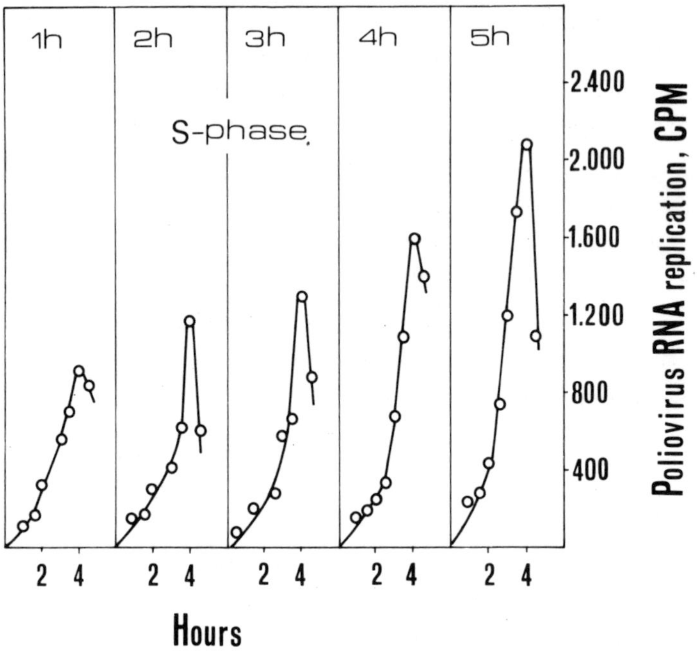

Fig. 4 - BIOSYNTHESIS OF POLIOVIRUS mRNA$^+$ DURING THE HELA S-PHASE. Cell samples were infected at 1-hr intervals with 80 p.f.u./cell. The labelling was followed for about 5 hrs with ^{3}H-uridine.

son for the block of *HeLa* cell protein biosynthesis in mitosis (Fig.3) could be attributed to a specific ribosome translational state (26) or to a possible presence of inhibitors of translation (12,26) in this cell cycle stage, since during the M-phase there are both free mRNAs and "frozen" polysomes (26,39) which probably store mRNAs "jumping" from one cell cycle to the next (61). As regards the poliovirus protein biosynthesis, in the M-phase (as well as in G_1 and G_2) the observed block (11,53) does not seem to depend upon the presence of inhibitors of virus translation or upon the inability of ribosomes to produce virus proteins (44), since during the M-phase (as well as in G_1 and G_2) the poliovirus $mRNA^+$, which is needed to yield virus proteins, is probably not present in abundance (53). Our laboratory has studied, on one hand, the distribution of the plus strand poliovirus RNA and, on the other, the poliovirus RNA^- and RNA^+ polymerase activities during the *HeLa* cell cycle, particularly during the S-phase and mitosis (12). The bulk of plus strand virus mRNA can be separated from the bulk of minus strand virus RNA on oligo-dT cellulose (39), since poliovirus $mRNA^+$ contains a "post-transcriptional" poly-A at the 3' end. The translating virus $mRNA^+$s are isolated instead from previously purified giant poliovirus polysomes (62). The purification of the two polymerases, of "poliovirus RNA^+ dependent-poliovirus RNA^- polymerase", which probably functions during the whole cell cycle, and of "poliovirus RNA^- dipendent-poliovirus RNA^+ polymerase", which probably functions during the S-phase only (12,54), is performed with minor modification of the methods known (63).

V. PROCESSING OF MACROMOLECULES

We have seen how cell and virus macromolecules are synthesized during the interphase (Figs.3 and 4). Let us look now at an inverse phenomenon, for instance, at the cell cycle dependence of the 45S rRNA processing (12,25). Maturation of this polymer to yield 18S and 28S rRNAs (Fig.5) is manifested in the following general circumstances of cell transcription.

The total RNA content per cell develops during the *Hela* mitotic cycle following an essentially S-shaped pattern (12,25,64), in accord with the known stability and accumulation of RNA with the cell aging (25). The partial concentration of the different RNA species in the bulk RNA is instead more or less constant during interkinesis (25). From these results comes the first indication that, in-

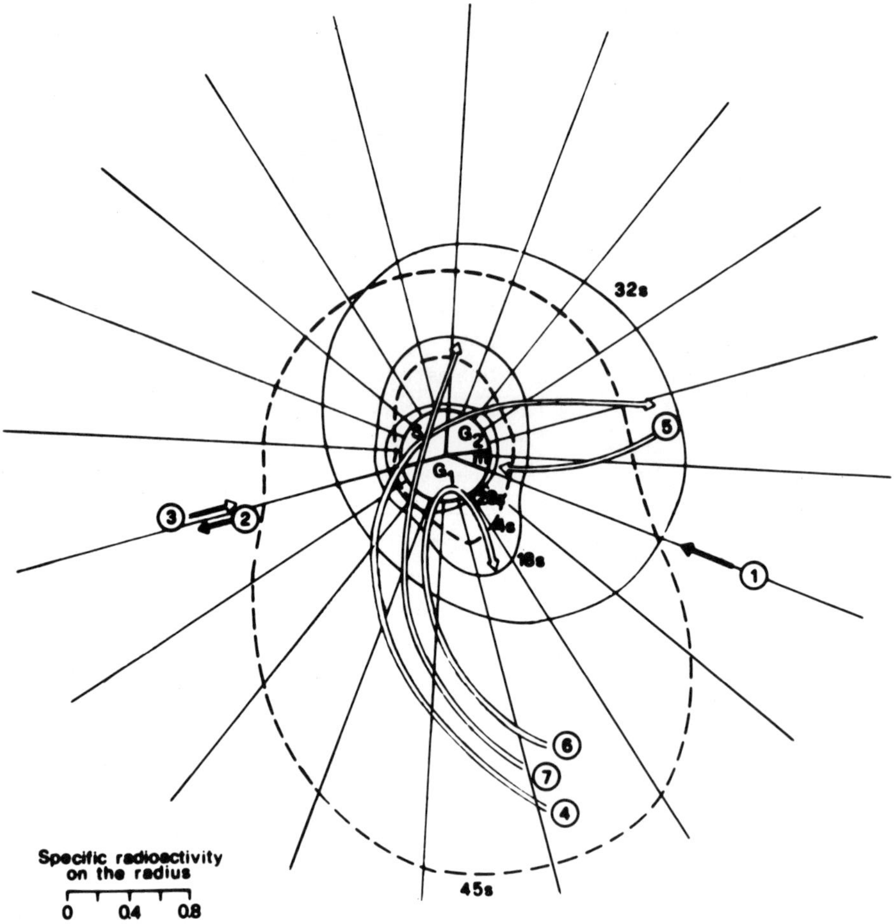

Fig. 5 - MODEL FOR "ORBITAL" REGULATION OF rRNA SYNTHESIS AND PROCESSING AROUND THE HELA CELLULAR CYCLE (G_1, S, G_2 and M). At 1-hr intervals during interkinesis and karyokinesis the cells were labelled for 50 min with ^{3}H-uridine, while tRNA and rRNAs were purified on sucrose gradients. The coordinates are those of Fig. 3 (the ordinate scale shows the rates of rRNA synthesis and processing and of tRNA synthesis). The arrows refer to the times during which should change: a - the regulation of 45S rRNA (1,2) and nDNA (3) polymerases; b - the regulation of 32S to 28S (4,5) and 20S to 18S (6,7) rRNA processing enzymes. "Orbits" from the central circular coordinate: 28S, 4S, 18S, 32S, and 45S (Ref. 12).

dependent of quantitative levels attained, transcription, processing and degradation should all co-operate in the establishment of a permanent equilibrium among the different RNA species - as required for translation during the cellular cycle (Fig.3). Thus, the rate of transcription of the total cell RNA shows a higher oscillation in G_1 and a lower oscillation in G_2, with minima in S and M (25). This is in accordance with the specific flow of pre-mRNA with attached poly-A which is continuous during the life of the cell (39) to yield proteins with maxima in G_1 and G_2 (Fig.3). The rate of 45S rRNA transcription shows a sharp maximum during the G_1-phase (Fig. 5). These observations pertain to the substrate for the RNA-polymerase system. Since with a 50 min pulse-labelling with ^{3}H-uridine the rate of transcription of total RNA (25), of pre-mRNA (39) and, most markedly, of rRNA (12,25) appears to occur to a higher extent during the G_1-phase, when compared with transcription occurring in G_2 (Figs.3 and 5), the so-called "gene-dosage effect" (65) in eukaryotes does not seem to be always a necessary event (12). That 45S rRNA transcription on the new nDNA chains, present in G_2, is partially suppressed is suggested by the finding that in this phase the rate of rRNA ^{3}H-uridine labelling is about twice as low as in G_1 instead of twice as high, as would be expected from a gene-dosage effect. To conclude, one should suspect that rDNA would be untranscriptable during some stages. Could it be that rDNA must interact with histones in some manner or just be modified before being able to serve as substrate for rDNA dependent-rRNA polymerase? Alternatively, could the amount or the activity of the RNA-polymerase vary strikingly with the different cell cycle stages?

The time correlation of the rates of synthesis and processing of rRNA during the *HeLa* cell cycle is illustrated as an "orbital" system (Fig.5). It is observed that transcription of the 45S rRNA precursor, as mentioned above, occurs at a higher rate in G_1. The rate of formation of the 32S rRNA intermediate is instead higher in S and G_2. 18S rRNA is produced at high rates in early G_1 and at the end of S. Therefore, the 45S rRNA transcriptase might be induced rapidly after mitosis (arrow 1) and be partially repressed when the cell enters the S-phase (arrow 2) and starts to duplicate its nDNA (arrow 3). S and G_2 might represent the stages of the maximal induction of the 32S rRNA forming enzyme (arrows 4 and 5). The 18S rRNA is formed from the 45S rRNA precursor via a 20S rRNA intermediate, so that it would be reasonable to believe this formation to follow kinetics not necessarily synchronized with the rate of formation of 28S rRNA (arrow 5). In fact, in figure 5, the specific labelling of the 18S rRNA shows two maxima. The early increase occurs at the beginning of the cell cycle, in parallel with the increase of transcription of the 45S rRNA (arrow 6). The late increase occurs at the end of S (arrow 7) after a partial damping in the rate of synthesis of 45S rRNA (the maximal production of tRNA takes place simultaneously with these two pathways). An important detail has to be stressed. The average interval needed to process pre-rRNA is on the order scale of 30 min from 45S rRNA to 18S rRNA, and of a few min longer from 45S rRNA to 28S rRNA (66). These time scales might seem too short for 45S rRNA transcription to be made in G_1 and 32S rRNA processing to be made in G_2. Of course there is no contradiction, since we measured, at 1-hr intervals throughout the cycle, the *rates* of synthesis and processing with ^{3}H-uridine pulse-labels of 50 minutes.

In summary, the main conclusion from figure 5 is that the rate of rRNA synthesis dominates over the rate of rRNA maturation when the cell is young, and that the rate of rRNA processing dominates over the rate of rRNA synthesis when the cell is old.

VI. MODIFICATION OF MACROMOLECULES

The significance of DNA, RNA and protein methylation is one of the mysteries of molecular biology. For instance, it is not known when in ribosomes only a few proteins are methylated (67). The participation of the rRNA methylation in the rRNA processing is *sub judice* (68). nDNA modification is, perhaps, the most intriguing phenomenon; its role has been connected with the defence against the nucleases (69), with the control of transcription (38), and with the control of genome replication (38). nDNA methylation was also considered as a factor of cell differentiation (70).

a - *nDNA methylation*. Essentially we shall observe "where" (38), "when" (37) and "how" (12) nuclear DNA is preferentially methylated in synchronized *HeLa* cells.

1. Let us look first at the "place" of methylation, namely at the genes which appear to be selectively modified (38). In *HeLa* cells, the hybrids of pre-mRNA with sheared methylated nDNA do contain 5-methylcytosines, whereas the hybrids between the same sheared methylated nDNA with mRNA purified from polysomes no longer contain any 5-methylcytosines (38). In accordance with the notion of the eukaryotic transcriptional unit (47), these results suggest that the regulatory, non-informative (proximal to the 5' nDNA end), genes are the ones preferentially methylated (38,71).

2. To study the "timing" of nDNA methylation, we followed in parallel biosynthesis and methylation of *HeLa* nDNA using ^{14}C-methyl-methionine as sole labelled precursor (37,38). In the whole *HeLa* cell, nDNA methylation parallels nDNA biosynthesis during the S-phase, while the extra-S-phase nDNA methylation is reduced to a minimum (37). Alternatively, in isolated nuclei, in the absence of nDNA biosynthesis (because of lack of trifosphonucleosides), nDNA methylation still goes on in correspondence of the S-phase (37).

3. The next aims in this laboratory were, on one hand, the elucidation of the temporal "order" of methylation of given nDNA sequences during their replication (77) and, on the other, the study of the language of gene modification in relation to gene replication (78) and organization (79). Obviously, both these topics are about the "how" of nDNA methylation (12).

HeLa nDNA can be separated in preparative Ag^+/Cs_2SO_4 gradient in two fractions (80). Sonication improves this separation (81). The heavy fraction hybridizes rRNA (12,81). As expected, temperature melting (81) and analytical ultracentrifugation in neutral CsCl (81) show that the smaller heavy fraction is GC-rich, while the light larger fraction is AT-rich (81). During the S-phase, these two nDNA populations are replicated early and late: the GC-rich nDNA, which contains genes for rRNA (81), is synthesized up to the 3rd hour of the S-phase; the AT-rich nDNA is instead synthesized later (77,78). At the first stage of S a higher specific methylation appears to precede the growth of the GC-rich nDNA; at the second stage of S specific methylation appears to precede the completion of the AT-rich nDNA replication (77,78). Therefore, the fact that genes (for instance those for rRNA) are methylated with a fixed temporal order near their replication time strongly suggests that nDNA methylation might be involved in some manner in the mechanism of the gene replication itself (77,78).

The differential methylation of the GC- and AT-rich nDNAs (81) and the nDNA methylase properties in early and late S (78) suggest that, probably, there are two languages for nDNA modification at these two S-phase stages (12,78). The methylase purified from *HeLa* nuclei at the 2nd hr of S *in vitro* appears to work with a higher specificity on the GC-rich nDNA substrate (78). On the contrary, the methylase purified from *HeLa* nuclei at the 4-5th hr of S shows a preference for the AT-rich nDNA (78). These facts imply the following generalizations. i - Supermethylation of the GC-rich nDNA sequences in early S should not be surprising because of its statistical character (12). Actually, with the increase in cytosine content, the probability of methylation should correspondingly increase (12). ii - Supermethylation of the AT-rich nDNA sequences, instead, would seem to occur in contradiction to the principle of the GC "target" size, since a decrease in cytosine content may correspond to an increase in methylation. As in this case the statistical rule cannot be taken into consideration, the existence in late S of specific recognition mechanisms for AT-rich sequences should be assumed (12). iii - Finally, it was postulated that cytosines to be methylated are in the sequence "purine-cytosine-purine" (82). Thus, in GC-rich nDNA, this sequence might be ...GCG ... or ...GCA..., while in AT-rich nDNA it might be ...ACA... or ...ACG... (12,78,81).

Lastly, we tried to purify at 1-hr intervals during S the supermethylated nDNA sequences according to their rate of reassociation (79). Figure 6 shows the specific methylation of the nDNA double strands reassociated in a large C_0t scale (79). At least 5 supermethylated nDNA families can be detected in *HeLa* cells (79). The 5-methylcytosine concentration increases on palindromes from 1 to 5 hrs during S. The early S-phase involves the accumulation of 5-methylcytosines on the nDNA sequences reassociating near a C_0t = 0.2. The late S involves the increment of these methylated bases on the sequences which reassociate at C_0t = 40. The unique sequences show a moderate methylation from 3 to 6 hrs of S. These facts (79) and those collected from the Ag^+/Cs_2SO_4 gradient analysis of the supermethylated GC- and AT-rich nDNA fraction synthesized early and late during S (77,78) give a coherent picture of the organization and structure modification of the *HeLa* genome during its replication.

One can return now to the general questions concerning the alternative roles of nDNA methylation in the gene expression of the eukaryotic cell.

The problem of the involvement of the nDNA methylation in the genetic regulation of transcription has to do with two main facts: the regulatory part of the transcriptional unit (47) is preferentially methylated (38); the supermethylated palindromes (Fig.6) preferentially hybridize pre-mRNA rather than mRNA purified from polysomes (unpublished). This suggests that some palindromes should be in the regulatory part of the transcriptional unit and that their supermethylation might be due to their inability to bind histones.

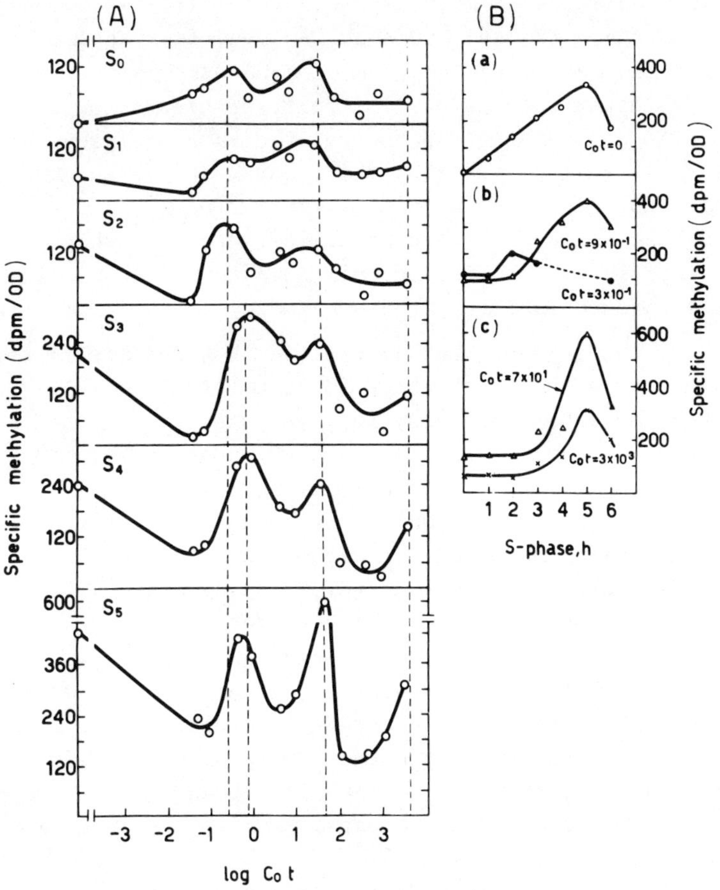

Fig. 6 - SPECIFIC METHYLATION AGAINST C_0t OF REASSOCIATED DOUBLE STRANDED HELA nDNAs PURIFIED AT 1-hr INTERVALS DURING THE S-PHASE (A). Insert B shows the development through S of specific 5-methylcytosine distribution on palindromes (a) and on sequences reassociating at low (b) and high (c) C_0t values (Ref.79).

The possibility that nDNA methylation is involved in the genetic regulation of gene replication was discussed in relation to the time shift observed in the ordering of gene methylation and gene replication during the S-phase (77,78). However, the meaning of the occurrence of nDNA methylation in isolated *Hela* nuclei, i.e. in the absence of nDNA synthesis (37), has to be further elucidated. A central point in nDNA methylation studies is the question about the occurrence of methylation on old (72,73), on new (74, 75) or on old and new (12) nDNA chains. The fact that nDNA methylation can proceed in the absence of nDNA synthesis in isolated nuclei (37) appears to favour the point of view according to which the old nDNA chains should be methylated to a preferential extent (76). This is an important remark, because the amount and the type of methylation detected on the eukaryotic "Okazaki" fragments are questionable, while there are preliminary results which appear to suggest that in the whole *HeLa* cell some inhibition occurs of nDNA synthesis through inhibition of nDNA methylation (unpublished).

Whatever the role of nDNA methylation may be, the fact that it occurs mainly during the S-phase (37), maintaining in G_1 and G_2 a low background level (37), may help in interpreting the known hypothesis about the correlation between the nDNA methylation and early sea urchin development (70). Obviously, development implies a succession of cellular cycles which after the egg fertilization proceeds with a high speed and then slows down, since the cells of many tissues enter the G_0-stage (51,83). Therefore, it should not be difficult to suppose that during the early stages of the development the succession of more new and new S-phases would lead to the accumulation in the embryo of a high amount of 5-methylcytosine. The differentiated tissues would regulate their level of methylation in accordance with their rate of cell proliferation. It is known, in fact, that rapidly-proliferating tissues are methylated to a higher extent, when compared with slowly-proliferating or non-proliferating tissues (84,85). In other words, the observed increase of nDNA methylation during the early stages of the sea urchin development should be interpreted as an epiphenomenon of differentiation, since it is probably correlated with gene replication (77,78). However, such interpretation, to be taken into consideration, requires the demonstration that methylated nDNA is not repaired (38,86). Recently, we have observed that the 5-methylcytosines acquired during the S-phase are stable on the nDNA-polymer (87). In fact, the methylated nDNA chains are semiconservatively replicated for at least 12 *HeLa* cell cycles (87). Hence, an arrest of the nDNA methylation when the cell cycles are stopped at the moment of entering the G_0-stage may coincide indeed with differentiation, although as epiphenomenon (78,79).

b - *rRNA methylation*. Since we have described in detail the investigation on synthesis and processing of ribosomal RNA (Fig.5), besides the interest for methylation of transfer (88) and messenger (89) RNAs, we want to pay some more attention here to methylation of the 45S rRNA (90). It was found that in this precursor me-

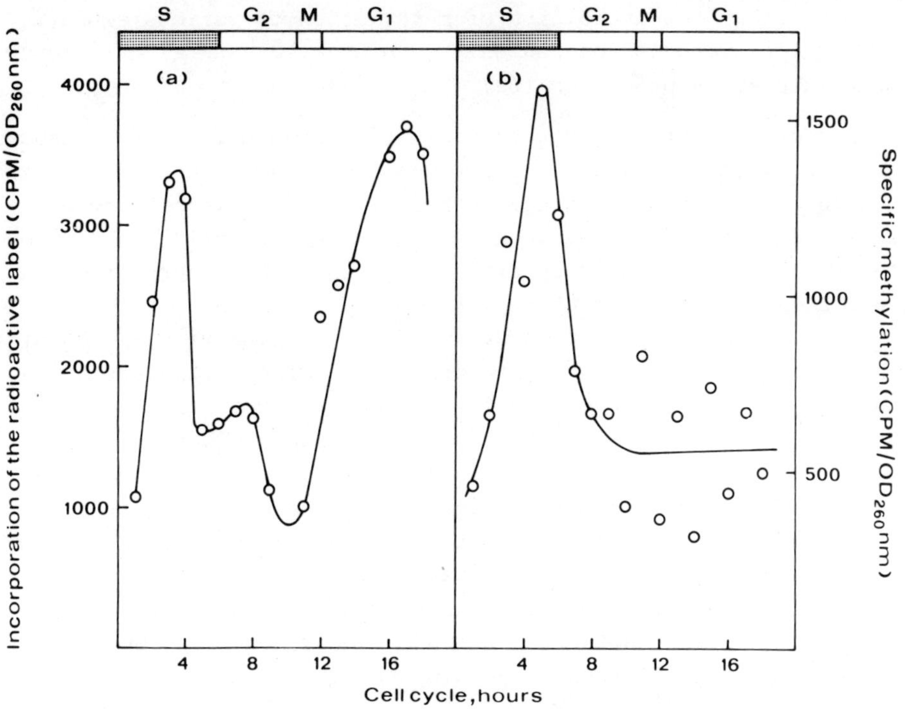

Fig. 7 - RATES OF 45S rRNA SYNTHESIS AND METHYLATION DURING THE HELA CELL CYCLE. At 1-hr intervals during interphase and mitosis, the cells were labelled with ^{14}C-methyl-methionine for 50 min, in the absence (a) and in the presence (b) of 5 μg/ml Actinomycin D.

thylation is confined to the sequences ultimately integrated into ribosomes (91). This and the fact that a limitation of methylation, by methionine starvation of cells, blocks ribosome production but allows 45S rRNA transcription to continue, suggested that methylation is involved in processing (92). Yet no close coupling is observed between methylation and transcription (68), for during an extensive blockage of methylation with cycloleucine, biosynthesis of pre-rRNA continues, although at a slightly reduced rate (68). Transcription and methylation can be temporarily uncoupled *in vivo* without significantly impairing the efficiency of the subsequent maturation of the transcript, which takes place when methylation is resumed (68). Thus, contrary to previous observations, an undermethylation of the ribosomal precursor does not block the maturation process at a definite stage. The stages of the processing pathway are affected by the inhibition of methylation to various extents. The degree of methylation can only modulate the general efficiency of the maturation processes (68).

An uncoupling of 45S rRNA transcription from methylation is shown in figure 7. In the absence of Actinomycin D, both the biosynthetic (through the C_1-chain) and the methylase pathways were going on (see the analogy with nDNA methylation in intact cells and in isolated nuclei). Under this condition, the 45S rRNA incorporates the labelled carbon atom coming from the methyl group of methionine with maxima during the phases G_1 and S (Fig.7a). Obviously the G_1-peak in figure 7a should be interpreted in terms of transcription, in accordance with figure 5. Alternatively, in the presence of the inhibitor of transcription, the G_1-peak disappears whereas the S-peak is still present (Fig.7b). Thus, the S-phase peak should be interpreted in terms of methylation. All this confirms that transcription of the 45S rRNA and its methylation may be uncoupled (68). In addition, the occurrence of a high rate of methylation during the S-phase, when combined with the picture of figure 5, would support the suggestion that methylation can only modulate the general efficiency of the 45S rRNA processing (68).

c - *Protein methylation.* Proteins form the most heterogeneous class of macromolecules in the cell. This implies that not all of them should be methylated and that the meaning of methylation should be certainly correlated with their functions, case by case. The non-random distribution of methyl groups on the different proteins is well documented. For instance, let us take the case of methylation of the ribosomal proteins. In prokaryotes, such as in *E. coli*, methylation occurs predominantly in the 50S subunit (67). In eukaryotes, such as in *HeLa* and *L* cells, both 40S and 60S subunits are methylated (67). However, in the population of the *HeLa* ribosomal proteins only seven macromolecules are definitively methylated (67).

We have investigated the timing of methylation of saline-soluble and SDS-soluble proteins during the whole *HeLa* interphase and karyokinesis (40). While the rate of methylation remains relatively low and essentially unchanged in the fraction of saline-soluble proteins, SDS-soluble proteins become methylated to a considerably larger extent, especially during the S-phase. This is a further demonstration that methylation falls in a heterogeneous fashion on proteins. In addition, the fact that methylation of the SDS-soluble proteins takes place at the highest rate during the S-phase (40) raises the question about the differential timing of protein modification in general, since during S the rate of the bulk protein biosynthesis finds one of the minima (Fig.3). It is likely that a significant fraction of protein methylation concerns the histones, whose synthesis also takes place in the S-phase (93). In this group of proteins, the delay between formation and modification should be reduced to minimum. On the other hand, if other proteins synthesized during the G_1 and G_2 phases are subjected to a similar process of methylation, the delay between synthesis and methylation would be expected to become of a few hours for macromolecules made during G_1 and of several hours for proteins synthesized during the G_2-Phase. In the latter case; proteins would pass through mitosis and undergo an almost full cell cycle before being methylated (40). Further work is needed to clarify this point, studying single proteins. In *HeLa* cells, synchronized with thymidine, methylation of the 40S ribosomal proteins occurs heavily at the threshold of the S-phase whereas methylation of the 60S ribosomal proteins is most pronounced in mid-S (67). This is in agreement with our results and can serve as a useful tool in the study of the correlation between synthesis and modification of the protein structure and of the stability of the CH_3 groups on the protein macromolecule.

In conclusion, methylation appears to be functionally heterogeneous on DNA, RNA and proteins, but its occurrence roughly coincides always with the S-phase. Thus, the S-phase might be characterized by a centralized mechanism for macromolecular modification. In relation to this, the turnover of methionine and S-adenosyl-methionine in S have to be investigated in detail. The correlation between the degree of methylation and nDNA repetition (Fig.6) is also of great interest, since nDNA repetition is correlated with the regulatory part of the eukaryotic transcriptional unit (47) and with the genome evolution (94). On the GC-rich nDNA, could the methylated ...GCG... or ...GCA... triplets be considered non-sense? What is the meaning of the methylated ...ACA... or ...ACG... triplets on the AT-rich nDNA? Another remark is about the fact that DNA methylation, at least as shown in bacteria, is needed for defence against the restriction enzymes (69) whereas, *viceversa*, 45S rRNA is believed to be methylated just to modulate (68) or facilitate (90-92) its processing by enzymes which are also nucleases.

VII. IMPLICATIONS OF CELL CYCLE METHODOLOGY

We have presented the flow of gene information during the *HeLa* cell cycle as a temporal system of "orbits" (Fig.3). Such "differential" time pathways show the periodicity of replication, transcription and translation, complementing the knowledge of "unidirectionality" implicit in the flow of gene information from the nucleus to the cytoplasm (Fig.1). The evaluation of gene expression as a function of the time coordinate introduces into the cell studies the idea of "relativity" of the interdependent processes. In other words, although the flow of gene information (Fig.1) has a "space" polarity (replication and transcription take place in the nucleus whereas translation takes place in the cytoplasm), its actual occurrence (which means life) shows a "time" polarity with respect to the M-phase, characterized by a block of gene expression (Fig.3).

Of course, by virtue of the concept itself of "relativity" of processes (50), one has to consider that the system of temporal "orbits", which reflects the flow of gene information during the cell life, cannot be strikingly fixed. Although the timing of the different steps of gene expression is believed to be genetically encoded (46), the temperature variations, for instance, severely affect the length of the G_1-phase in human amnion cells (96), of the G_1- and S-phases in L5178Y cells (97), of the M-phase in *HeLa* cells (98). The pH changes alter the G_1-phase in pig kidney cells (99). Such a regulation of the length of the cell cycle stages by physico-chemical means, as a consequence, implies a modulation of the timing of the different steps of gene expression. Therefore, one has to consider the model of figure 3 as a potentially variable feature even in the same cell type.

The important fact is that the analysis of the cancer *HeLa* cell cycle has provided a key to the understanding of the time correlations among the different steps of gene expression in general.

a - *Cell differentiation*. The implication of the control of the cell cycle biochemical events in differentiation is obvious. According to new concepts, the cycling cells should arrest in three points of the cell cycle: in early G_1, in late G_1, and in late G_2 (51). Thus, in a tissue *in vivo*, there should be four major classes of cells (51): cycling cells, non-cycling G_1-blocked cells (liver, mammary gland, ear epidermis etc.), non-cycling G_2-blocked cells (kidney epithelium, salivary gland, duodenum etc.), and non-cycling G_0-blocked cells (tongue, hepatocytes etc.). What is the nature of, and how and when the "chalones"/"triggers" (45) needed to arrest or to move the cell cycle in one of these stages? How, and how much of,the cell saves in the differentiated tissue the biochemical condition which was typical or prominent for that stage at which its mitotic cycle was arrested?

As regards the sequential expression of the "liver" and "kidney" ornithine-transaminase forms (100, 101) during the *HeLa* generation time (102), it has been shown that the "liver" form of the enzyme appears in the S-phase, while the "kidney" form of the enzyme appears in G_1 (35). This fact suggested that only in G_1 the enzyme is subjected to the same type of regulation as in kidney. Such an analysis should be extended to all proteins which characterize a given tissue or are prominent in it. Alternatively, one cannot consider nDNA methylation as a typical or prominent event occurring in the non-cycling G_1-, G_2- or G_0-blocked cells, since it appears to be rather correlated with the regulation of gene expression, in S (37). Rapidly-proliferating tissues do have a high rate of nDNA methylation, while slowly-proliferating tissues are characterized by a low rate of nDNA methylation (84,85).

b - *Cell transformation*. Figure 3 raised the question about the probability that a virus has to infect the cell in a favourable point of the mitotic cycle (11).

As regards the infection with poliovirus (which does not lead to the cell transformation but to the cell death), we have seen that its initiation is always possible during the *HeLa* cellular cycle (53,60) whereas the progeny virus $mRNA^+$ is made during the S-phase only (Fig.4).

Let us pay further attention to the infection by oncogenic viruses (which leads not to the cell death but to cell transformation). The interest of this research is linked not only with the struggle against the tumors induced by natural oncoviruses, but also with the utility or danger of artificial genetic manipulations, as they are equally possible in animals and plants (12,103).

Our starting hypothesis was that an exogeneous DNA could not become covalently bound to the host nDNA when this is strongly packed or covered by histones or some other proteins. We thought that any exogeneous DNA, to be inserted into the cell genome, should find it in a relaxed and naked fashion. The cell cycle stages would provide such different conditions, since nDNA is tightly packed during mitosis, in any case strongly combined with histones and other proteins during G_1 and G_2, and in the process of relaxation during the S-phase (104).

The preliminary results suggest that the oncogenes are not integrated into the host cell genome during the M-phase but instead are preferentially inserted into it during the S-phase (105). We investigated this in two systems: in the first, there is an interaction between the normal 3T3 cells and the SV40 (which is a DNA virus); in the second, there is an interaction between the normal NRK cells and the RSV (which is an RNA virus). The integration of the virus genome into the host cell genome occurs directly in the system 3T3/SV40, and via the inverse transcriptase in the system NRK/RSV. Therefore, it is possible that the inverse transcriptase mechanism is working at an optimal rate during the S-phase, since the integration finds this phase as a favourable "weak" point of the cell cycle (105). Figure 8 shows the different growth of NRK and 3T3 cells, before and after their transformation by RSV and SV40. The transformed 3T3 cells accelerate their growth to a higher extent when compared with the transformed NRK cells. A pecular phenomenon, which parallels transformation of both 3T3 and NRK cells, is the decrease in them of the g = 2.003 EPR signal (45).

The suggestion that the insertion of an exogeneous DNA into the host cell genome may be successful only when this is in a favourable state during the mitotic cycle (12,105) appears to be important for genetic research in general, since it shows the way to facilitate or prevent the integration by arresting the cells in S or in other phases, respectively. The hope regards first of all the possibility of arresting the propagation of the transforming action by oncoviruses through a local cell synchronization directly in the sick animal.

The study of the cell cycle favourable point in cell-oncovirus interactions (105) provided a stimulus for studying the place on nDNA ("weak" point) where viral genome is inserted (38).

The normal cells of several animal species contain a large number of nDNA sequences which are homologous to oncoviral genes (106). Such a homology should serve for recognition of the nDNA target by the oncoviral genome.

On this basis, we exploited the preferential methylation of regulatory genes at the level of the transcriptional unit (38) to

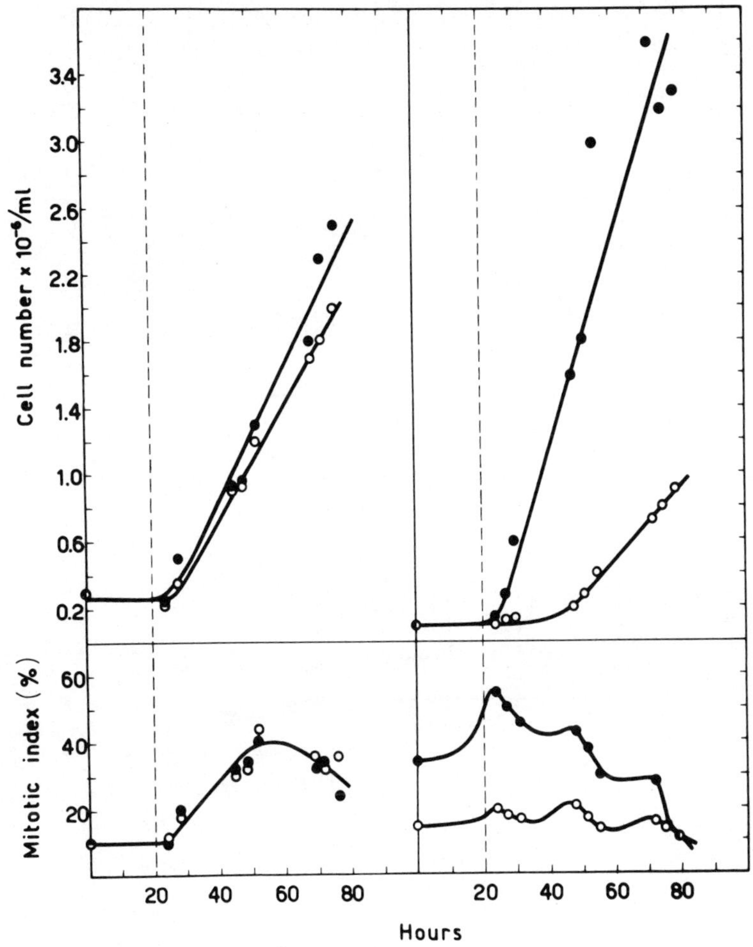

Fig. 8 - GROWTH CURVES AND MITOTIC INDEX OF NRK AND 3T3 CELLS, BEFORE AND AFTER THEIR TRANSFORMATION WITH RSV AND SV40. The cell growth is shown above; the mitotic index is shown below. The NRK transformation with RSV is shown on the left; the 3T3 transformation with SV40 is shown on the right. O——O, Normal cells; ●——●, transformed cells.

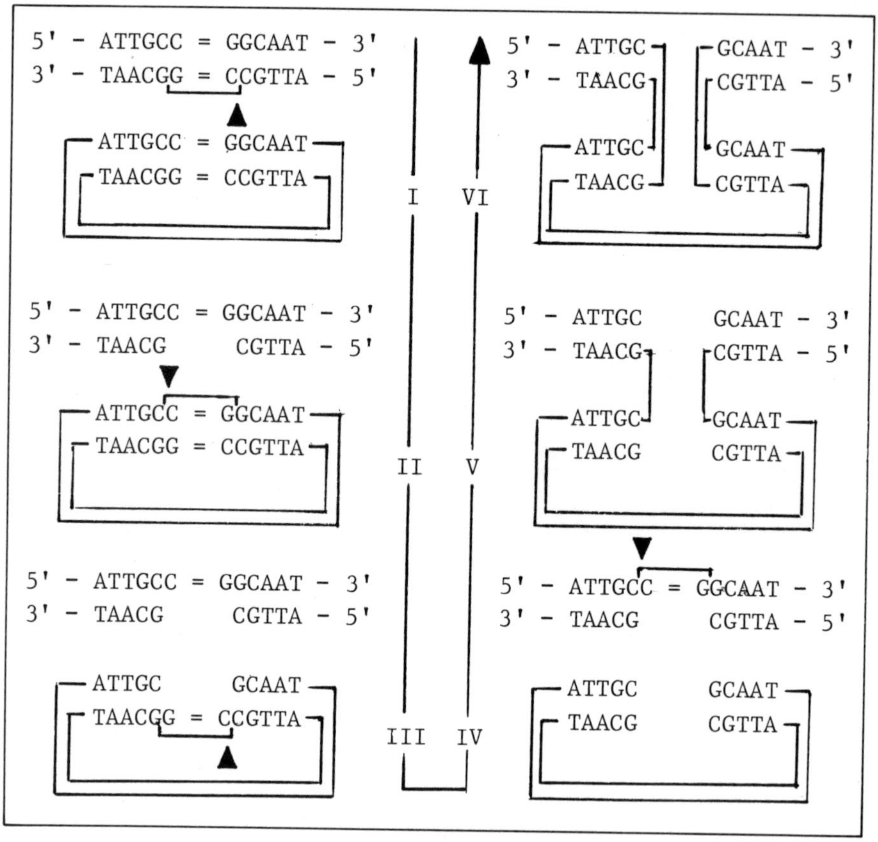

Fig. 9 - SCHEMA FOR ONCOVIRUS GENOME INTEGRATION INTO THE HOST CELL GENOME AT THE LEVEL OF THE REGULATORY ZONE OF THE TRANSCRIPTIONAL UNIT CONTAINING "NAKED" PALINDROMES. The schema shows four possible alternate cuts by the same nuclease (I-IV) and the lygase bindings (V-VI).

get an idea about the possible location of the site on host nDNA for recognition of the oncogenes (107).

The preliminary work shows that the hybrids between the sheared methylated nDNA from rat liver and the Rous sarcoma virus RNA does contain 5-methylcytosines (38,107), while the virus oncogenes do not hybridize the population of unique nDNA sequences (103,107) in which the bulk of the structural genes should be. This is supported by the experiments in the systems NRK/RSV and 3T3/SV40 (Fig.8).

Thus, our working hypothesis is that the repetitive sequences of the regulatory zone of the transcriptional unit (proximal to the 5' end) might represent a target for the oncovirus genome input (107) when the nDNA is relaxed in the S-phase (105).

It is worth mentioning that the supermethylated palindromes (Fig.6) appear to be located in the regulatory part of the transcriptional unit (unpublished) and appear to hybridize the oncogenes in the system 3T3/SV40 (Fig.8). On this basis, we propose that palindromes might provide the "weak" point, i.e. the homology needed as a start signal for oncovirus integration in S. This is in agreement with two main prerequisites: i - Probably, the homologous zone on nDNA to be recognized by the oncogenes should be an inverted sequence, considering that palindromes are present also in DNA of SV40 (110) and polyoma (111) viruses; ii - the inverted sequences should be "naked" points (without histones or other proteins).

The first prerequisite conforms with the schema below (Fig.9).

The second prerequisite is instead suggested by a recent finding showing that on pre-mRNA (and perhaps on nDNA) a high percentage of inverted sequences does not bind the proteins at all (108). Therefore, when nDNA is relaxed during the S-phase, palindromes might become an easy target for oncogenes.

We have discussed elsewhere (12) the assumption of two possible mechanisms of cell transformation, one involving a complete loss of cell regulation, the other involving a partial loss of cell regulation (in the first case, the oncogenes would be integrated between the regulatory and structural zones of the transcriptional unit; in the second case, the virus oncogenes would be integrated in the regulatory zone only). To the point, the virus specific mRNA appears to be transcribed during the S-phase (48), while it may be linked with a cell messenger to produce both virus and cell proteins (109). This led us to consider the bloc "oncogene-cell gene" as an "involution" of the eukaryotic transcriptional unit, since in the transformed cell the bloc "oncogene-cell gene" would produce again, as in bacteria (4), polycistronic messages (103). Nevertheless, how can we combine the assumption that the oncogenes are integrated at the level of the transcriptional unit (47) during the S-phase (105) with the knowledge that, on average, 1-7 oncogenes can be integrated per cell genome? The possible answer to this should come from the knowledge of repetition on nDNA (94), particularly from that of regulatory genes.

Acknowledgements. The skilful assistance of Mr. C. Buono is gratefully appreciated. This work was partly supported by the CNR-Programma Finalizzato Virus.

REFERENCES

1. Crick, F.H.S., Symp. Soc. Exptl. Biol., 12, 138 (1958).
2. Baltimore, D., Nature, 226, 1209 (1970).
3. Watson, J.D., Molecular Biology of the Gene. Benjamin, New York (1970).
4. Jacob, F., and Monod, J., J. Mol. Biol., 3, 318 (1961).
5. Gurdon, J.B., Control of Gene Expression in Animal Development. Harvard University Press (1975).
6. Howard, A., and Pelc, S.R., Heredity, 6, 261 (1953).
7. Brachet, J., Introduction to Molecular Embryology. Heidelberg Science Library (1973).
8. Mitchison, J.M., The Biology of the Cell Cycle. Cambridge University Press (1971).
9. Epifanova, O.I., The Cell Cycle. Nauka, Moscow (1973).
10. Prescott, D.M., Advances Genet., 18, 100 (1976).
11. Eremenko, T., Benedetto, A., and Volpe, P., Nature New Biol., 237, 114 (1972).
12. Volpe, P., Horizons Biochem. Biophys., 2, 285 (1976).
13. Gey, G.O., Coffman, W.D., and Kubicek, M.T., Cancer Res., 12, 264 (1952).
14. Swanson, C.P., The Cell. Prentice-Hall, Englewood Cliffs, N.J., (1969).
15. Riley, D.M., and Keller, J.M., J.Cell Sci., 29, 139 (1978).
16. Mauron, A., and Spohr, G., Eur. J. Biochem., 82, 619 (1978).
17. Daniell, E., and Mullenbach, T., J. Virol., 26, 61 (1978).
18. Vannestrom, B., and Philipson, L., J. Virol., 22, 290 (1977).
19. Darnell, J.E., Girard, M., Baltimore, D., Summers, D.F., and Maizel, J.V., in The Molecular Biology of Viruses. Academic Press, New York, pp. 375-401 (1967).
20. Puck, T.T., and Yamada, M., Radiat. Res., 16, 589 (1962).
21. Volpe, P., and Eremenko, T., Exptl. Cell Res., 60, 456 (1970).
22. Puck, T.T., Cold Spring Harb. Symp. Quant. Biol., 29, 167 (1964).
23. Eremenko, T., Menna, T., and Volpe, P., Micro Biol., in press (II).

24. Volpe, P., and Eremenko, T., Europ. J. Biochem., 32, 227 (1973).
25. Volpe, P., Menna, T., and Eremenko, T., Bull. Mol. Biol. Med., 1, 18 (1976).
26. Eremenko, T., and Volpe, P., Eur. J. Biochem., 52, 203 (1975).
27. Weber, J., Jelenek, W., and Darnell, J.E., Cell, 10, 611 (1977).
28. Goldberg, S., Weber, J., and Darnell, J.E., Cell, 10, 617 (1977).
29. Amalric, F., Merkel, C., Gelfand, R., and Attardi, G., J. Mol. Biol., 118, 1 (1978).
30. Fraser, J.M.K., and Huberman, J.A., J. Mol. Biol., 117, 249 (1977).
31. Volpe, P., and Eremenko, T., in Methods in Cell Biology. Academic Press, New York, Vol. 2, pp. 113-126 (1973).
32. Ooka, T., Girgis, A., and Daillie, J., Exptl. Cell Res., 81, 207 (1973).
33. Tupper, J.T., Mills, B., and Zorgniotti, F., J. Cell Physiol., 88, 77 (1976).
34. Suarez, M., Contreras, G., and Fridlender, B., J. Virol., 16, 1337 (1975).
35. Volpe, P., Menna, T., and Pagano, G., Eur. J. Biochem., 44, 455 (1974).
36. Eremenko, T., Menna, T., and Volpe, P., Micro Biol., in press (III).
37. Geraci, D., Eremenko, T., Cocchiara, R., Granieri, A., Scarano, E., and Volpe, P., Bioch. Biophys. Res. Commun., 57, 853 (1974).
38. Volpe, P., and Eremenko, T., FEBS-Letters, 44, 121 (1974).
39. Eremenko, T., Menna, T., and Volpe, P., 12th FEBS-Meeting, Abstr., 355 (1978).
40. Eremenko, T., Cimarra, P., Giuditta, A., and Volpe, P., submitted.
41. Spalding, J., and Kajwara, K., Proc. Natl. Acad. Sci., USA, 56, 1535 (1966).
42. Alberghina, L., J. Theor. Biol., 69, 633 (1977).
43. Neskovic,B., Intern. Rev. Cytol., 24, 71 (1968).
44. Salb, J.M., and Marcus, P.I., Proc. Natl. Acad. Sci., USA, 54, 1353 (1965).
45. Segre, A.L., Benedetto, A., Eremenko, T., Volpe, P., Di Nola, A., and Conti, F., Biochim. Biophys. Acta, 197, 615 (1977)
46. Defendi, V., and Manson, L.A., Nature, 198, 359 (1963).
47. Georgiev, G.P., J. Theor. Biol., 25, 473 (1969).

48. Hoffman, P., and Darnell, J.E., J. Virol., 15, 806 (1975).
49. Eremenko, T., and Volpe, P., in The Cell Nucleus. Nauka, Moscow, pp. 256-258 (1972).
50. Wiener, N., Cybernetics. The Technology Press, N.Y. (1948).
51. Gelfant, S., Cancer Res., 37, 3845 (1977).
52. Zalmanzon, E.S., and Lobareva, L.S., Acta Virol., 10, 301 (1966).
53. Eremenko, T., Benedetto, A., and Volpe, P., J. Gen Virol., 16, 61 (1972).
54. Koch, A.S., Eremenko, T., Benedetto, A., and Volpe, P., Intervirology, 4, 221 (1975).
55. Bienz, K., Egger, D., and Wolff, D.A., J. Virol., 11, 565 (1973).
56. Koch, A.S., and Feher, G., J. Gen. Virol., 18, 319 (1973).
57. Pages, J., Mantenil, S., Stehelin, D., Fiszeman, M., Marx, M., and Girard, M., J. Virol., 12, 99 (1973).
58. Hampar, B., Tanaka, A., Nonoyama, M., and Derge, J.G., Proc. Natl. Acad. Sci., USA, 71, 631 (1974).
59. Kaplan, J.C., Keinman, L.F., and Black, P.H., Virol., 68, 215 (1975).
60. Koch, A.S., and Volpe, P., in International Virology. Karger, Basel, Vol. 2, pp. 277-282 (1972).
61. Hodge, L.D., Robbins, E., and Scharff, M.D., J. Cell Biol., 40, 497 (1969).
62. Darnell, J.E., Perspect. Virol., 4, 16 (1965).
63. Baltimore, D., in The Biochemistry of Viruses. Dekken, New York, pp. 101-176 (1969).
64. Neskovic, B., Polic, D., Babin, J., and Aidaric, Z., Jug. Phsysiol. Pharm. Acta, 2, 163 (1967).
65. Pfeiffer, S.E., and Tolmach, L.J., J. Cell Physiol., 71, 77 (1968).
66. Perry, R.P., Ann. Rev. Biochem., 45, 605 (1976).
67. Chang, F.N., Navickas, I.J., Au, C., and Budzilowicz, C., Bioch. Biophys. Acta, 518, 89 (1978).
68. Caboche, M., and Bachellerie, J.P., Eur. J. Biochem., 74. 19 (1977).
69. Harber, W., and Linn, S., Ann. Rev. Biochem., 38, 467 (1969).
70. Scarano, E., Ann. Embryol. Morphol., suppl., 1, 51 (1969).
71. Romanov, G.A., Biokhimiya, 41, 1038 (1976).
72. Adams, R.L.P., Biochim. Biophys. Acta, 254, 205 (1971).
73. Evans, H.H., Evans, T.E., and Littman, S., J. Mol. Biol., 74, 563 (1973).

74. Turkington, R.W., and Spielvogel, R.L., J. Biol. Chem., 246, 3835 (1971).
75. Schneiderman, M.H., and Billen, D., Biochim. Biophys. Acta, 308, 352 (1973).
76. Adams, R.L.P., Biochim. Biophys. Acta, 335, 365 (1973).
77. Eremenko, T., Granieri, A., and Volpe, P., 10th Intern. Congr. Biochem., Abstr., 01-1-086 (1976).
78. Volpe, P., and Eremenko, T., 14th Intern. Congr. Genet., Abstr., 1, 194 (1978).
79. Eremenko, T., Timofeeva, M.Y., Granieri, A., and Volpe, P., 14th Intern. Congr. Genet., Abstr., 1, 191 (1978).
80. Eremenko, T., Granieri, A., Scarano, E., and Volpe, P., 10th FEBS-Meeting, Abstr., 130 (1975).
81. Eremenko, T., Granieri, A., and Volpe, P., Mol. Biol. Rep., in press.
82. Vanyushin, B.F., Masharina, L.V., and Belozersky, A.N., Proc. Acad. Sci., USSR, 147, 145 (1962).
83. Epifanova, O.I., and Terskikh, V.V., Cell Tissue Kin., 2, 75 (1969).
84. Vanyushin B.F., Mazin, A.L., Vasyliev, V.K., and Belozersky, A.N., Biochim. Biophys. Acta, 299, 397 (1973).
85. Turnbull, J.F., and Adams, R.L.P., Nucl. Ac. Res., 3, 677 (1976).
86. Volpe, P., and Eremenko, T., 10th Intern. Congr. Biochem., Abstr., 01-1-082 (1976).
87. Eremenko, T., and Volpe, P., 13th Congr. Ital. Soc. Mol. Biol. Biophys., Abstr., 23 (1976).
88. Fleissner, E., and Borek, E., Biochem., 2, 1093 (1963).
89. Furuichi, Y., Morgan, M., Shatkin, A.J., Jelenek, W., Salditt-Georgieff, M., and Darnell, J.E., Proc. Natl. Acad. Sci., USA, 72, 1904 (1975).
90. Greenberg, H., and Penman, S., J. Mol. Biol., 21, 527 (1966).
91. Wagner, E., Penman, S., and Ingram, V., J. Mol. Biol., 29, 371 (1967).
92. Vaughan, M.H., Soeiro, R., Warner, J.R., and Darnell, J.E., Proc. Natl. Acad. Sci., USA, 58, 1527 (1967).
93. Stein, G.S., and Borun, T.W., J. Cell Biol., 52, 292 (1972).
94. Britten, R.J., and Davidson, E.H., Quart. Rev. Biol., 48, 111 (1971).
95. Davidson, E.H., and Britten, R.J., Cancer Res., 34, 2034 (1974).
96. Sisken, J.E., Exptl. Cell Res., 40, 436 (1965).

97. Watanabe, I., and Okada, S., J. Cell Biol., 32, 309 (1967).
98. Rao, P.N., and Engelberg, J., Science, 148, 1092 (1965).
99. Sisken, J.L., in Cinemicrography in Cell Biology. Academic Press, New York, pp. 143-168 (1963).
100. Volpe, P., Sawamura, R., and Strecker, H.J., J. Biol. Chem., 244, 719 (1969).
101. Strecker, H.J., Hammar, U., and Volpe, P., J. Biol. Chem., 245, 3328 (1970).
102. Volpe, P., and Eremenko, T., Advances Cytopharmacol., 1, 257 (1971).
103. Volpe, P., Biochimica del Ciclo Cellulare. Clu, Napoli (1977).
104. Caplan, A., Ord, M.G., and Stocken, L.A., Biochem. J., 174, 475 (1978).
105. Eremenko, T., Menna, T., and Volpe, P., 4th Intern. Congr. Virol., Abstr., P14, 250 (1978).
106. Temin, H., Proc. Natl. Acad. Sci., USA, 52, 323 (1964).
107. Volpe, P., and Eremenko, T., 3rd Intern. Congr. Virol., Abstr., C, 116 (1975).
108. Calvet, J.P., and Pederson, T., J. Mol. Biol., 122, 361 (1978).
109. Green, M., Ann. Rev. Biochem., 39, 701 (1970).
110. Subramanian, K.N., Dhar, R., Weissman, S.M., J. Biol. Chem., 252, 355 (1977).
111. Soeda, E., Miura, K., Nakaso, A., and Kimura, G., FEBS-Letters, 79, 383 (1977).

PART II

MACROMOLECULE STRUCTURE AND FUNCTION

EUKARYOTIC TRANSLATION FACTORS AND RNA-BINDING PROTEINS

L. P. Ovchinnikov, T. N. Vlasik, S. P. Domogatsky, T. A. Seryakova and A. S. Spirin

Institute of Protein Research, Academy of Sciences of the USSR, Poustchino, Moscow Region U.S.S.R.

INTRODUCTION

Messenger RNA in animal (1-5), higher plant (5,6) and lower eukaryotic (7) cells is always complexed with protein and exists in the form of ribonucleoprotein particles with an RNA to protein weight ratio of 1 to 3 (buoyant density in CsCl is about 1.4 g/cm^3). The term "informosomes" was suggested to denote this type of particles (1,2). Three classes of such particles are distinguished: nuclear ribonucleoproteins (8-10), free cytoplasmic informosomes (1-5) and polyribosomal messenger ribonucleoproteins (11-13). A hypothesis was proposed that informosomal proteins can serve for the transport of mRNA from the nucleus to the cytoplasm (1,2,9), for the stabilization and protection of mRNA in the cell (2) and for regulation of protein biosynthesis (1,2,4).

Together with informosomes a special class of proteins with a high affinity to RNA is present in the cytoplasm (2,14-16) and in the nucleus (17). These RNA-binding proteins are capable of interacting with RNA, forming artificial ribonucleoprotein particles resembling informosomes in sedimentation coefficients and in buoyant density value in CsCl.

Figure 1 gives an example of equilibrium density distribution of such artificial informosome-like particles in the CsCl gradient. In this case, particles

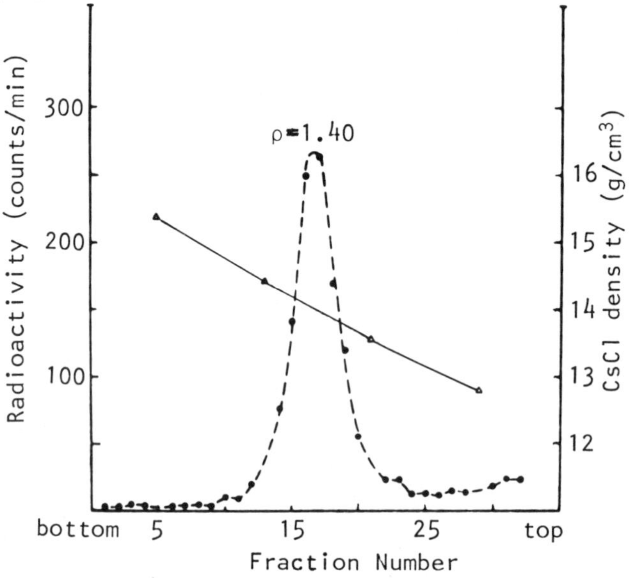

Figure 1. Density distribution in the CsCl gradient of the informosome-like particles formed by exogenous [^{14}C]RNA in the cytoplasmic extract of loach embryos. 23S + 16S [^{14}C]RNA of E. coli was mixed with the cytoplasmic extract of loach embryos. The mixture was incubated for 30 min at +4°. The products of the interaction were fixed with formaldehyde and isolated by sucrose gradient centrifugation. Centrifugation was done in a CsCl gradient in the presence of 4% formaldehyde in a SW-39 rotor of the Spinco L centrifuge at 35,000 rev/min for 22 hours. ●---●, radioactivity; △—△, CsCl density. (Taken from [21]).

were formed with ribosomal 23S + 16S [^{14}C]RNA of *Escherichia coli* in the cytoplasmic extract of loach embryos. These particles appear very homogeneous according to the density and are distributed in the narrow region of the CsCl gradient with a density of 1.4 g/cm^3.

The characteristic feature of the RNA-binding proteins of the cytoplasm of all the eukaryotic cells studied is their existence predominantly in the form of relatively high molecular weight aggregates with sedimentation coefficients from 6 to 10S (2,14,16,18-20). Figure 2 presents an example of sedimentation distribution of RNA-binding proteins from rat liver cytoplasmic extract after centrifugation in the sucrose gradient. One can see that the RNA-binding activity sediments faster than the bulk of the cell proteins.

RNA-binding proteins are capable of forming stoichiometric complexes both with messenger and ribosomal RNA. But the complexes formed with messenger RNA are characterized by a higher stability (21). This means that RNA-binding proteins are specific to mRNA and

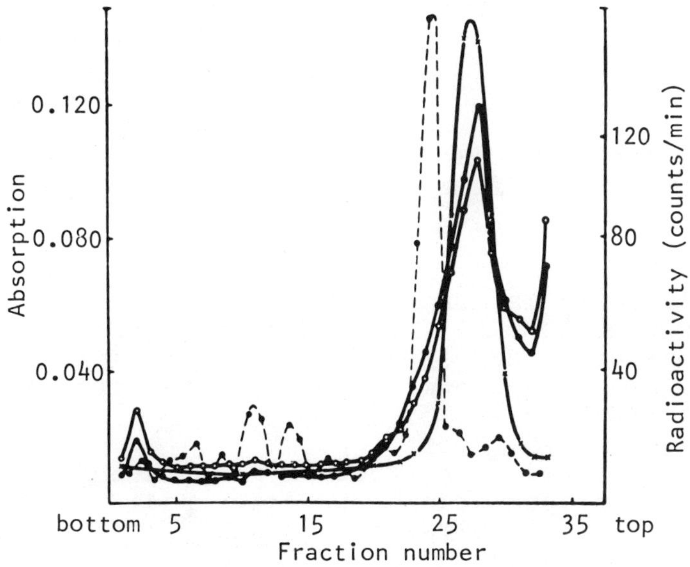

Figure 2. Sedimentation distribution in the sucrose gradient of the RNA-binding proteins of rat liver cytoplasmic extract. Centrifugation was done in a SW-39 rotor of the Spinco L centrifuge at 36,000 rev/min for 10 hours. After centrifugation [^{14}C]RNA was added to all the odd fractions, the samples were filtered through membrane nitrocellulose filters and the radioactivity retained on the filters was counted as a measure of the RNA-binding activity. Absorption at 260 nm, 280 nm and 410 nm was measured in all the even fractions. o——o, absorption at 260 nm; •——•, absorption at 280 nm; x——x, absorption at 410 nm (hemoglobin as a marker); •---•, radioactivity as a measure of the RNA-binding activity. (Taken from [14]).

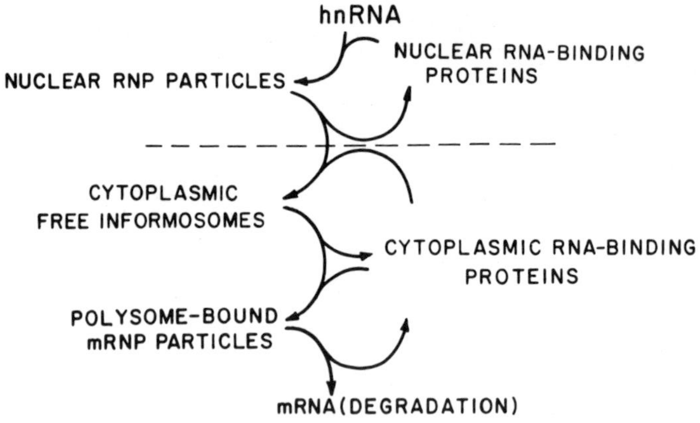

Figure 3. Scheme of possible interrelation between messenger ribonucleoproteins (informosomes) and RNA-binding proteins of diverse localization in the eukaryotic cell (Taken from [5]).

this specificity is manifested in a higher affinity of these proteins to mRNA (21). It is interesting to note that RNA-protein complexes formed at physiological temperatures are more stable in comparison with those formed at +4° (22).

The described properties of RNA-binding proteins permitted to propose that they are a pool involved in the formation of informosomes <u>in vivo</u>, representing free informosome-forming proteins (2,14).

The hypothetic interrelation between the three classes of informosomes and RNA-binding proteins are represented schematically in figure 3 (5).

ISOLATION OF RNA-BINDING PROTEINS

The high equilibrium constants of binding of RNA-binding proteins with RNA (in the range from $10^7\ M^{-1}$ to $10^{13}\ M^{-1}$ for ribosomal RNA at +4° (23)) permitted to use the technique of their affinity adsorption on free RNA (20,24,25), or on RNA covalently bound to Sepharose (26-28) for their selective removal from the extract and for their very effective purification from all other proteins.

Figure 4 shows the result of fractionation of proteins of the rabbit reticulocyte ribosome-free extract on the RNA-Sepharose column (28). It is seen that the bulk of proteins of the extract passes through the column without retention. On the other hand, from 80 to 95% of the RNA-binding activity measured by the binding of RNA on nitrocellulose filters (14,16) is retained on the column. The RNA-binding activity adsorbed on the column is completely eluted with a buffer containing 1 M KCl. As a result the specific RNA-binding activity of the protein washed from the column with 1 M KCl increases about 150 times as compared with the specific RNA-binding activity of the proteins of the original extract.

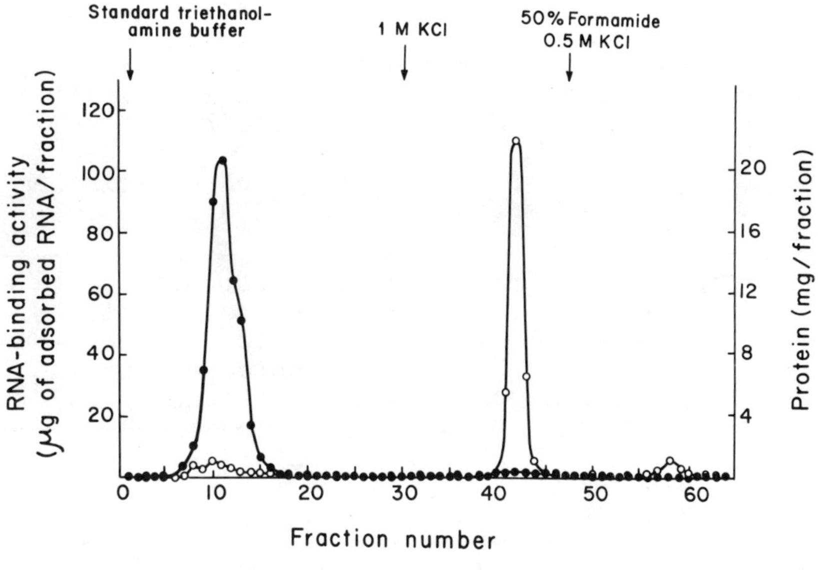

Figure 4. Chromatography of rabbit reticulocyte ribosome-free extract on the RNA-Sepharose 4B column. RNA-Sepharose was equilibrated with the standard triethanolamine buffer (10 mM triethanolamine•HCl, pH 7.8, 10 mM KCl, 1.5 mM $MgCl_2$). Stepwise elution of the adsorbed protein was done with 1 M KCl, then with 50% formamide containing 0.5 M KCl. ●——●, protein; o——o, RNA-binding activity. (Taken from [28]).

RNA-BINDING PROTEINS OF WHEAT EMBRYO EXTRACT CONTAIN TRANSLATION FACTORS

Using the method of selective separation of RNA-binding proteins, we put the question: how did RNA-binding proteins affect the functioning of the cell-free translation system? To answer this question we used a wheat embryo cell-free system for the first experiments (29).

The control cell-free system contained: washed 80S ribosomes from wheat embryos, tobacco mosaic virus (TMV) RNA, [^{14}C]aminoacyl-tRNA; as a preparation of the total initiation and elongation factors the system also contained postribosomal supernatant of wheat embryos, freed from tRNA by a DEAE-cellulose column (complete S100-DEAE fraction).

In the preliminary experiments, we noticed that the passing of the preparation of S100-DEAE fraction through the RNA-Sepharose column resulted in the loss of its translation factor activity (Table 1, second line). Consequently, either the matrix-bound RNA inactivated some of the translation factors, or at least one or some of the factors had a special affinity to RNA, i.e. behaved as RNA-binding proteins.

Table 1. Translation factor activity of RNA-binding proteins (taken from [29])

Presence of protein fractions in the cell-free system			Incorporation of [^{14}C] amino acids into trichloroacetic acid-insoluble polypeptide (counts/min)	
Complete S100-DEAE fraction	S100-DEAE fraction deprived of RNA-binding proteins	Fraction of RNA-binding proteins	Exp. 1	Exp. 2
+	–	–	4400	
–	+	–	510	530
–	+	+	3380	4200
–	–	+		3350

The cell-free system contained, in addition to the protein fractions indicated in the table: 80S ribosomes, [^{14}C]Aa-tRNA, TMV RNA, GTP, ATP and a GTP-ATP-regenerating system.

To test these two alternatives, we eluted the fraction of RNA-binding proteins from the RNA-Sepharose column with the high ionic strength buffer and added it into the cell-free system with the inactive S100-DEAE fraction passed through RNA-Sepharose. The activity of the system was significantly restored (Table 1, experiment 1). Thus, the factors were not inactivated by the RNA-Sepharose column, and the second alternative proved to be true: at least one or some of the translation factors were adsorbed on RNA as RNA-binding proteins.

In the next experiment we decided to test the translation factors activity of the RNA-binding protein preparation itself, without introducing either of the S100-DEAE fractions into the cell-free system. The result shown in Table 1, experiment 2, was unexpected: the preparation of RNA-binding proteins ensured the active working of the system, thus substituting for all the totality of the initiation and elongation factors. Consequently, RNA-binding proteins of the wheat embryo cytoplasmic extract include all (or almost all) the initiation and elongation factors of the protein-synthesizing system.

On the other hand, this experiment indicates that the proteins, being initiation and elongation factors, at the same time possess yet another function, namely that of binding to RNA with a high affinity.

RNA-BINDING PROTEINS OF RABBIT RETICULOCYTES CONTAIN THE TWO ELONGATION FACTORS AND SOME OF THE INITIATION FACTORS OF TRANSLATION

In following experiments together with Dr. B.Erni and Prof. T.Staehelin from the Basel Institute for Immunology, we tested the ability of RNA-binding proteins from rabbit reticulocytes to replace individual initiation and elongation factors in a purified eukaryotic cell-free translation system (30).

Table 2 shows the elongation factor activities in the preparation of RNA-binding proteins, as determined in the system with poly(U) as a template. The residual polyphenylalanine synthesis in the absence of EF-1 is about 2% and in the absence of EF-2 about 15 to 17% of the complete system. The addition of RNA-binding proteins to assays without EF-1 and EF-2 results in a recovery of the original activity, i.e. RNA-binding pro-

Table 2. Stimulation of [^{3}H]phenylalanine incorporation by the preparation of RNA-binding proteins with a high affinity to RNA in the poly(U)-dependent cell-free system (taken from [30])

Composition of cell-free system	Incorporation of [^{3}H]phenylalanine into the acid-insoluble product (pmol)
Complete	2.02
- EF-1	0.04
- EF-1 + RNA-binding protein (2 μg)	2.23
- EF-1 + RNA-binding protein (4 μg)	2.24
- EF-1 + RNA-binding protein (6 μg)	2.36
- EF-2	0.35
- EF-2 + RNA-binding protein (2 μg)	2.13
- EF-2 + RNA-binding protein (4 μg)	2.33
- EF-2 + RNA-binding protein (6 μg)	2.52
Complete + RNA-binding protein (6 μg)	2.42

teins completely substitute for EF-1 and EF-2. As little as 2 μg of the preparation of RNA-binding proteins fully compensate for the absence of EF-1 or EF-2 in the system. The addition of the RNA-binding proteins to the complete system does not give a significant stimulation.

The initiation factor activities in the preparation of RNA-binding proteins from rabbit reticulocytes were determined in the globin mRNA-dependent translation system described as the system "b" in the paper (31). The results presented in Table 3 show that the RNA-binding proteins are able to compensate entirely for the absence of the (eIF-4C + eIF-5) preparation in the system. In addition, 6 μg of preparation of RNA-binding proteins seems to substitute partially also for eIF-1 and to a lesser extent for eIF-3 and eIF-4B. On the other hand, the preparations of RNA-binding proteins from reticulocytes contain practically no eIF-2 and eIF-4A activities.

Table 3. Stimulation of [^{14}C]leucine incorporation by the preparation of RNA-binding proteins with a high affinity to RNA in the globin mRNA-dependent cell-free system (taken from [30])

Composition of cell-free system	Incorporation of [^{14}C] leucine into the acid-insoluble product (pmol)	
	Exp. 1	Exp. 2
Complete	50.8	48.2
- eIF-1	13.0	15.1
- eIF-1 + RNA-binding protein (6 μg)	30.6	27.1
Complete	46.0	41.2
- eIF-2	4.0	3.2
- eIF-2 + RNA-binding protein (6 μg)	9.0	5.2
Complete	57.8	39.2
- eIF-3	10.6	9.2
- eIF-3 + RNA-binding protein (6 μg)	22.0	15.1
Complete	60.5	47.1
- eIF-4A	5.0	5.0
- eIF-4A + RNA-binding protein (6 μg)	6.6	6.1
Complete	47.6	40.8
- eIF-4B	12.5	11.5
- eIF-4B + RNA-binding protein (6 μg)	17.2	20.5
Complete	45.4	44.4
- (eIF-4C + eIF-5)	5.9	3.8
- (eIF-4C + eIF-5) + RNA-binding protein (6 μg)	45.9	28.1

Thus we showed that the preparation of RNA-binding proteins of rabbit reticulocytes compensated completely the absence of the elongation factors EF-1 and EF-2, and the absence of at least the two initiation factors eIF-4C and eIF-5.

Figure 5 shows a densitogram of Coomassie blue stained polyacrylamide gel after electrophoresis of the preparation of rabbit reticulocyte RNA-binding proteins in the presence of sodium dodecyl sulfate. The preparation of RNA-binding proteins from rabbit reticulocytes contained three major polypeptide chains with molecular ratios of about 95 000, 49 000 and 36 000, and also a large number of less prominent components. Co-electrophoresis of the elongation factors EF-1 and EF-2 of reticulocytes and the RNA-binding proteins shows a complete coincidence of their position in the gel with the main polypeptides of the RNA-binding proteins: the main component of the factor EF-1 coincides with the 49,000 Mr polypeptide and EF-2 - with the 95,000 Mr polypeptide (33). The high speci-

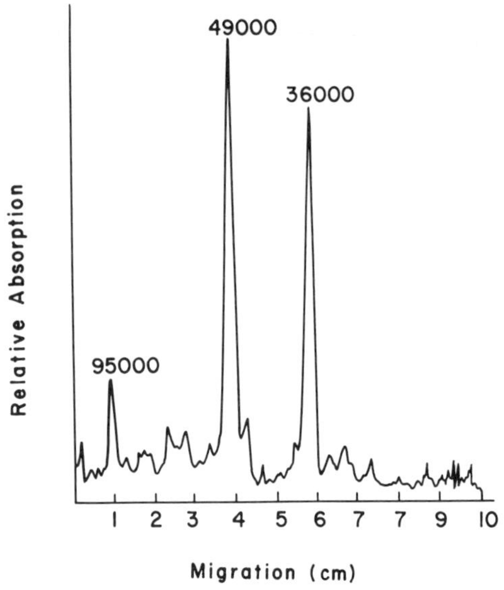

Figure 5. Result of polyacrylamide gel scanning after electrophoresis of the preparation of RNA-binding proteins of rabbit reticulocytes. Electrophoresis was done in the presence of sodium dodecyl sulfate as in (28). (Taken from [30]).

fic EF-1 activity of the RNA-binding protein preparation suggests that the 49,000 Mr component of RNA-binding proteins represents EF-1.

The factors eIF-4C (Mr~19,000), eIF-5 (Mr~160,000) and eIF-1 (Mr~15,000) (31) can correspond to minor components in the preparation of RNA-binding proteins from this source.

A NEW CONCEPT CONCERNING RNA-BINDING AND INFORMOSOMAL RPOTEINS

The discovery of RNA-binding activity of eukaryotic translation factors led to the hypothesis explaining why eukaryotic mRNA is complexed with protein ("Omnia mea mecum porto") (32). It was proposed that: 1) RNA-binding activity is generally characteristic of many eukaryotic proteins having something to do with RNA and RNA-dependent processes. 2) The protein moiety of messenger ribonucleoproteins and informosomes consists of RNA-binding proteins of this kind.

Thus, according to the concept proposed, mRNA in eukaryotic cells carries on itself the proteins which are required for its own biogenesis, existence and functioning. In connection with this, it can be thought that the protein moiety of free cytoplasmic informosomes and polyribosomal messenger ribonucleoproteins consists of proteins serving translation, including both the initiation, elongation and termination factors themselves and various regulators. Thus, in addition to the set of the translation factors, free informosomes can contain special protein components which repress and mask mRNA. It is likely that some enzymes responsible for modifications of the translation factors, mRNA itself and bound regulatory proteins can also be included in the protein moiety of free informosomes or polyribosomal messenger ribonucleoproteins.

In the light of this concept, nuclear ribonucleoproteins of the informosome type must have another set of proteins. If the above-mentioned principle is obeyed, proteins of the nuclear particles must ensure modifications of the newly synthesized mRNA precursors, their processing and mRNA transport from the nucleus.

Since informosomes were discovered only in eukaryotic cells, it was proposed (32) that the capability to bind with RNA is an additional evolutionary acquisition of many eukaryotic proteins which serve protein biosynthesis and other RNA-involving processes to prevent their uniform dilution and to ensure at least their partial compartmentation (due to formation of mRNA-protein complexes) in the large eukaryotic cell. If this is true, then the corresponding prokaryotic proteins, being analogous in function, must not possess the affinity to RNA.

EUKARYOTIC ELONGATION FACTORS (EF-1 AND EF-2) ARE RNA-BINDING PROTEINS WHEREAS THEIR PROKARYOTIC ANALOGS (EF-T AND EF-G) ARE NOT

The following experiments (33) lend direct support to the above hypothesis. We compared the two eukaryotic elongation factors, EF-1 and EF-2, with their prokaryotic equivalents, EF-T_u, EF-$T_u \cdot T_s$, and EF-G, all in a highly purified state.

The eukaryotic elongation factors, EF-1 and EF-2 were isolated from rabbit reticulocytes according to the procedure described in detail by Merrick et al. (34,35). The bacterial elongation factors, EF-T_u, EF-T_s and EF-G, were prepared from Escherichia coli MRE 600 as described by Kaziro et al. (36,37).

The RNA-binding activity of the proteins was measured by two methods: 1) by retention of [^{14}C]RNA on nitrocellulose filters in the presence of the proteins under study (14,16) and 2) by direct adsorption of the proteins on the Sepharose-coupled RNA column (28). Figure 6 shows the dependence of the [^{14}C]RNA$\cdot$protein complex formation measured by the nitrocellulose filter retention technique on the amount of the tested protein added to the incubation mixture containing 0.2 μg of [^{14}C]RNA. First of all it is seen that the eukaryotic elongation factors do have a significant affinity to RNA. The highest affinity of protein to RNA is observed in the case of EF-1; the titration curve is steep, thus indicating a high binding constant. EF-2 is also found to be capable of forming complexes with RNA, though its affinity to RNA seems to be lower than that of EF-1; here, binding of the same amounts of RNA requires four times more protein by weight than in the case of EF-1.

It is striking that no form of the prokaryotic elongation factors tested displays any visible RNA-

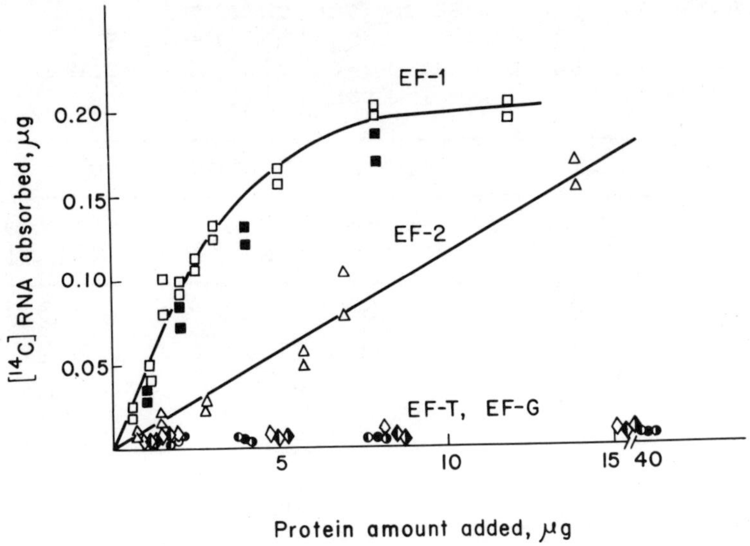

Figure 6. Amount of [^{14}C]RNA retained (in the form of RNA•protein complexes) on nitrocellulose filters versus the amount of the protein added (eukaryotic or prokaryotic elongation factors). The protein was added to the constant 0.2 μg amount of [^{14}C]RNA (4,000 cpm) in a total volume of 100 μl. □, EF-1_L; ■, EF-1_H; △, EF-2; ◐, EF-T_u•GDP; ◑, EF-T_u•GTP; ●, EF-T_u•T_s; ◇, EF-G; ◆, EF-G•GDP; ◆, EF-G•GTP. (Taken from [33]).

binding activity: neither EF-T_u, nor EF-T_u•T_s, nor EF-G, with and without nucleotides, give any retention of radioactive RNA on the filter (Fig. 6).

The other method of assaying the RNA-binding activity of the proteins was used to exclude the possibility that complexes of prokaryotic elongation factors with high molecular weight RNA for some reason could be incapable of being adsorbed on nitrocellulose filters of the Millipore type. Table 4 shows the results of passing the eukaryotic and prokaryotic elongation factors through the Sepharose-coupled RNA column. In all the cases at the given ionic conditions (10 mM Tris-HCl, 10 mM KCl and 1 mM $MgCl_2$) the eukaryotic elongation factors, EF-1 and EF-2, were completely retained in the column; at 1 M KCl they were eluted. On the other hand, neither of the prokaryotic elongation fact-

Table 4. Results of passing the eukaryotic and prokaryotic elongation factors through the Sepharose-coupled RNA column (taken from [33])

Elongation factor	Amount of protein (μg)		
	Applied to the column	Determined in 10 mM KCl wash	Determined in 1 M KCl eluate
$EF\text{-}1_L$	2.5	<0.1	2.3
$EF\text{-}1_H$	3.0	<0.1	3.2
EF-2	2.1	<0.1	1.9
$EF\text{-}T_u$	2.0	2.3	< 0.01
EF-G	2.0	2.1	< 0.05

ors displayed the slightest affinity to RNA in the column, even under the used conditions of a relatively low ionic strength (Table 4).

In our special experiments the total ribosome-free supernatant of the rabbit reticulocyte lysate was passed through the Sepharose-coupled RNA column. The result was that at least 97% of the EF-1 activity and 99% of the EF-2 activity were adsorbed in the column. On the contrary, the activities of $EF\text{-}T_u$ and EF-G were not retained at all in analogous experiments with the total ribosome-free E. coli supernatant, though some proteins of the bacterial extract were adsorbed.

ARE THE ELONGATION FACTORS THE CONSTITUENTS OF INFORMOSOMES?

The demonstration of the RNA-binding activity of the eukaryotic elongation factors suggests that they can complex with mRNA in the cytoplasm. From this, it is likely that the elongation factors, in particular the firmly binding EF-1, are one of the constituents of free cytoplasmic informosomes and polyribosomal messenger ribonucleoproteins.

The result of one-dimensional polyacrylamide gel co-electrophoresis of RNA-binding proteins and the proteins of polyribosomal messenger ribonucleoproteins of rabbit reticulocytes in the presence of sodium

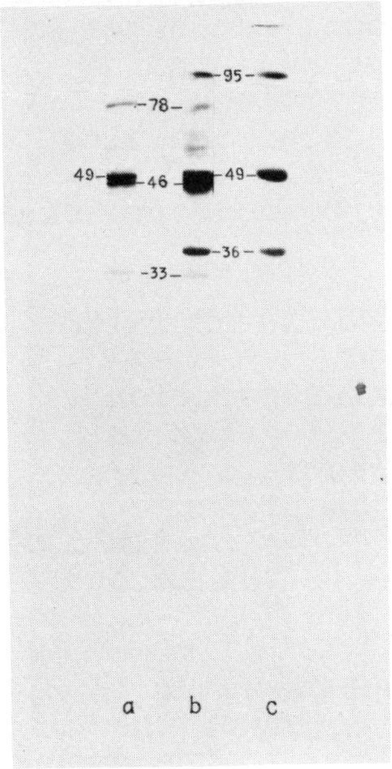

Figure 7. Electrophoretic distribution of proteins of rabbit reticulocyte polyribosomal messenger ribonucleoproteins and RNA-binding proteins in polyacrylamide gel with sodium dodecyl sulfate. Gel a - proteins of 13S polyribosomal ribonucleoproteins; gel c - RNA-binding proteins; gel b - mixture of a and c; numerals indicate the molecular weight (x 10^{-3}) of the corresponding polypeptide chains. (Taken from [38]).

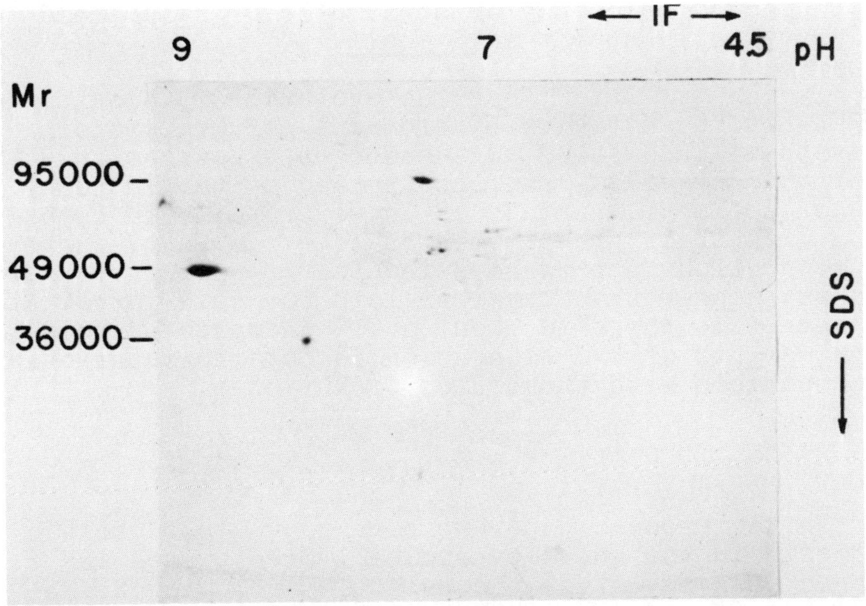

Figure 8. Two-dimensional separation of RNA-binding proteins of rabbit reticulocytes according to the O'Farrel procedure (39). (Performed by V. B. Minikh).

dodecyl sulfate (38) lend support to the suggestion on the presence of EF-1 in the polyribosomal messenger ribonuceloproteins (Fig. 7). One of the RNA-binding proteins of rabbit reticulocytes with a molecular ratio of 49,000 coinciding with EF-1, coincides also in electrophoretic mobility with one of the polypeptides of the polyribosomal messenger ribonucleoproteins.

TWO-DIMENSIONAL SEPARATION OF RNA-BINDING PROTEINS

Figure 8 shows the two-dimensional separation of RNA-binding proteins of rabbit reticulocytes in polyacrylamide gel according to the O'Farrell (39) procedure with isoelectrofocusing in the presence of urea in the first dimension and electrophoresis in the presence of sodium dodecyl sulfate in the second dimension.

A considerable part of RNA-binding proteins of rabbit reticulocytes is found in the region of the pH gradient from 7 to 9, i.e. they belong to the group of moderate basic proteins. Such moderate basic proteins, according to the data of O'Farrell et al. (40) are characteristic only of eukaryotic organisms and are practically absent in E. coli.

It can be thought that the evolutionary acquisition of RNA-binding activity by proteins serving protein biosynthesis and other RNA-dependent processes in eukaryotic organisms was accompanied by the increase of the isoelectric points of these proteins in comparison with their prokaryotic analogs. In fact, eukaryotic elongation factors have isoelectric points 1.5-2.5 pH units higher than their analogs in prokaryotes. At the same time there is about a 20% increase of molecular weights of the elongation factors in eukaryotes in comparison with those from prokaryotes.

SUMMARY

In the present paper we have shown that protein synthesis in the wheat embryo cell-free system was stopped as a result of the selective removal of RNA-binding proteins from the system. On the other hand, the total preparation of RNA-binding proteins from wheat embryos compensated completely the absence of initiation and elongation factors in the system.

The preparations of rabbit reticulocyte RNA-binding proteins were also shown to fully restore the activity of the cell-free system deprived of individual elongation factors EF-1 and EF-2 and initiation factors eIF-4C and eIF-5. Partial compensation of the activity of the system deprived of some other initiation factors was also observed. A conclusion was drawn that the eukaryotic initiation and elongation factors can be RNA-binding proteins.

A direct corroboration of this conclusion was obtained in experiments with pure rabbit reticulocyte elongation factors, EF-1 and EF-2. On the other hand, bacterial elongation factors, EF-T and EF-G, were found to display no RNA-binding activity.

The result obtained is in agreement with the hypothesis (Spirin, A. S. (1978) FEBS Lett. 88, 15-17) that the RNA-binding activity of eukaryotic translation factors is an evolutionary acquisition connected with the increase of the volume of cells and the necessity of compartmentation of the molecules serving RNA and RNA-dependent processes.

The acquisition of RNA-binding activity by eukaryotic proteins was accompanied by some increase of the isoelectric points and the molecular weights in comparison with corresponding prokaryotic proteins.

REFERENCES

1. Spirin, A. S., Belitsina, N. V. and Ajtkhozhin, M. A. (1964) Zh. Obshch. Biol. 25, 321-327 (English translation: Fed. Proc. 24, T907-T915, 1965).
2. Spirin, A. S. (1969) Eur. J. Biochem. 10, 20-35.
3. Ovchinnikov, L. P. and Spirin, A. S. (1970) Naturwiss. 57, 514-521.
4. Spirin, A. S. (1972) in: The Mechanism of Protein Synthesis and its Regulation (Bosch, L., ed.), pp. 515-537. North-Holland Publ. Co., Amsterdam-London.
5. Preobrazhensky, A. A. and Spirin, A. S. (1978) in: Progress in Nucleic Acid Research and Molecular Biology (Cohn, W., ed.), vol. 21, pp. 1-37. Acad. Press, New York.
6. Ajtkhozhin, M. A., Akhanov, A. U. and Doschanov, Kh. I. (1973) FEBS Letters 31, 104-106.
7. Radjabov, Kh. M. and Ovchinnikov, L. P. (1977) Dokl. Akad. Nauk SSSR 237, 732-734.

8. Samarina, O. P., Krichevskaya, A. A. and Georgiev, G. P. (1966) Nature (london) 210, 1319-1322.
9. Samarina, O. P., Lukanidin, E. M., Molnar, J. and Georgiev, G. P. (1968) J. Mol. Biol. 33, 251-263.
10. Samarina, O. P., Lukanidin, E. M. and Georgiev, G. P. (1973) Karolinska Symposia on Research Methods on Reproductive Endocrinology, 6th Symposium. Protein Synthesis in Reproductive Tissue, pp. 130-160. Bogtry Forum, Copenhagen.
11. Henshaw, E. C. (1968) J. Mol. Biol. 36, 401-411.
12. Perry, R. P. and Kelley, D. E. (1968) J. Mol. Biol. 35,37-59.
13. Cartouzou, G., Attali, J. C. and Lissitzky, S. (1968) Eur. J. Biochem. 4, 41-54.
14. Ovchinnikov, L. P., Voronina, A. S., Stepanov, A. S., Belitsina, N. V. and Spirin, A. S. (1968) Molekul. Biol. (USSR) 2, 752-763.
15. Baltimore, D. and Huang, A. S. (1970) J. Mol. Biol. 47, 263-273.
16. Stepanov, A. S., Voronina, A. S., Ovchinnikov, L. P. and Spirin, A. S. (1971) FEBS Letters 18, 13-18.
17. Voronina, A. S. (1973) FEBS Letters 32, 310-312.
18. Stepanov, A. S., Voronina, A. S., Ovchinnikov, L. P. and Spirin, A. S. (1972) Biokhimiya 37, 3-9.
19. Preobrazhensky, A. A. and Ovchinnikov, L. P. (1974) Dokl. Akad. Nauk SSSR 214, 951-954.
20. Preobrazhensky, A. A. and Ovchinnikov, L. P. (1974) FEBS Letters 41, 233-237.
21. Ovchinnikov, L. P. and Avanesov, A. Ts. (1969) Molekul. Biol. (USSR) 3, 893-899.
22. Stepanov, A. S. and Voronina, A. S. (1972) Dokl. Akad. Nauk SSSR 203, 1418-1421.
23. Voronina, A. S. and Stepanov, A. S. (1972) Biokhimiya 37, 437-442.
24. Voronina, A. S., Stepanov, A. S. and Ovchinnikov, L. P. (1972) Biokhimiya 37, 10-15.
25. Avanesov, A. Ts., Tentsov, Yu. Yu. and Ovchinnikov, L. P. (1977) Dokl. Akad. Nauk SSSR 234, 1205-1208.
26. Preobrazhensky, A. A. and Elizarov, S. M. (1975) Bioorg. Khim. 1, 1633-1638.
27. Liautard, J. P., Setyono, B., Spindler, E. and Köhler, K. (1976) Biochim. Biophys. Acta 425, 373-383.
28. Ovchinnikov, L. P., Seriakova, T. A., Avanesov, A. Ts., Alzhanova, A. T., Radzhabov, Kh. M. and Spirin, A. S. (1977) Eur. J. Biochem., submitted for publication.
29. Vlasik, T. N., Ovchinnikov, L. P., Radjabov, Kh. M. and Spirin, A. S. (1978) FEBS Letters 88, 18-20.

30. Ovchinnikov, L. P., Spirin, A. S., Erni, B. and Staehelin, T. (1978) FEBS Letters 88, 21-26.
31. Schreier, M. H., Erni, B. and Staehelin, T. (1977) J. Mol. Biol. 116, 727-753.
32. Spirin, A. S. (1978) FEBS Letters 88, 15-17.
33. Spirin, A. S., Domogatsky, S. P., Vlasik, T. N., Seriakova, T. A. and Ovchinnikov, L. P. (1978) submitted for publication.
34. Merrick, W. C., Kemper, W. M., Kantor, J. A. and Anderson, W. F. (1975) J. Biol. Chem. 150, 2620-2625.
35. Kemper, W. M., Merrick, W. C., Redfield, B., Lin, Ch.-K. and Weissbach, H. (1976) Arch. Biochem. Biophys. 174, 603-612.
36. Kaziro, Y., Inoue-Yokosawa, N. and Kawakita, M. (1972) J. Biochem. (Japan) 72, 853-863.
37. Arai, K., Kawakita, M. and Kaziro, Y. (1972) J. Biol. Chem. 247, 7029-7037.
38. Ovchinnikov, L. P., Avanesov, A. Ts., Seryakova, T. A., Alzhanova, A. T. and Radzhabov, Kh. M. (1977) Eur. J. Biochem., submitted for publication.
39. O'Farrell, P. H. (1975) J. Biol. Chem. 250, 4007-4021.
40. O'Farrell, P. Z., Goodman, H. M. and O'Farrell, P. H. (1977) Cell 12, 1133-1142.
41. Nagata, S., Iwasaki, K. and Kaziro, Y. (1977) J. Biochem. 82, 1633-1646.

METHYLATION OF TRANSFER RIBONUCLEIC ACID

F. Cimino, C. Traboni, P. Izzo, and F. Salvatore

Istituto di Chimica Biologica
2 Facoltà di Medicina e Chirurgia
Università di Napoli, Via S. Pansini 5
80131 Napoli, Italy

The central role of transfer ribonucleic acid (tRNA) in the multistep mechanism of cellular protein biosynthesis justifies by itself the existence of a variety of specific molecules, all of which appear to have a spatial structure basically similar to each other. Moreover, in recent years several other roles of tRNA, besides the structural ones (i.e., those related to the formation of a protein), have been shown to be related to some regulatory cellular processes (1,2). Therefore, a variety of tRNA molecules exists within a single cell (up to 50-60 species), which is due to the different primary sequence of the four *major* nucleosides, as well as to a series of modified, or *minor* nucleosides which are formed at posttrancriptional level. Among these modifications, methylation is the most frequent, both in prokaryote and in eukaryote cells, allowing the formation of a variety of methylated nucleosides, which represent up to 50 per cent of the total number of the modified nucleosides (more than 50 have been identified so far) (3).

SELECTED ASPECTS ON THE STRUCTURE AND THE BIOSYNTHESIS OF tRNA

Our knowledge about the biosynthesis of tRNA is scant compared with the plethora of information about its structure and functions. In the last few years much information became available both on the primary sequence and the three-dimensional configuration of tRNA

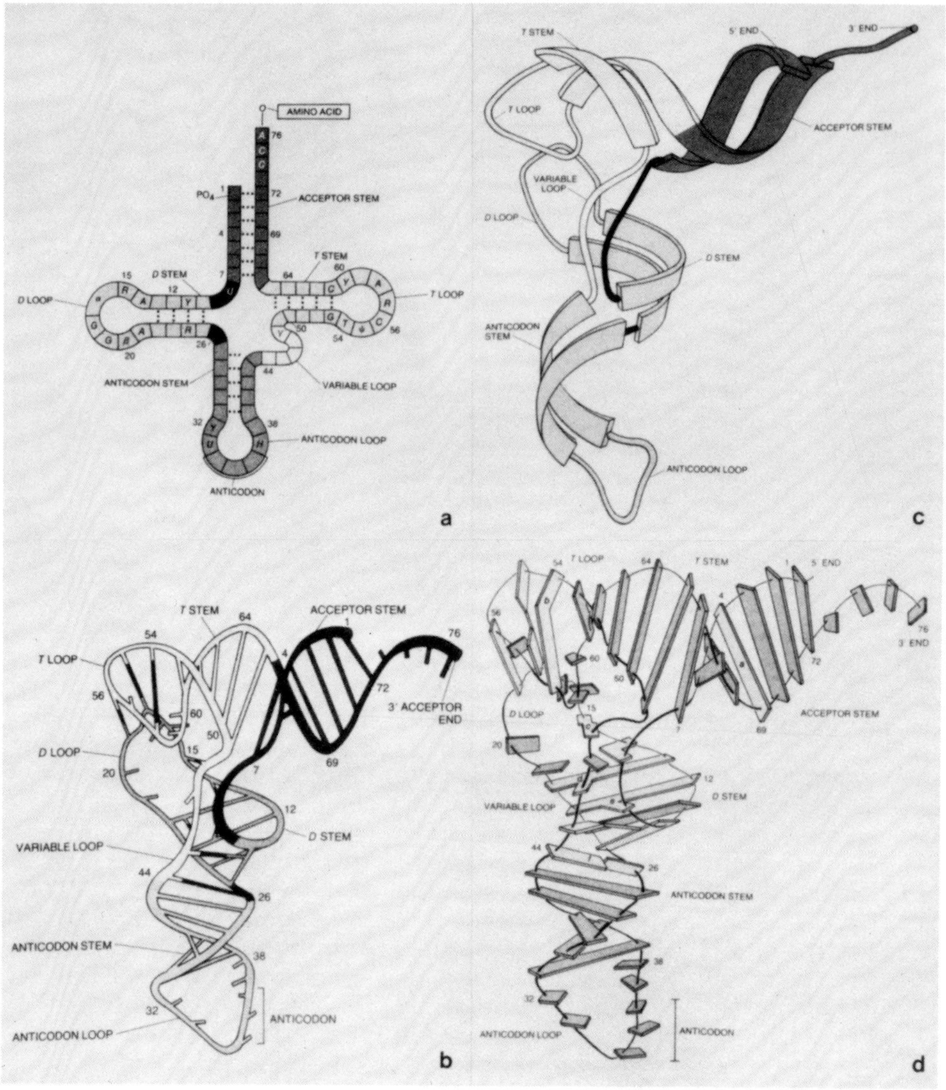

Fig.1. Two- and three-dimensional schematic representations of tRNA structure. **a** Cloverleaf diagram of tRNA, mostly deduced from sequence analysis of more than 80 tRNA species. The number of nucleotides is generally constant except than in α and β portion of D loop, and in the variable loop (up to 21 nucleotides). A, C, G, U, T, Ψ indicate *conserved* bases in all tRNA; R (adenosine or guanosine), Y (cytidine or uridine), H (modified aden-

molecules: the sophisticated X-ray diffraction analysis up to 2.5 Å resolution, has been instrumental to this latter aim (4,5). *Fig.1* shows, on the top, the general cloverleaf diagram of tRNA molecule, and the helical segments of the molecule corresponding to the four stems of the cloverleaf, as they are arranged in the three-dimensional conformation, showing the typical L-shaped structure. On bottom of the same *Figure*, the folding pattern of the polynucleotide chain and the orientation of the bases in yeast phenylalanine tRNA are represented. Phenylalanine tRNA structure has been worked out at very high resolution and the main results deriving from these crystallographic studies (see Ref.4) are: *(i)* the molecule is organized in two base-stacked columns of nucleotides positioned at right angles to each other. This extensive stacking interaction functions as a very strong force in stabilizing the three-dimensional structure of the molecule; *(ii)* the molecule is characterized by the presence of the so-called tertiary interactions formed by hydrogen bonds not of conventional *Watson-Crick type*. These peculiar hydrogen bonds, mostly present among bases called *conserved* and *semi-conserved* (see *Fig.1a*), determine the architectural framework of tRNA molecules.

The structure of tRNA derives from a very complex process of biosynthesis (see 6,7) which is usually described as occurring in two parts (see *Fig.2*). First, tRNA genes are transcribed through the action of RNA-polymerase into a polynucleotide chain which is longer than the mature-size tRNA, and is very often devoid of modified nucleotides. Second, the original transcript is then converted into a mature molecule by an assortment of specific enzymatic reactions, which include: *(i)* nucleolytic tayloring of the longer precursor; *(ii)* 3'- and 5'-end refinement (not always necessary); *(iii)* for-

<

osine or guanosine) indicate *semi-conserved* positions, where one of the bases belonging to the same group (R or Y or H) is present in all tRNA. **b** Helical segments (ribbon-like drawings) of tRNA molecule corresponding to the four stems of the cloverleaf diagram: the four loops are represented as rod-like drawings. **c** Folding pattern of polynucleotide chain of yeast phenylalanine tRNA. **d** Orientation of bases in tRNA, with longer boards representing base pairs (taken from Ref.5 with permission of Scientific American).

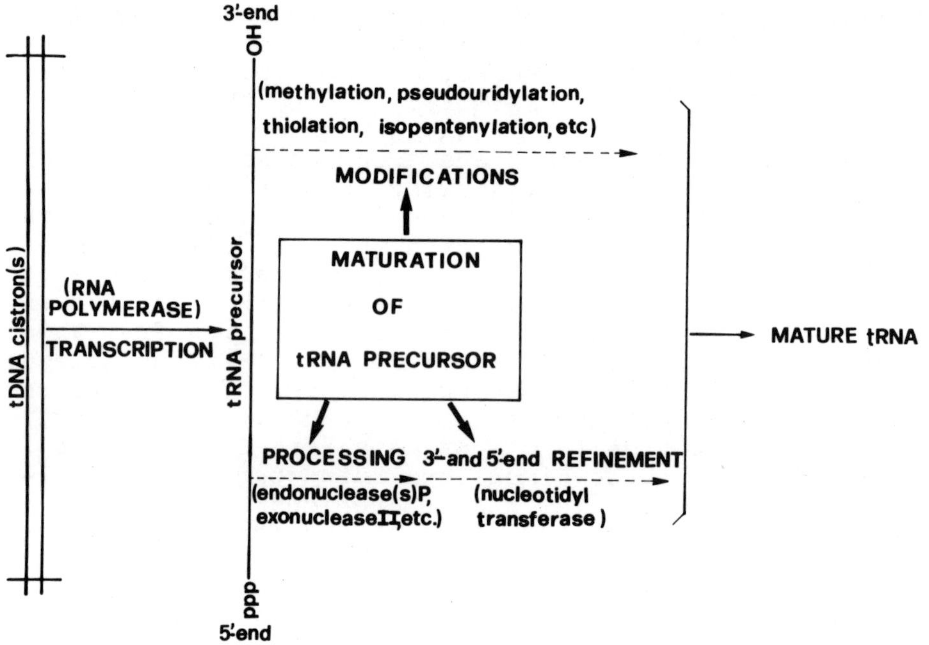

Fig.2. Sequence of events occurring in the biosynthesis of tRNA (derived from Ref.7).

mation of a variety of modified nucleosides up to 16 per cent of the total nucleotide content. As far as the timing of events occurring in tRNA maturation, results obtained suggest that modification reactions take place in some cases prior to, and, in some other, after the trimming process. Furthermore, the three-dimensional structure of yeast phenylalanine tRNA shows that all base modifications occur at the surface of the molecule, thus suggesting that no conformational change of tRNA is required for the modification to take place, and that base modifications are unlikely to play a major role in building up the unrefined spatial configuration (4). Again, the methylation at 2'-O site of the ribose moiety always occurs at specific positions which are among the most exposed regions of the molecule, thus suggesting that methyl groups may well be there for a protection from ribonuclease attack (4).

METHYLATED NUCLEOSIDES IN tRNA

Methyl groups may be attached as single or multiple modification to either purine or pyrimidine base, or to ribose; they are also present within the structure of hypermodified nucleosides, which result from the attachment of a more complex side chain to the four *major* nucleosides (3,7). More than 30 methylated nucleosides have been so far identified in tRNA from different sources. *Fig.3* shows the structure of methylated nucleosides found in tRNA from prokaryote cells, even if present in trace amount. Only three or four of the structures shown in *Fig.3* constitute the bulk of all methylated nucleosides present in each type of prokaryote cell. On the basis of this observation they have been called *major methylated nucleosides* (7), indicating that in each cell type a specific set of tRNA methyl transfer enzymes predominates. This molecular aspect could be of relevant importance in defining the species differences among prokaryote organisms: this hypothesis might be amenable to investigation. Moreover, it seems worthwhile noting that the average number of methylated nucleosides present in a population of tRNA molecules varies from lower to higher organisms: less than 3 per cent in *Mycoplasma*, about 3 per cent in prokaryotes, and 10 per cent in eukaryotes (3). Not only does the absolute number of methylated nucleosides increase from lower to higher organisms, but their variety increases as well, at least as far as the major methylated nucleosides are concerned (see above).

In the last ten years evidences have been presented which indicate variations in patterns of tRNA species in differentiating and malignant cells, often associated with modulation of activity of the enzymes which methylate tRNA. We refer to extensive reviews for a detailed discussion on this matter (8,9): here we shall only mention a recent finding particularly interesting since it shows a qualitative specific change of a malignant cell. In fact, Kuchino and Borek (10) have found that the isoaccepting phenylalanine tRNAs from Novikoff hepatoma and Ehrlich ascites cells both contain two supernumerary methylated bases: one of them, 1-methylguanine, is absent in phenylalanine tRNA from normal rat, calf, mouse and rabbit liver.

A number of studies have also shown an important role of methylated nucleosides in some of the molecular functions of tRNA in protein biosynthesis (amino acid acceptance , codon response, ribosomal attachment, etc.).

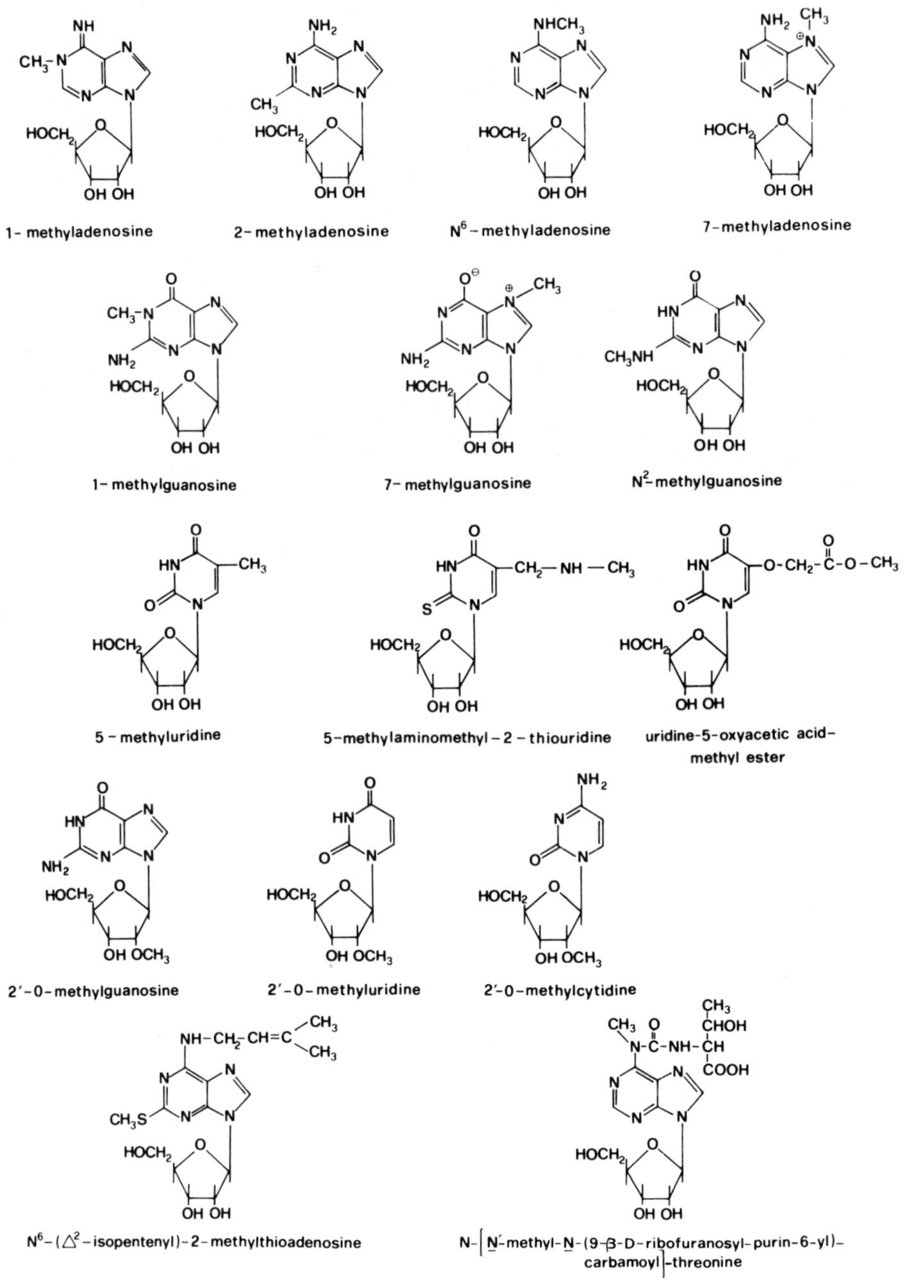

Fig.3. Structures of methylated nucleosides found in tRNA from prokaryote cells.

For a detailed discussion on these aspects we refer to previous reviews in this field (3,7,11,12); we shall limit our discussion only to very recent and relevant results. Pope *et al.*(13) reported that a *supK* strain of *S.typhimurium*, which is able to suppress a non-sense mutation, is defective in a tRNA methylating enzyme which produces the methyl ester of uridine-5-oxyacetic acid in two tRNA species, i.e. alanine tRNA and serine tRNA. This methylated nucleoside is normally present at the first position of the anticodon, thus indicating that such modified nucleoside should be critical for codon-anticodon recognition. It seems also worth mentioning the recent result obtained by Freundlich *et al.* (14), who demonstrated that conditions which *in vivo* produce undermethylated valine tRNA, isoleucine tRNA and leucine tRNA concomitantly produce a ten-fold derepression of valine and isoleucine biosynthetic enzymes, even in the presence of excess of both these amino acids. The derepression observed may be possibly caused by the inability of methyl deficient tRNAs to adeguately participate in normal regulatory functions.

Though these data do indicate a direct involvement of methylated nucleosides in some tRNA functions, it is not yet possible to draw definite conclusions about the biological functions of methylated nucleosides in tRNA. These difficulties have sometime led to propose that tRNA might be a substrate used by the cell to eliminate excess of methyl groups: the so-called *sink function* of tRNA for methyl group. However, most of the Authors in the field (see Ref.3) are reluctant to accept that the evolutionary conservation of a very specific and energy-consuming process would be totally functionless.

tRNA METHYLATING ENZYMES

A) General Properties and Difficulties in Their Study.

S-adenosylmethionine tRNA methyltransferases have been shown to occur in all kind of cells so far studied, both prokaryote and eukaryote (see also Ref.3,7). In eukaryote cell they are mostly localized in cytosol and in mitochondria; a very recent result (15) has shown the presence of these enzymes in nuclei of mouse L-cells. The only enzyme so far known, which uses a methyl donor different from S-adenosylmethionine is 5-mU tRNA methyltransferase from *B.subtilis*, for which the methyl donor is N^5-methyltetrahydrofolic acid (16,17).

This class of enzymes bears high specificity toward the tRNA substrate. It should be pointed out (see 7) that three requirements must be fulfilled in order to achieve the enzymatic attachment of a methyl group to tRNA. In fact, each enzyme must recognize: *(i)* the proper moiety along the polynucleotide chain (either the specific base or the ribose); *(ii)* the position of the modification at the purine, pyrimidine or furanosic ring; *(iii)* the localized nucleotide sequence and the spatial locus of the three dimensional configuration in which the *methylatable* nucleoside is positioned. These three requirements allow to define three different types of specificity for tRNA methyltransferase, namely: *moiety* specificity, *ring-atom* specificity, and *site* specificity, respectively (see 7).

Though some recent success has been obtained in the purification and in the study of molecular properties of some methylating enzymes (18-21), the general picture which emerges from the extensive studies carried out so far is that the enzymology and the physical chemistry of these molecules are still poorly known. Most likely, this lack of knowledge contributes to hamper the understanding of the role of methylation in the functions of tRNA, both structural and regulatory.

The major difficulties faced in the purification of tRNA methyltransferases are: *(i)* enzyme multiplicity, which makes cumbersome to isolate each specific reaction product, and then difficult to evaluate specific activity and extent of purification; *(ii)* enzyme instability; *(iii)* frequent contamination by ribonuclease activity; *(iv)* possible presence of complexes with endogenous tRNA; *(v)* unavailability of a proper tRNA substrate, due to difficulty in preparing suitable amounts of single undermethylated or precursor tRNA species. All these factors would very likely cause missing or masking of one or more enzymes (see 7 for a detailed discussion on this point).

B) tRNA Methyltransferases from *S.typhimurium*.

In order to contribute to the studies on the role(s) of methylated nucleosides in the functions of tRNA, as well as to investigate the timing of the methylation reactions during the multistep process of maturation of tRNA, we have started to study the pattern of tRNA methyltransferases in *S.typhimurium*.

The first approach was to investigate, by using a *met*⁻ strain of *S.typhimurium* grown in presence of $^{14}CH_3$-methionine, the tRNA methyltransferases functioning *in vivo* during cell growth. Isolation of tRNA and analysis of radioactive nucleosides, derived from enzymatic hydrolysis of the polynucleotide chain, has been performed to this aim. *Fig.4a* shows at least ten spots of methylated nucleosides which have been definitely identified. *Fig.4b* reports the pattern of methylated nucleosides obtained after enzymatic hydrolysis of an undermethylated tRNA, remethylated *in vitro* by a crude extract of *S.typhimurium* and $^{14}CH_3$-adenosylmethionine: the undermethylated tRNA was isolated by a *rel*⁻ *met*⁻ mutant of *S.typhimurium*. A comparison of the results obtained *in vivo* and *in vitro* is summarized in *Fig.4c*: the enzymes which produce m^2A, Gm, Um, ms^2i^6A are missed or masked in the crude extract of *S.typhimurium*. Furthermore the quantitative ratios of radioactivity *in vivo* and *in vitro* appear to be different. Among the factors that could explain such discrepancies, have to be taken into account: enzyme instability, different enzyme affinity for one or both substrates, need for particular cofactors which are present in the *in vivo* conditions, and, most of all, the use *in vitro* of a crude undermethylated preparation of tRNA. Such preparation contains a heterogenous population of undermethylated tRNA with an uneven distribution of methyl groups and different extent of methylation.

Purification of the various tRNA methyltransferases present in the crude extract of *S.typhimurium* was attempted by the use of phosphocellulose chromatography (phosphocellulose resembles the polyanionic backbone of a nucleic acid)(7). A typical fractionation is shown in *Fig.5*, where four distinct peaks of activity are separated. Each of these activities appears to be highly purified since about 90 per cent of the proteins are eluted in the void volume. The product specificity of the four enzymes is indicated in the legend of *Fig.5*: only one enzyme produces four different methylated products, thus indicating that such enzyme preparation is contaminated by the two adjacent peaks and by another activity producing a not yet identified methylated nucleoside . The specificity for m^5U of the activities present in the flow-through and in the third peak was unambiguously proved by the use of a tRNA extracted from an *E. coli* mutant defective in ribothymidine methyltransferase (22). This result supports the hypothesis that the two activities belong to the same protein molecule. Activities present in peaks A, B and D were uneffective when

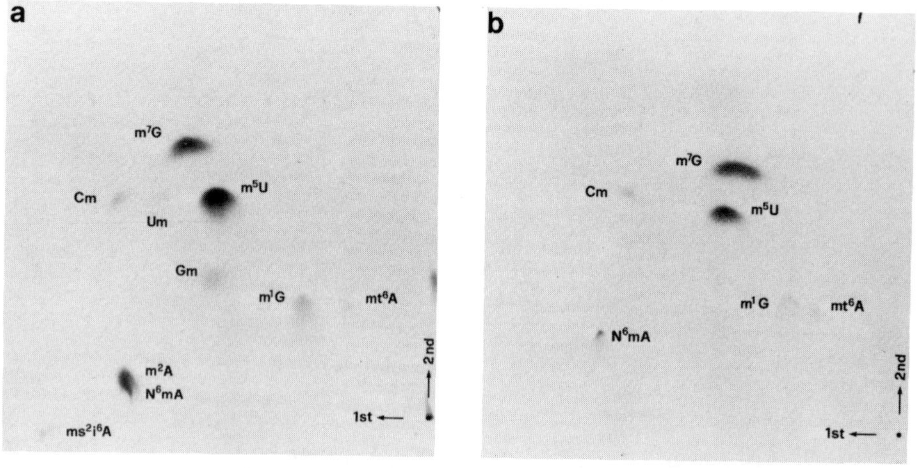

c

NUCLEOSIDE NAME AND ABBREVIATION	RELATIVE AMOUNT*	
	IN VIVO	IN VITRO
5-methyluridine (m^5U)	100	63
7-methylguanosine (m^7G)	35.80	100
2-methyladenosine (m^2A)	16.42	---
2'-O-methylguanosine (Gm)	15.05	---
1-methylguanosine (m^1G)	8.51	13
N^6-methyladenosine (N^6mA)	7.58	15
2'-O-methyluridine (Um)	6.83	---
2'-O-methylcytidine (Cm)	5.41	18
N-[$\underline{N}^1$-methyl-$\underline{N}$-(9-β-D-ribofuranosylpurin-6-yl)-carbamoyl]-threonine (mt^6A)	4.31	7
N^6-(Δ^2-Isopentenyl)-2-methylthioadenosine (ms^2i^6A)	4.31	---
3-methylcytidine (?) (m^3C)	2	19

*taken equal to 100 the radioactivity measured in the most abundant nucleoside

Fig.4. Patterns of methylated nucleosides in tRNA of *S.typhimurium* obtained in experiments performed *in vivo* and *in vitro*, as described in Ref.21. **a.** *in vivo* experiment showing methylated nucleosides obtained from tRNA extracted and digested as reported in Ref.21; cells of *met*⁻ strain of *S.typhimurium* were grown in presence of

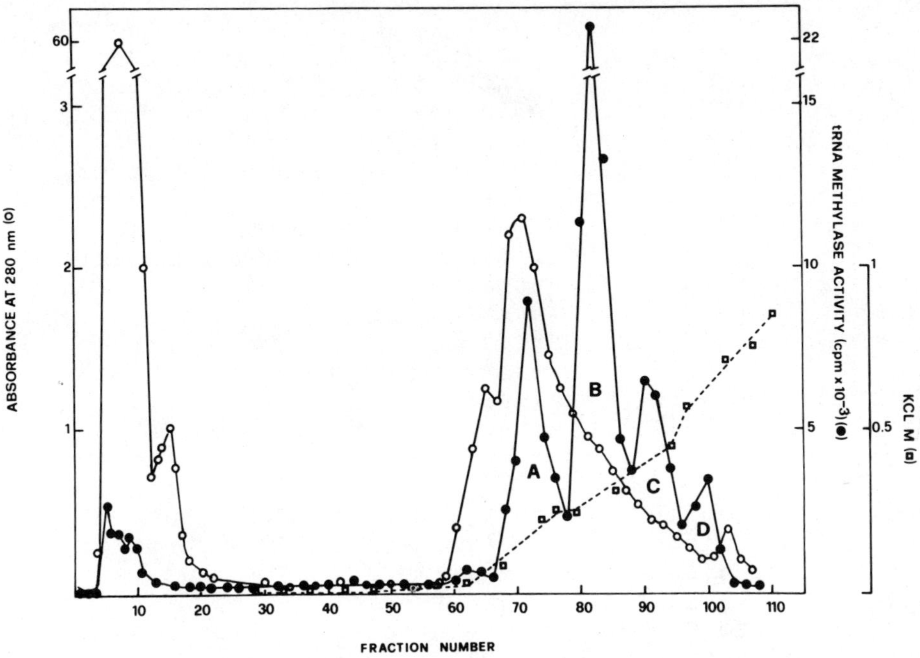

Fig.5. Phosphocellulose chromatography profile of tRNA methyltransferases from a crude extract of *S.typhimurium* (taken from Ref.21; see also Ref.7). Product specificity of the four peaks was the following: mt^6A for peak A; m^1G for peak B; m^5U, m^1G, m^7G, m^3C(?) for peak C; m^7G for peak D; m^5U for the activity present in the flow-through.

←

$^{14}CH_3$-methionine. **b**. *In vitro* experiment showing methylated nucleosides obtained from an undermethylated tRNA, methylated *in vitro* with $^{14}CH_3$-adenosylmethionine and crude extract of *S.typhimurium*. **c**. Quantitative comparison between the experiments reported in **a** and **b**.

tested against tRNA specifically lacking m^5U.

Table I shows the apparent K_M for adenosylmethionine and crude undermethylated tRNA of the four enzymatic activities separated by phosphocellulose chromatography. While the apparent affinity constants for adenosylmethionine are very similar, those for undermethylated tRNA appear to be scattered in a wider range. This could imply that the site for adenosylmethionine is similar in all the different enzymes, whereas interaction with tRNA implies the presence of enzyme sites which are more typical for specific methylatable sites. It must be noted, however, that the measure of K_M for tRNA can be strongly influenced by the use of a non suitable substrate, like crude undermethylated tRNA, in which it is difficult to assess precisely what proportion of the entire preparation is the true substrate.

The molecular size of the four methylase activities was studied by comparing their sedimentation profile with that of proteins of known molecular weight (see *Fig.6*). Peak A is resolved in two activities: whether they belong to the same protein molecule deserves further investigation. A general interesting feature, which appears from the results of *Fig.6*, is that the four tRNA methylases (except the larger fraction present in peak A) have a M.W. ranging from 25,000 to 65,000 daltons, which is lower than those of other tRNA methyltransferases so far studied (19,23-25).

Table II reports preliminary experiments (performed in collaboration with Dr. Zappia's group) on the effect of adenosylhomocysteine and some of its analogs on the activities of tRNA-(guanine-7)-methyltransferase from *S.typhimurium* (peak D of *Fig.5*). These data indicate that also for this enzyme activity adenosylhomocysteine is a powerful inhibitor as for other methyl transfer reactions. Furthermore, the analysis of the results obtained with the last three analogs listed in Table II indicates that the inhibitory effect depends on the presence of the carboxyl group and of the amino group of the amino acid chain, regardless of the length of carbon chain attached to the sulfur. Finally, the inhibitory effect is significantly reduced when the amino acid moiety is shortened.

Table I. Apparent affinity constants (K_M) for adenosylmethionine and undermethylated tRNA of tRNA methyltransferases from *S.typhimurium* (from Cimino *et al.*, 21).

Methylase Activity *	Apparent K_M	
	Adenosylmethionine	Undermethylated tRNA
Peak A	1.5×10^{-5}M	3.1×10^{-4}M
Peak B	3.2×10^{-5}M	1.2×10^{-4}M
Peak C	2.2×10^{-5}M	6.3×10^{-5}M
Peak D	1.5×10^{-5}M	3.1×10^{-5}M

* from the phosphocellulose column (see *Fig.5*).

Table II. Effect of adenosylhomocysteine and its analogs on the activity of tRNA-(guanine-7)-methyltransferase from *S.typhimurium*.

Compound	Relative activity*		
	A	B	C
None	100	100	100
L-Adenosylhomocysteine	20	7	2
D-Adenosylhomocysteine	75	71	21
DL-Adenosylhomocysteine	51	17	8
L-Adenosylcysteine	71	67	38
Methylthioadenosine	80	84	28
Thioethanoladenosine	105	127	86
n-Butylthioadenosine	109	112	78
Isobutylthioadenosine	112	114	103

* In columns A, B and C the results obtained by using concentration of analog corresponding respectively to 1,10 and 60 fold the K_M for adenosylmethionine (1.5×10^{-5}M) are reported; for each column the activity measured in absence of inhibitor is taken equal to 100.

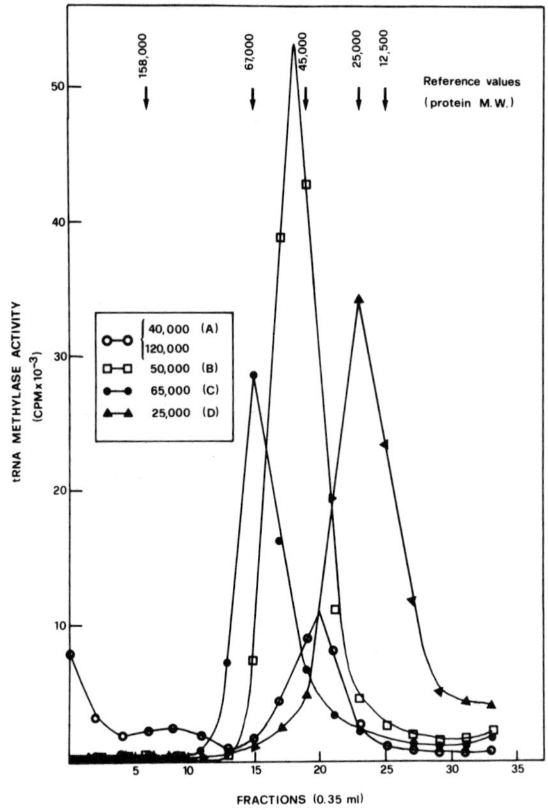

Fig.6. Rate-zonal sedimentation profiles of tRNA methyltransferases in glycerol gradient (taken from Ref.21). (A), (B), (C) and (D) refer to the four peaks of enzyme activities spearated by phosphocellulose chromatography (see previous *Figure*).

CONCLUDING REMARKS

Modified nucleosides account up to 16 per cent of the total nucleoside content of tRNA. The study of their function has been hampered by several difficulties and only very few definite data are available so far. The well established role of pseudouridine in the anticodon region of several tRNAs, for their regulatory functions (26) at the level of the operon of metabolic pathways for the biosynthesis of several amino acids, is the most impressive evidence of the role of a modified nucleoside in tRNA functions. As far as tRNA methylation is concerned, some of the most recent and interesting results (10,13,14; see also 7,11,12) have been discussed before. However, we may say that the study of the function of tRNA methylation is still at its beginning. The use of bacterial mutants is certainly an important tool for approaching such a problem, though the production of these mutants as well as their characterization is a difficult task. Progress could be achieved also by using an *in vitro* cell-free protein synthesis system, tRNA dependent, or even better if dependent on a single species of tRNA. In this ideal system the efficiency of undermethylated species of tRNA might be tested and compared to that of normal species. The several functions connected with the structural role of tRNA, i.e. amino acid adaptor in protein synthesis, could be then studied in a very controlled system.

Another important point is the clarification of the enzymology and timing of the several steps involved in tRNA maturation.To this aim the availability of tRNA precursors, or even better of single species of precursor tRNA, would be instrumental. In fact, these precursors, together with the entire battery of enzymes and factors necessary to achieve the maturation of tRNA, might be used for the study of tRNA maturation in a controlled *in vitro* system.

ACKNOWLEDGEMENT

The experimental work carried out in the Authors' Laboratory has been supported by grants from CNR (Rome) and by an International Science Cooperative Program (CNR, Italy-National Science Foundation, USA).

REFERENCES

1. Clark, B.F.C., *Prog.Nucleic Acid Res.Mol.Biol.*, 20, 1 (1977).

2. Cortese, R.,in *Biological Regulation and Development*, R.F. Goldberger (Ed), vol.I (Plenum Press, New York and London, 1978), p.401.

3. Nau, F., *Biochimie*, 58, 629 (1976).

4. Kim, S.H., and Sussman, J.L., *Horizons Biochem.Biophys.*, 4, 159 (1977).

5. Rich, A., and Kim, S.H., *Scientific American*, vol. **238**, n.1, p.52 (1978).

6. Smith, J.D., *Prog.Nucleic Acid Res.Mol.Biol.*, **16**, 25 (1976).

7. Salvatore, F., and Cimino, F., in *The Biochemistry of Adenosylmethionine*, F. Salvatore, *et al.*, (Eds.), (Columbia University Press, New York, 1977), p.187.

8. Borek, E., and Kerr, S.J., *Adv.Cancer Res.*, **15**, 163 (1972).

9. Kerr, S.J., *Adv.Enzyme Regul.*, **13**, 379 (1974).

10. Kuchino, Y., and Borek, E., *Nature*, **271**, 126 (1978).

11. Littauer, U.Z., and Inouye, H., *Ann.Rev.Biochem.*, **42**, 439 (1973).

12. Kerr, S.J., and Borek, E., in *The Enzymes*, P.D. Boyer (Ed.), vol.9, pt.B, 3d ed. (Academic Press, New York, 1973), p.167.

13. Pope, W.T., Brown, A., and Reeves, R.H., *Nucleic Acid Res.*, **5**, 1041 (1978).

14. Rizzino, R., Mastanduno, M., and Freundlich, M., *Biochim.Biophys.Acta*, **475**, 267 (1977).

15. Colonna, A., and Kerr, S.J., *Cell*, in press (1978).

16. Kersten, H., Sandig, L., and Arnold, H.H., *FEBS Letters*, **55**, 57 (1975).

17. Delk, A.S., and Rabinowitz, J.C., *Proc.Natl.Acad. Sci. U.S.A.*, **72**, 528 (1975).

18. Wierzbicka, H., Jakubowski, H., and Pawelkiewicz, J., *Nucleic Acid Res.*, **2**, 101, (1975).

19. Glick, J.M., and Leboy, P.S., *J.Biol.Chem.*, **252**, 4790 (1977).

20. Aschoff, H.J., and Kersten, W., in *The Biochemistry of Adenosylmethionine*, F. Salvatore, *et al.*, (Eds.), (Columbia University Press, New York 1977) p.231.

21. Cimino, F., Traboni, C., Colonna, A., Izzo, P., and

Salvatore, F., submitted for publication.

22. Björk, G.R., and Isaksson, L.A., *J.Mol.Biol.*, 51, 83, (1970).

23. Aschoff, H.J., Elten, H. Arnold, H.H., Mahal, G., Kersten, W., and Kersten, H., *Nucleic Acids Res.*, 3109 (1976).

24. Björk, G.R., and Kjellin-Stråby, K., in *The Biochemistry of Adenosylmethionine*, F. Salvatore *et al.* (Eds.), (Columbia University Press, New York, 1977), p.216.

25. Izzo, P., and Gantt, R.R., *Biochemistry*, 16, 3576, 1977.

26. Singer, C.E., Smith, G.R., Cortese, R., and Ames, B.N., *Nature New Biology*, 238, 72 (1972).

INTERACTION OF *E.COLI* RNA POLYMERASE WITH SUBSTRATES DURING INITIATION OF RNA SYNTHESIS AT DIFFERENT PROMOTERS

E.D.Sverdlov, S.A.Tsarev, T.L.Levitan, V.M.Lipkin, N.N.Modyanov, M.A.Grachev, E.F.Saychikov*, A.G.Pletnev*, Yu.A.Ovchinnikov*

Shemyakin Institute of Bioorganic Chemistry, USSR Academy of Sciences, Moscow, USSR

*Institute of Organic Chemistry, Siberian Division of the Academy of Sciences, Novosibirsk, USSR

Escherichia coli RNA polymerase is an object of intensive study. However, even the part played by its various subunits at the individual stages of its functioning is not unequivocally known.

We attempted to attack this problem using photoaffinity modification by analogs of substrates and template. The present report describes the first results obtained with analogs of nucleoside-5'-triphosphates having photoreactive groups either at the γ-phosphate residue, or at the heterocyclic nucleus and also with photoreactive analogs of template.

A. INTERACTION OF RNA POLYMERASE WITH NEWLY SYNTHESIZED RNA

ATP γ-anilide (compound 1 in Fig. 1) is known to be a substrate of *E.coli* RNA polymerase [1]. In the course of the present studies we found that γ-azidoanilides of ATP and GTP (Fig. 1, compounds II and III, respectively) are also substrates of the enzyme. Therefore these analogs may be used for photoaffinity labeling of RNA polymerase within transcribing complexes. The γ-azidoaniline residue is present at the 5'-terminus of nascent RNA. Formation of new phosphodiester bonds leads to a shift of the photoreactive group along the complex. Hence, illumination of the complex at different moments of time in the course of RNA synthesis makes it possible to label different sites of RNA polymerase forming the "corridor" along which the product leaves the transcriptional complex.

In the present studies we used promoter-containing templates which were

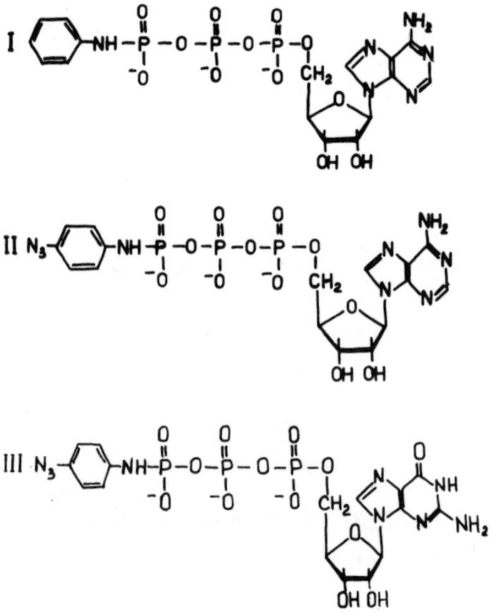

Fig. 1. The various RNA polymerase substrates

individual restriction fragments of phage DNA's. At present primary structures of many promoter recognized by *E.coli* RNA polymerase are known. All these structures appeared to be different [2]. In our opinion, attempts to find common features of these sequences [2,7] are not very successful. It may well happen that the modes of interaction of different promoters with RNA polymerase are very different. The reasons for such diversity may be the necessity for RNA polymerase to be operated by various regulatory peptides or proteins. Therefore it is possible that similar photoaffinity reagents will label different subunits of RNA polymerase or different sites of their polypeptide chains.

One of the promoter containing fragments was obtained from DNA of bacteriophage λi^{434}. Fig. 2. shows the partial genetic and physical map of this genome. The heavy line shows the fragment which is 1150 b.p. long and contains the complete gene of oop-RNA [3], including promoter and terminator. This fragment is obtained by cleavage of λi^{434} DNA by restriction nuclease EcoRI and can be prepared in large amounts [4].

The 5'-terminal sequence of oop-RNA is pppGUUGAUAGAUC... [3,5]. The first four residues of this sequence are only composed of two base residues G and U. The first ten residues are a combination of three base residues G, U, A. The first C residue is at position 11. Therefore, it may be expected

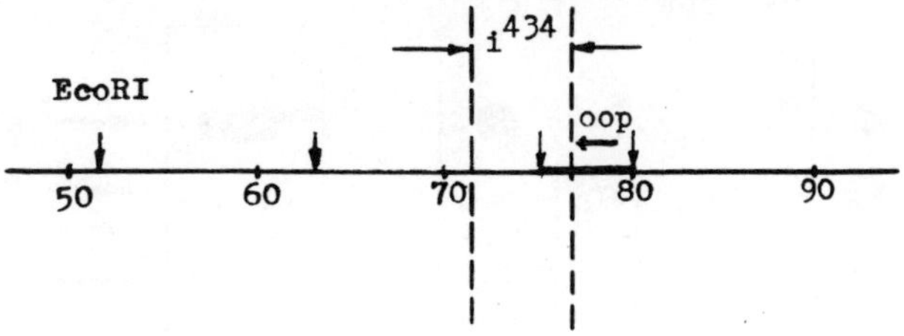

Fig. 2. Physical and genetic map of the immunity and ori regions of phage λi^{434}. The positions on the map (%λ) are reffered to the left terminus as 0%λ. The 1%λ unit corresponds to 465 b.p. The positions of genes N, CI and CII are indicated on the map. The vertical dashed lines determine the immunity region of the phage 434. The positions of the EcoRI cut are indicated by arrows. The heavy line indicates the fragment containing the gene of the oop-RNA. The map is not drawn precisely to scale.

that the tetranucleotide pppGUUG will be synthesized with an uncomplete combination of substrates RpppG + UTP and the decanucleotide RpppGUUGAUAGAU - with the combination RpppG + UTP + ATP, where R stands for the γ-azidoaniline residue.

Fig. 3a shows the result of an affinity labeling experiment with a combination of substrates RpppG + [α - ^{32}P]- UTP. Obviously, any radioactivity covalently bound by RNA polymerase in such an experiment on illumination at λ > 290 nm may be only due to oligonucleotides synthesized in the system, rather than to any of the starting substrates, because the one with the photoreactive group is nonradioactive, whereas the one with the radioactive label is not photoreactive. After synthesis and illumination, the reaction mixture was treated with SDS* to dissociate RNA polymerase into subunits which were separated subsequently by gel-electrophoresis in SDS. The radioactivity was traced by autoradiography, and the subunits band were stained by Coomassie. It is seen that only β,β'- and σ-subunits become radioactive after photoaffinity labeling.

There is no doubt, that the selectivity of the labeling is very high: (1) It

---

* Abbreviation used: SDS, sodium dodecyl sulfate; IU, 5-iodouridine; IUTP, 5-iodouridine triphosphate; BSA, bovine serum albumine.

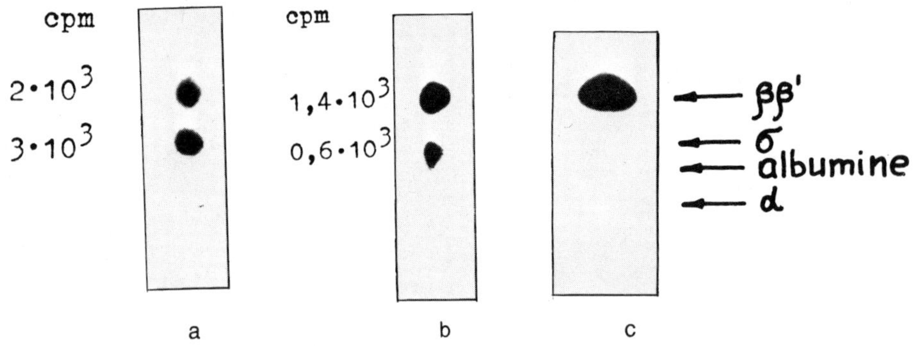

Fig. 3. Crosslinking of RNA polymerase with photoreactive oligonucleotides in transcribing complex. Standard reaction mixture (20 μl) contained 4.5 μg of RNA polymerase, 0.1 μg of λi434 DNA fragment, 0.2 mM γ-azidoanilide of GTP, 8μM [α-^{32}P]-UTP (350 Ci/mmol), 10 mM $MgCl_2$, 20 mM Tris-HCl, pH 7.9, 50 mM NaCl, 4.5 μg of BSA. After 7 min incubation at 30°C, the mixture was irradiated for 1 - 2 min using SVD-12A mercury lamp and cut-off filter transmitting wave lengh greater than 290 nm. RNA polymerase was dissociated to subunits by adding 20 μl of SDS buffer (3% SDS, 5% 2-mercaptoethanol, 10% glycerol and 0.063 M Tris-HCl, pH 6.8) and heating to 90°C for 2 min. The subunits were subjected to electrophoresis on SDS-polyacrylamide gradient gel (3- 30%). The radioactivity was detected by autoradiography and the positions of subunits - by staining with Coomassie. The positions of the RNA polymerase subunits are indicated by arrows. The figures to the left of the autoradiographs represent the Cerencov radiation of the bands excised from the gel.

a, standard mixture; b, standard mixture supplemented with 0.2 mM ATP; c, standard mixture with denaturated DNA fragment.

was mentioned that the covalently bound radioactivity may only be due to nascent oligonucleotides. (2) Albumine which was present in all the reaction mixtures was never modified, if illumination was performed in a short time after addition of substrates. (3) Subunits remain unlabeled, if GTP was present instead of RpppG. (4) Subunits remain unlabeled, if the complex consisting of RNA polymerase, template and nascent oligonucleotide is destroyed before illumination. (5) Subunits also remain unlabeled, if a mixture of substrates and enzyme is illuminated without template. (6) Labeling does not occur without illumination.

Subunits are also labeled selectively with the combination of substrates

RpppG, [α-^{32}P]-UTP and ATP (Fig. 3b). However, radioactivity predominates in this case in β,β'-subunits. Presumably, this reflects shift of the 5'-terminus of nascent oligonucleotides within the transcribing complex when this lenght increases continously up to the decanucleotide RpppGUUGAUAGAU.

Fig. 3c shows the result of affinity labeling obtained with the fragment, which was denaturated before the experiment. It is seen, that with such denaturated template labeled are only β,β'-subunits, and no label enters the σ-subunit. The same result is obtained, if a few nicks are introduced into the fragment, e.g., by mild treatment with DNAse. Hence, labeling of the σ-subunit is observed only under specific initiation conditions. Probably, the role of the σ-subunit is not limited by recognition of the promoter, it may also directly participate in RNA synthesis initiation.

A similar study was performed with a fragment of DNA of bacteriophage T7, which contains the three early promoters A_1, A_2 and A_3. The fragment was obtained by cleavage of T7 DNA by restriction endonuclease BsuI [6]. Qualitatively, the results obtained with this fragment and combinations of substrates, which should result in oligonucleotides RpppGC (promoter A_2 [7]) and and RpppAUC (promoter A_1 [8]) are similar to those described above - the label appears in the β,β'- and σ-subunits. However, the combination of substrates which should finally give RpppAUGAAA (promoter A_3 [7]) provides also labeling of the α-subunit. Two alternative explanations may be proposed to this end. It may happen, that contact of the α-subunits with the triphosphate residue of the nascent oligonucleotides is the case only at its length equal to six monomer units. On the other hand, the result might also reflect the different topography of transcribing complexes with different promoters.

Finally, β,β'-, σ- and, in some cases, α-subunits contact with the 5'-triphosphate group of growing RNA at different steps of the transcription.

In order to find out which subunits of the enzyme contact with the 3'-terminus of nascent RNA, we used the reaction of photoaffinity labeling by means of a photoreactive NTP in combination with an initiating dinucleoside monophosphate. These experiments were made using the λi^{434} fragment as template. Table 1 shows the results of a study of the efficiency of initiation by different dinucleoside monophosphates.

It is seen that only three dinucleoside monophosphates UpG, GpU and CpU are active in initiation. Comparison of this evidence with the primery structure of the promoter region for oop-RNA reveals that the three dinucleoside monophosphates correspond to the site of oop-RNA initiation (Fig. 4).

TABLE 1. Stimulation of RNA Synthesis by Dinucleoside Monophosphates with the Fragment of DNA as a Template* λi^{434}.

NpN used for initiation	acid insoluble radioactivity (cpm)	±m
GpA	39.5	48.3
ApU	20	50.4
UpG	818.5	52.5
GpU	311.5	11.9
CpU	739.5	83.3
UpG	28.5	109
ApC	122	112
UpU	17	9.8

* The reaction mixture (20 μl) contained: polymerase -1μg, dinucleoside monophosphate - 0.32 mM, λi^{434} DNA fragment - 0.6 μg. ATP, GTP, CTP and ^{14}C-UTP (6·10^8 cpm/mmol) - 8 μM of each compound. Acid insoluble radioactivity was determined in 10 min at 37° C.

On of these oligonucleotides, GpU, was used as the initiating one in the photoaffinity labeling experiments. In combination with this compound, we used IUTP as a photoreactive substrate. In order to render the labeling oligonucleotide radioactive, [α-^{32}P]-GTP was added to the reaction mixture. The combination GpU, IUTP and [α - ^{32}P]-GTP, according to the structure of oop-RNA, should yield the oligonucleotide G-p-U-p-IU-^{32}p-G with the photoreactive IU residue nearly the 3'-terminus. Similary to the experiments above described, only the nascent oligonucleotide combines radioactivity and photoreactivity.

```
                  oop-RNA
                 <--------
          CUAGAUAGUUGppp

5'   TCTTACTGGATCTATCAACAGG   3'
3'   AGAATGACCTAGATAGTTGTCC   5'
```

Fig. 4. The sequence of the oop-RNA* promoter region. oop-RNA 5'-end sequence is shown at the top of the picture. The scores below indicate the dinucleotides capable to initiate the RNA synthesis.

* This sequence has been established by Scherer et al. [5] and confirmed also by our group from Shemyakin Institute of Bioorganic Chemistry of the Academy of Sciences in cooperation with the group of Dr. K.Skryabin of the laboratory of Prof. A.Baev (Institute of Molecular Biology of the Academy of Sciences of USSR, Moscow).

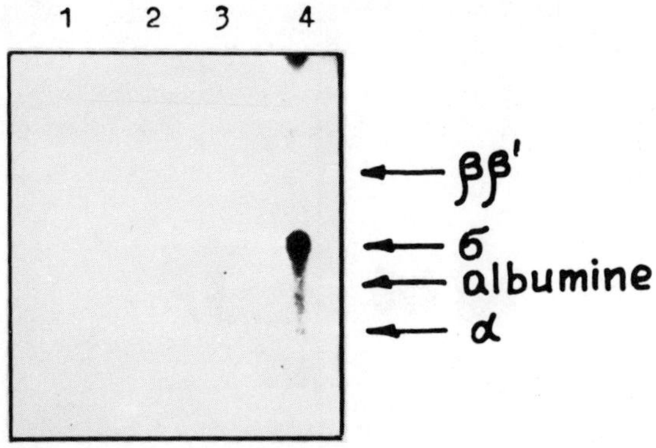

Fig. 5. Crosslinking of RNA polymerase subunits with photoreactive oligonucleotide synthesized using GpU, IUTP, and [α-^{32}P]-GTP as substrates and the λi^{434} DNA fragment as the template. The reaction mixture (20 μl) contained: 4.5 μg RNA polymerase, 0.2 μg of DNA fragment, 0.13 mM IUTP, 0.2 mM GpU, 9 μM [α-^{32}P]-GTP (350 Ci/mM), 4.5μg of BSA. After 10 min incubation at 37° C the mixture was irradiated for 10 min and subjected to treatment as described in the legend to Fig. 1.

1, mixture withour IUTP; 2, reaction with UTP insted of IUTP; 3, reaction mixture without the DNA fragment; 4, reaction mixture.

After the synthesis, the reaction mixture was illuminated by light of λ>290 nm, the subunits of RNA polymerase were separated by electrophoresis in SDS.

Fig. 5 shows the autoradiograph of the gel after the separation. The position of subunits were found by means of staining with Coomassie blue. It is seen that the only subunit which is labeled under such conditions is σ. This labeling is specific, because: (i) Albumine present in the same mixture is not labeled. (ii) RNA polymerase is not labeled in the absence of 5-iodoridine-5'-triphosphate (columns 1 and 2), or in the absence of template (column 3). This result is one more indication to the immediate participation of the σ-subunit in initiation of RNA.

B. INTERACTION OF RNA POLYMERASE WITH DNA

Mechanism of the interaction of proteins with nucleic acids is a problem the solution of which seems to be quite far. There are no general approaches

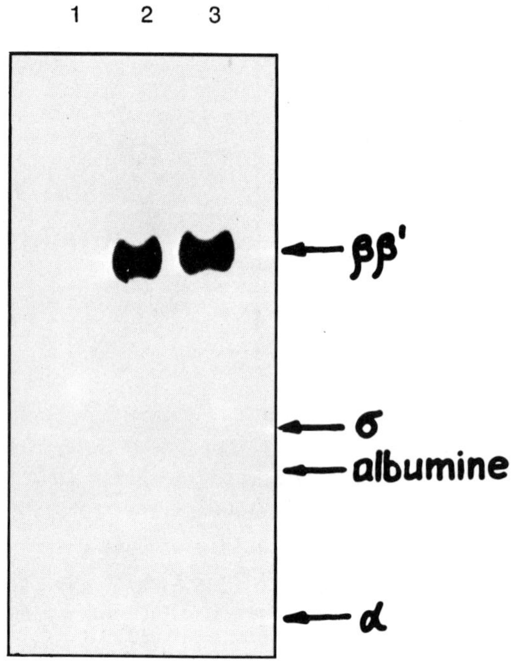

Fig. 6. Crosslinking of RNA polymerase with poly-d(A-IU) by irradiation. The reaction mixture (10 μl) contained 28 pmoles (10^6 cpm) of polynucleotide, 4.5 μg of RNA polymerase, 40 mM Tris-HCl, pH 7.9, 10 mM $MgCl_2$, 4.5 μg of BSA. The irradiation and subsequent treatment were carried out as described in the legend to Fig. 2

1, Control without irradiation; 2, The mixture is treated as described above; 3, After irradiation the mixture was supplemented with DNAase (10 μg/ml) and incubated for 10 min at 37° C.

to it. However, some information on the closest contacts between a nucleic acid and a protein may be obtained by means of covalent crosslinkage of the contacting grouping. In the case of RNA polymerase, attempts were made to introduced such crosslinks by UV-irradiation of a complex of RNA polymerase with poly-d(AT) [9]. We also observed covalent linkage between the enzyme and DNA when complexes of RNA polymerase and T7 or λ DNA were irradiated by UV-light, but, similary to previous workers [9], failed to identify the participating subunits. Presumbly, this happened because UV-light destroyed, at least partly, the RNA polymerase.

In order to overcome this difficulty, we synthesized a copolymer poly-d(A-IU) [10] which contains photoreactive 5-iodouridine residues. This polymer appeared to bind quantitatively RNA polymerase giving firm complexes; it is active as a template in the synthesis of poly- (AU). Complexes of [^{32}P]-labeled poly-d (A-IU) with RNA polymerase were illuminated at $\lambda > 290$ nm, and the subunits of RNA polymerase were separated in the usual way. Autoradiography showed, that the photoaffinity label is covalently bound by β- and (or) β'-subunits, (Fig. 6).

C. CONCLUSION

1. An approach has been developed to identify the sites of RNA polymerase involved in contacts with the 5'- and the 3'-termini of nascent RNA in the transcribing complex at different steps of the process.

2. With different promoters, the 5'-terminus of nascent RNA contacts with β, β'-, and σ-subunits. There is one example when the contact with α-subunit takes place. Presumably, this divergency is due to the different topography of the transcribing complex induced by different promoters.

3. The σ-subunit interacts with the 3'-terminus of nascent RNA during initiation of RNA synthesis. Presumably, this is due to direct participation of the σ-subunit in initiation.

4. Poly-d(A-IU) interacts with the β- and (or) the β'-subunits of RNA polymerase.

Therefore, the same subunits of the enzyme contain sites of interaction both with substrates and with the templates. Determination of the peptides which participate in these contacts may give information on the topography of the corresponding active centers.

REFERENCES

1. *Grachev M.A., Zaychikov E.F.* (1974) FEBS Letters, *49*, 163 - 166

2. *Gilbert W.* (1976) in: RNA Polymerase (Losick R., Chamberlin M. eds) pp 193 - 205, Cold Spring Harbor Laboratory, New York

3. *Dahlberg J.E., Blattner F.R.* (1973) in: Virus Research (Fox C.F., Robinson W.S. eds) pp 533 - 543, Academic Press, New York

4. *Sverdlov E.D., Monastyrskaya G.S., Rostapshov V.M.* (1978), Bioorganicheskaya Khimiya, USSR, *4*, 894 - 900

5. *Scherer G., Hobom G., Kössel H.* (1977), Nature, *265*, 117 - 121

6. *Grachev M.A., Zaychikov E.F., Kravchenko V.V., Pletnev A.G.* (1978), Doklady Akademii Nauk SSSR , USSR, *239*, 475 - 478

7. *Pribnov D.* (1975), J. Mol. Biol., *99*, 419 - 443

8. *Iida Y., Matsukage A.* (1974), Molec. Gen. Genet., *129*, 27 - 35

9. *Strniste G.F., Smith D.A.* (1974), Biochemistry, *13*, 485 - 493

10. *Dale R.M.K., Ward D.C., Livingston D.C., Martin E.* (1975), Nucl. Acid Res., *2*, 915 - 930

DNA-DEPENDENT ATPASES AND VICEVERSA

E. P. Whitehead

Biology Department, Commission of the European Communities

F. Palitti, G. Cerio-Ventura, A. Vellante, P.M. Fassella

Istituto di Chimica Biologica, Università di Roma

SUMMARY

DNA-dependent ATPases, some of which are ATP-dependent DNAases are involved in many of the vital processes in the molecular biology of the gene such as genetic recombination, DNA replication, restriction and termination of transcription. The enzymological properties and in vivo functions so far known are reviewed, with emphasis on the exonuclease V's of bacteria, involved in recombination.

1. Introduction

In 1964 Tsuda and Strauss (1) reported the partial purification and characterisation of an enzyme that catalysed the curious-seeming reaction

$$\text{DNA} + \text{ATP} \longrightarrow \text{small nucleotides} + \text{ADP} + P_i$$

DNA was not hydrolysed in the absence of ATP, nor was ATP in the absence of DNA, and the enzyme thus earns the names of ATP-dependent DNAase and DNA-dependent ATPase.

This reaction was curious, firstly because ATP hydrolysis is not usually coupled in this way to reactions where the equilibrium would strongly favour products, as in the hydrolysis of nucleic acids (and of course there are many DNAases not coupled to ATP hydrolysis) secondly because it was quite unknown what the biological role of the enzyme might be.

We now know many nucleic acid-dependent ATPases, many, but not all, of which are ATP-dependent nucleases. In outline at least, neither of the above puzzles are complete mysteries any more. In brief, ATP hydrolysis serves in enzyme mechanisms that unwind DNA, or change its conformation in other thermodynamically unfavoured ways. And these enzymes serve in most of the important biochemical processes involving genes, (Fig. 1), at least genes in prokaryotes. Only relatively recently was a nucleic acid dependent ATPase from a eukaryotic reported (mouse myeloma (2) and two from an animal virus (3,4)) and an enzyme similar to the unwindases of section 3.2. below has recently been found in a human cell-line (A. Falaschi personal communication); their function is unknown. Indeed a good three-quarters of what we know concerns enzymes from _Escherichia coli_ and its phages.

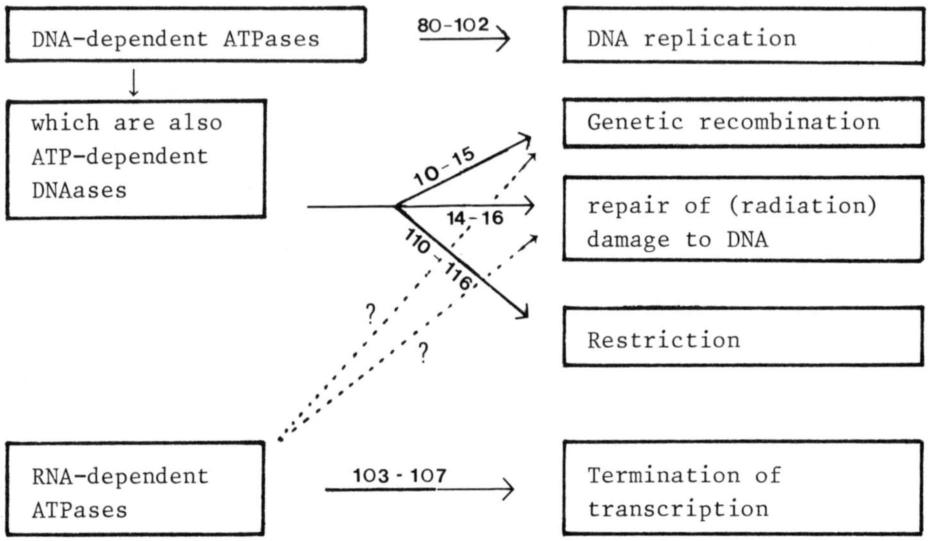

Fig. 1 - Some functions of nucleic acid-dependent ATPases. The numbers on the arrows are references.

In this essay we shall review the enzymology of nucleic-acid dependent ATPases relevant to each of the processes indicated in Fig. 1, and we hope to make it apparent that mechanistic information about the enzymes is more-than-usually relevant to their in vivo role. Another important aspect is their composition in subunits (and their genetic determinants) and their interactions with other proteins - two things that turn out to be not completely distinct. Some related reviews are references (5-9).

Two terminology-related but substantial points are that the reactions catalysed by these enzymes require magnesium ions as well as ATP, like other enzyme-catalysed reactions of ATP, and this is to be understood at all points below when we speak of 'ATP-dependence' or 'presence of ATP' etc. Other nucleoside 5'-triphosphates can substitute ATP in the reactions of most of the enzymes mentioned, but in all cases ATP is the best substrate/cofactor, and usually the second-best is dATP.

2. Exonuclease V

In this section we consider a group of enzymes of which the original M. luteus enzyme of Tsuda & Strauss (1) is a typical example, and the enzyme controlled by the recB and recC genes of E.coli is the best known. The latter enzyme became known as E.coli exonuclease V (29). More recently Wilcox & Smith (30) proposed that the name be adopted for similar enzymes from other bacteria, and we adopt this nomenclature here.

The characteristics of the enzymes listed in Table 1 are broadly so similar that it seems justifiable to consider them a family. Their most obvious characteristic is that they can degrade DNA to small acid-soluble fragments, and this is the basis of the usual assay for them (under some circumstances, however, they can give rise to rather large products, see below). The product nucleotides always have 3'-OH and 5'-phosphate ends (23, 17, 22, 29, 26, 31). A further common point is that several exonuclease V's have been shown to participate in genetic recombination processes, and in the repair of radiation damage to DNA. This is shown by the deficiency in these processes of strains with mutations that result in the absence, or alteration of their exonuclease V (see Table 1). The exonuclease molecules of Table 1 are all molecules of molecular weight in the region of 300,000 (17, 22, 20, 32, 12).

TABLE 1

Bacterium	Recombination	Repair	Purification
E. coli	(10) (11) (12)	(16) (11)	(17) (18)
B. subtilis	(13)	-	(19)(20)
Diplococcus pneumoniae	(14)	(14)	(21)
Hemophilus influenzae	(15)	(15)	(22)
Micrococcus luteus	-	-	(23)(24)(25)
Mycobacterium smegmatis	-	-	(26)
Alcaligenes faecalis	-	-	(27)
Bacillus laterosporus	-	-	(28)

2.1. Specificity and the processive mechanism

The enzymes of the family are all fairly similar in specificity for the type of DNA attacked. None of those tested attacks double stranded circular DNA (29, 31, 33, 27, 17, 34, 35). All degrade linear double-stranded DNA (33, 29, 19, 28, 17, 26, 21, 13, 27, 1, 23, 32). All degrade single-stranded linear DNA, but this activity is in all cases weaker than the action on double-stranded. Degradation of single strands ranges from about 50% to 5% of that of double strands depending on conditions of assay (pH, ATP concentration & c.) (22, 23, 36, 27, 21, 29, 17, 19, 28, 26). In all cases double-stranded DNA is a better cofactor than single-stranded for the ATPase activity (36, 17, 19, 27). All these statements refer to the action in the presence of ATP, an action believed to be exonucleolytic (see below). Goldmark & Linn reported (17) that in the absence of ATP E.coli exo V makes limited endonucleolytic attacks on single stranded but not double stranded DNA. This was shown by its conversion of circular single stranded molecules into a form which could be degraded by exonuclease I. Even this activity, however, was stimulated about sevenfold by ATP. The H.influenzae and M. luteus enzymes at least also seem to have endonucleolytic activity towards single strands (33, 23, 34). It has been concluded that the significant activity in vivo is the double-stranded exonucleolytic one, since temperature-sensitive recB and recC mutants of E.coli were thermolabile, relative to the wild type, only in this activity (37, 38). Benziger et.al. (39) reach similar conclusions from the transfectivity of various type of phage DNA in spheroplasts of rec E.coli but with the reservation that things could be different in normal cells.

The mechanism of action of all the exonuclease V's on double-stranded linear DNA may be categorised as essentially end-attachment exonucleolytic processive, and is summarised in Fig. 2. In this mechanism the enzyme attaches tightly to the extremities of a DNA molecule, and proceeds along it, digesting, without dissociating until it reaches the other end, or possibly another enzyme molecule coming the other way.

Evidence for end attachment is the incapacity to attack circular double-stranded molecules mentioned above. The enzymes can form complexes with linear double-stranded DNA, and these complexes are very stable. They have been evidenced by cosedimentation of the enzyme along with the DNA (32, 40) retention of enzyme-bound DNA on filters which allow uncomplexed DNA to pass through (41, 30) distri-

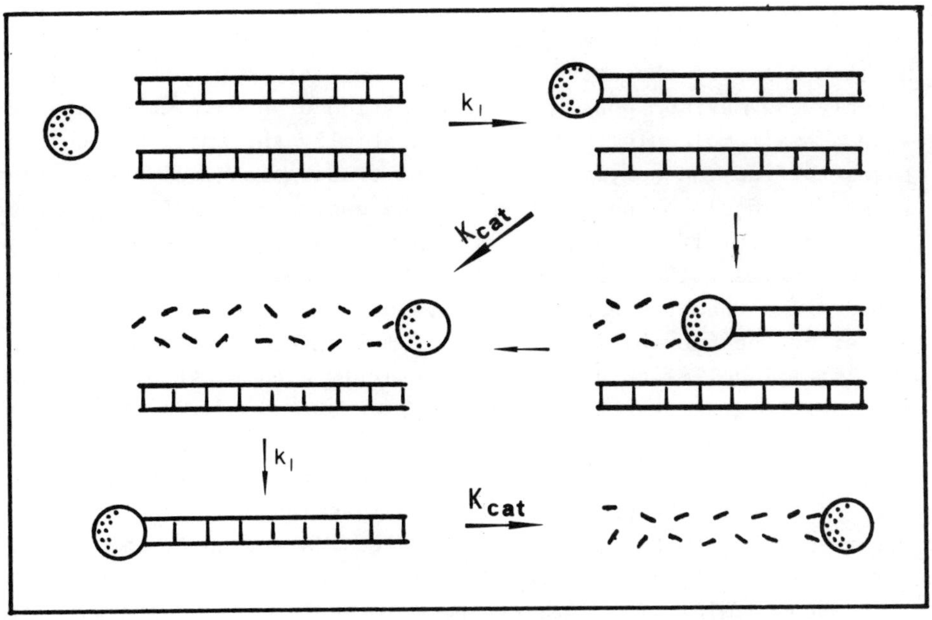

Fig. 2 - The processive (one-by-one) degradation of DNA by exonuclease V. The k_{cat} arrow does not represent a simple step: it is supposed that the degradation of the molecule can be represented by a single first order rate constant k_{cat}.

bution in a phase-partition experiment (32) and the demonstration in a kinetic experiment that in a preincubation without ATP the DNA becomes saturated with a certain quantity of enzyme and adding more enzyme does not further increase the velocity on subsequent addition of ATP (32, 47). The stoichiometry of this binding has been measured, and it has always been found that the molecular ratio enzyme:DNA is of the order 2:1 justifying the idea that attachment is to the ends, and the mechanism exonucleolytic. Further support for exonucleolytic action is that double-stranded molecules are more slowly attacked when they have short single strands at the ends (32, 33). Such molecules are, however, degraded, and in conformity with the fact that the enzyme does not absolutely require flush ends it degrades double stranded circular DNA with single-strand gaps of about 5 or more residues (43).

Smith et.al. (30) have shown how stable the attachment of H. influenzae exo V to DNA is, by measuring the rate constants for attachment and dissociation of the complex. The latter is 3.5×10^{-4} s^{-1}, indicating that the mean lifetime of the complex in their conditions is about 30 minutes. The attachment rate constant is around $10^{9} M^{-1} s^{-1}$, which gives a dissociation equilibrium constant of 5×10^{-13} M. This value of the attachment rate constant is remarkably high, roughly the maximum possible for formation of a complex of an enzyme with a small molecule (42) where essentially the small molecule diffuses to find the enzyme. In our case translational diffusion of the substrate should be negligible and the enzyme must diffuse to find the substrate. If the DNA is a rigid molecule, the production encounter of enzyme with ends should be at least one or two orders of magnitude less than the experimental value. This must be explained either by the flexibility of the DNA molecule, or by supposing that the initial collision does not necessarily need to be with the ends, in order to give rise to the complex, or both.

The evidence for a processive action (that is that the enzyme molecule does not leave the DNA molecule until it has traversed it all) is that in the course of normal digestion no products of size intermediate between intact linear molecules and quite small product are observed (24, 31, 23, 22, 21, 17, 26).

When termini of double-stranded molecules are marked with a different radioactive label than their internal regions, the ratio of the two labels appearing as acid-soluble products is constant with time, in incubations at low enzyme concentration (43, 31). If the action were econucleolytic but not processive the terminal label would be released earlier. This does happen when the DNA is saturated by enzyme as expected for any exonucleolytic action (43) (there was no preference for release of 3' or 5' label). With single stranded DNA, attack of the E.coli enzyme seems to be preferentially from the 3' end (29).

If unlabelled DNA is added after the start of digestion of labelled DNA by the H.influenzae enzyme the appearance of labelled products is not inhibited until about 1.25 minutes after the start of the reaction, showing that during this period the enzyme cannot switch from one (labelled) DNA molecule to another (unlabelled). Qualitatively similar observations have been made with the M. luteus (32, 47) and M.smegmatis (41) exo V's. After this 1.25 minute period the diminuition of appearance of labelled products is quite abrupt, as the processive mechanism would predict if the procession rate is fairly uniform so that all the enzyme molecules arrive at the end at about the same time.

We have observed in the steady-state kinetics of digestion of T7 DNA, a Michaelian dependence of reaction velocity on DNA concentration. The Michaelis constant is of the order 1.2 μg/ml, which means a concentration of DNA ends of 10^{-10}M. This is much higher than the binding constant mentioned above measured by Smith et.al. but we believe that it is not an equilibrium dissociation constant, but an extreme kind of Briggs-Haldane constant. In the steady-state of the mechanism of Fig. 1 the Michaelis (Briggs-Haldane) constant is k_{cat}/k_{+1}. In support of this idea we observe that the Michaelis constant expressed in terms of μg/ml is rather little affected when the T7 molecule split into about 17 pieces by a restriction enzyme is used as substrate.

At constant μg/ml the rate of association should increase in proportion to the number of ends available, whereas dissociation should not be affected. This would result in a dramatic decrease of an equilibrium K_m expressed in μg/ml, in proportion to the number of molecules. On the other hand, if the dissociation in the steady state really represents degradation of the substrate DNA molecule, encounters of enzyme and substrate at constant μg/ml should be greater with the restricted than with the intact DNA, but the lifetime of each complex should be diminished in exactly the same proportion so that the ratio k_{cat}/k_{+1} should be unaffected, which is approximately what we observe.

2.2. Products and the unwinding action

As mentioned above exo V's give small acid-soluble products which is the basis of their assay. These are a mixture of small oligonucleotides of different lengths, for instance after complete digestion the E.coli enzyme gives products of average length 4.5 (17). However under some conditions (such as high ATP concentrations, low or high ionic strength) quite high molecular weight acid-insoluble products are formed. Strikingly these contain a considerable proportion of long single-stranded material (44, 45, 43, 46, 18, 48). Simple single-stranded polynucleotides several hundred residues long, and partially double-stranded structures with single-stranded tails at one or both ends are found. These single-stranded tails are from several hundred to four thousand or more long, and are probably 3'-OH and 5'-phosphate ended in roughly equal proportions (45, 44). The length of some of the tailed molecules observed by the various workers is a considerable fraction of that of the entire substrate molecule, and there are probably substrate molecules in which the processively acting exonuclease has only travelled some of the way along the DNA molecule. The enzyme appears to release single-stranded and partially double-stranded products, which are then re-attacked in preference to intact double stranded substrate to give smaller products (44, 35) somewhat complicating the scheme of Fig. 2.

These single-stranded products need to be explained by a mechanism. They are not due to the enzyme simply digesting only one strand to small products leaving the other unpaired, as in many of the experiments much more product was in the form of long single-strands than in the form of small acid-soluble products.

An attractive mechanism was proposed by MacKay and Linn (44). The essential idea is illustrated in Fig. 3 and is that the ends of the DNA are attached to a site on the enzyme so that as it travels it unwinds the strands. This suggests a role for the ATPase activity of the enzyme.

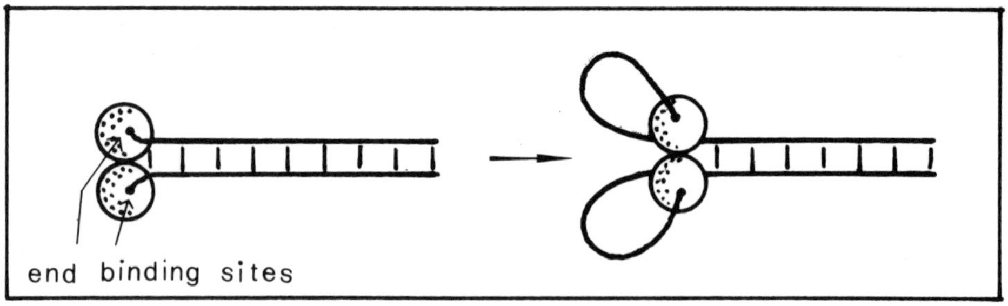

Fig. 3 - Principle of mechanism of DNA unwinding by exo V proposed by MacKay and Linn (44). The enzyme binds strongly to ends and proceeds unidirectionally along the duplex.

2.3. ATPase and stoichiometry

It is sometimes suggested that ATP hydrolysis provides energy for an enzyme to travel along DNA. Although this directs attention to an important feature of this class of enzymes, it may be a biochemical misuse of thermodynamic concepts (50) since an enzyme would not 'need energy' to diffuse along, its free energy being essentially the same at any position along the DNA. But the thermodynamically unfavourable unwinding of strands in the mechanism of Fig. 3 does need to be coupled to some free-energy degrading reaction; without this the tendency of the single strands to renature would just pull the enzyme back towards its starting point. This however, will not happen if the mechanism of this reversal requires a large amount of ATP synthesis from ADP and phosphate.

This brings us to the question of the stoichiometry of ATP hydrolysis relative to DNAase action, a subject still not satisfactorily clarified. It can be seen from Table 2 that ATP hydrolysis is quite disproportionate to phosphodiester bond hydrolysis. The ratio of ATP hydrolysed to nucleotides or nucleotide-pairs rendered acid-soluble is more reasonable, but is still not a reasonable simple Daltonian ratio. In any simple and obvious conception of the processive mechanism, one would expect a molecule of ATP to be hydrolysed at every step of the enzyme molecule from one nucleotide residue to the next, or from one pair to the next, and thus either a 1:1 or 2:1 ratio for (residues liberated: ATP hydrolysed). However it is easy to think of reasons, both trivial and fundamental why this might not be observed. Probably in most of the experiments summarised in Table 2 not all the nucleotide products were acid soluble, we have already mentioned that the enzyme can attack its own initial products, and possibilities of the enzyme slipping back along the DNA have also been invoked (51, 53). Possibly some of these complications are avoided when the single strand products are stabilised by tetraethylammonium bromide, and it was observed that this reduces the ratio form in experiments with the *A.faecalis* exo V about 3:1 to 1:1 (27).

It may be mentioned here that when exo V attacks artificially crosslinked DNA it hydrolyses up to the first crosslink, and then stays there, hydrolysing ATP but unable to progress further and hydrolyse DNA (52, 53). Also, although DNA-RNA complexes are cofactors for ATP hydrolysis neither strand of such a complex is hydrolysed to any extent (43, 36).

Bacterium	(ATP hydrolysed)/(phosphodiester bonds hydrolysed)	(ATP hydrolysed)/(acid soluble nucleotide)	(Average length of acid-soluble products)
E.coli	10 - 20 (29) 8 - 9 (36) 23 (17)	1,2 - 2,2 (29) 3 (36) 5,1 (°)	7 - 9 (°) 2,8 (36) 4,5 (17)
E.coli(β subunit)		100 (55)	
H.influenzae	31 - 39 (78)	5 (46)	5,9 (78)
M.luteus	3 (79)	0,5 (°)	5,5 (79)
M.smegmatis	3 (26)	1 (26)	3 (26)
B.laterosporus	3 (28)		
B.subtilis	30 (19)		
D.pneumoniae		2 (21)	
A.faecalis		1 - 3,5 (27)	

Table 2. ATPase/DNAase stoichiometry of exonuclease V's. Bracketed figures are reference numbers. (°) calculated from other two columns).

2.5. Subunit structure and protein-protein interactions.

Exonuclease V of E.coli is controlled by two known genes, recB and recC. Inactive crude extracts from $recB^-$ or C^- strains produce active enzyme which appears identical to the wild type enzyme when mixed in certain conditions, and the effect is probably due to recombination of undamaged subunits from each strain (54). The native enzyme has a molecular weight of 270,000 and in sodium dodecyl sulfate dissociates into two different subunits of approximately 140,000 and 128,000 daltons (17). Lieberman and Oishi (55) were able to dissociate native enzyme into an inactive α subunit of approximately 60,000 daltons and a β subunit, with slight residual DNAase and ATPase activities, of about 170,000 daltons. The two types of subunit could be assayed by their mutual complementation. The β subunit could complement extracts from $recB^-$, rec C^- and rec BC^- strains, while the α subunit could complement none of these. The authors concluded that the recB and probably recC genes code for the β subunit and the α subunit must be coded for by some as yet unknown gene. The exonuclease V of B.subtilis appears to contain five polypeptide chains. When its activity is lost due to a mutation in the recE gene one of these is found to be altered in molecular weight which is 70,000 in the wild type and 56,500 in the recE mutant (20).

Exo V of E.coli forms a physical and functional complex with polymerase I (and possibly other polymerases too) which is more effective than polymerase alone in making new strands: polymerisation by the complex is stimulated by ATP (56). The double stranded structures mentioned above as intermediates of degradation of double-stranded DNA by exo V are suitable primer/templates for action of polymerases, as has been shown experimentally (18).

It is reasonable to suppose that the complex is significant in vivo. Other protein-protein interactions that are undoubtedly significant in vivo are those between the exo V of E.coli and specific inhibitory proteins synthesised during phage infection, and coded by genes of phages such as λ, T4, T7 (59, 60, 61). These inhibit the exo V of the host; this appears essential for a late stage of phage replication. The inhibitory protein ("gamma") of λ has been purified (59). It appears to form a complex, possibly reversibly, with exo V of E.coli in which all of the latter's activities are inhibited, but it cannot inhibit the enzyme when it is already 'working' at DNA digestion.

2.6. Miscellaneous

The following facts concerning the effects of potassium ions and ATP seem worth mentioning because there is some analogy with the behaviour of other DNA-dependent ATPases. In the absence of K^+, DNA hydrolysis by E.coli exo V stops after two or three minutes, and the enzyme is no longer bound to DNA (57). Activity can be restored by K^+. The 2-3 minutes is probably of the order of the time required to digest a substrate mole DNA molecule in these experiments and it is tempting to think that perhaps the enzyme leaves the DNA after digestion in an inactive conformation, and that restoration of the active form requires K^+. K^+ is also an activator of another DNA-dependent ATPase (90) and of DNA gyrase (90). ATP in the absence of DNA inactivates exo V (48). A similar phenomenon has been seen with an ATP-dependent restriction enzyme (109) and RNA polymerase (58).

2.7. In vivo functions of exonuclease V's in recombination and repair.

The references of Table 1 and the cited reviews (6-8, 64-66) must be consulted for more detail of the behaviour of mutants lacking exo V than we can give here.

There have always been some reservations in relating exonuclease V (in particular that of E.coli) specifically to recombination and repair functions, due to the poor viability of cells with the recB$^-$ or recC$^-$ mutations. Various evidence reviewed by Radding (7) and Clark (6) suggest that such cells can form some kind of recombinant structure capable of being transcribed and translated into the recombinant phenotype, but that they are unable to execute some late step, cutting and trimming perhaps involved in separation and segregation of sister duplexes.

It is an amusing paradox that the exonuclease V of E.coli which serves in general in promoting genetic recombination, actually helps prevent one sort of recombination: transformation. It was for many years impossible to achieve transformation of this bacterium, until it was found that this could be done using recB$^-$ or recC$^-$ strains (62,63). Presumably this is because the exo V degrades the transforming DNA that penetrates the cell. (It has a similar effect on restricted phage DNA, see below). For transformation, the bacterium must necessarily have a functioning recombination mechanism, and this can be provided by a phenotypic suppressor of recB called sbc. On the other hand exo V serves in promoting transformation in other bacteria.

Besides recombination in the usual sense of redistribution of genotype, exonuclease V is also involved in a recombinational process of repair of damage to DNA by ultraviolet light, X-rays and other agents to which the recB$^-$ and recC$^-$ strains are sensitive. The cell has several pathways of repair available. The best known is excision repair (64-66) of uv-induced thymine dimers which depends ultimately on a polymerase restoring a correct base sequence by using the undamaged strand as template. But this is not possible if both strands are damaged and have lost genetic information at the same locus. Examples of such damage are crosslinking of the DNA.Another arises from the fact that normal DNA replication bypasses any uv-induced thymine dimers leaving gaps of about a thousand nucleotides on the newly synthesised strand (67) opposite the damage. In these cases repair has to use the undamaged sequence on the sister molecule,and a recombinational event involving physical exchange between sister molecules takes place (68-70). The recombinational repair process leads to long stretches of newly synthesised DNA (long repair patches as opposed to other short repair patches of excision repair). RecB$^-$ strains of E.coli have very limited ability to perform this resynthesis (71), and are also unable to rejoin the X-ray induced single stranded breaks (72). In the repair of damage by crosslinking agents, endonucleases make nicks on both sides of a damaged residue on one strand so again causing a single-strand gap (73). The eventual rejoining of the cut DNA is dependent on a functioning recombinational pathway (74).

Watson and Cricks' 1953 paper announcing the double-helical DNA structure ends with the famous sentence 'It has not escaped our notice that the specific pairing we have postulated immediately suggests a possible copying mechanism for the genetic material'. As much could be said about genetic recombination. Practically all models assume that formation of a (hetero) duplex structure from complementary single strands, one from each parent duplex, is a key step in recombination, but various ways can be imagined for this to happen. The activities of various enzymes can be invoked in models for the formation of such complexes and their conversion to separate recombinant DNA molecules, but there doesn't seem to be any agreed and proved pathway for general recombination yet. Several suggestions have been made for the possible role of exo V.

One (46) is that it may make available single-strands for pairing, from its ability to act on duplexes to produce single-stranded molecules. By renaturing the fragments produced from exo V's action on T7 DNA, it was possible to produce molecules as long as the original substrate or longer. The action of polymerase and ligase can then be imagined to produce an intact, possibly recombinant, duplex molecule for this DNA. The fact, mentioned above, that some sort of recombinant duplexes are produced in cells lacking exo V, seems rather against a role for exo V in formation of heteroduplexes in vivo. A mysterious 'pairing' of DNA molecules induced by exo V may also be relevant (77).

Another idea is that exo V could extend the region of heteropairing in a structure such as that illustrated in Fig. 4. The imagined DNA configurations are consistent with the branched electron-microscopically observed T4 recombination intermediate structures (76). Any exonuclease acting on double-stranded DNA might be imagined to achieve the same thing, but some kinds of topological constraint of the parent molecules could impose a requirement for coupling to ATPase.

Another idea for the consolidation of heteroduplex structure by exo V, which is consistent with its action on model structures (75) is illustrated in Fig.5. Step 1 can be executed by an endonuclease, step 2 is favoured when the 'receptor' duplex has supertwists, as it ~~releives~~ the tension and the region of heteropairing could be extended by a gyrase screwing up the tension (see below) (cellular

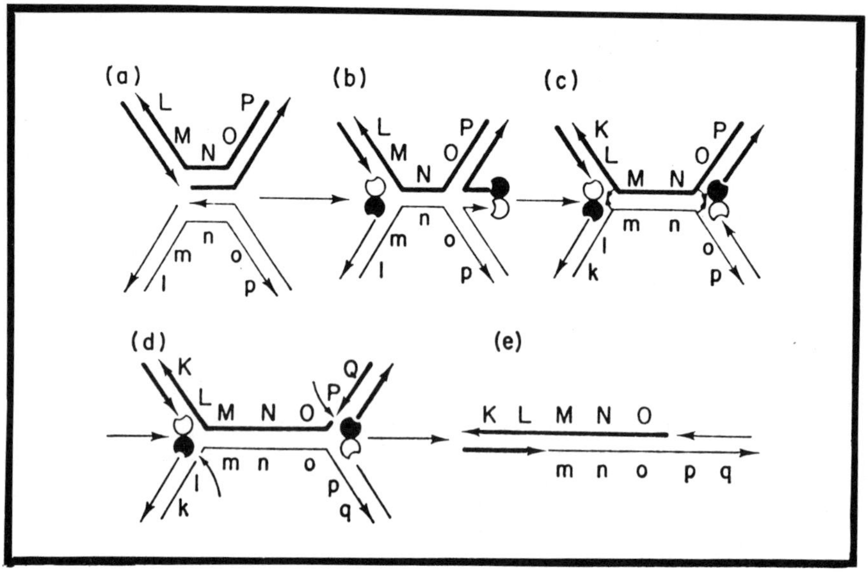

Fig. 4 - Suggestion of Broker and Lehman (76) for a role of exo V (the black-and-white dimer) in extending pairing regions in recombination.

chromosomal DNA appears to be supercoiled). The real possibility of this 'single-stranded agression'is illustrated by its demonstration with a model system, where the supercoiled receptor duplex is a circular DNA molecule (75). With the same model system it was shown that exo V can attack the 'D-loop' presumably by its endo and exonuclease activity towards single strands to generate the structure shown in Fig. 5 . On a blackboard the invading strand can then be joined to the resident broken strand by ligase, which however fails to do this in an in vitro experiment (75) although the gap between resident and invading strands are very short. Perhaps it would be useful for the exo V to carry polymerase with it, as we have seen it can, to solve this problem. This model does not seem to use the exo V's double strand exo activity.

Altogether we see that the knowledge of the enzymology of exo V and other enzymes is suggestive, and a necessary but not a sufficient condition for being able to deduce a pathway for genetic recombination.

3. DNA replication

Genetic evidence and in vitro experiments have shown that DNA-dependent ATPases participate in DNA replication.

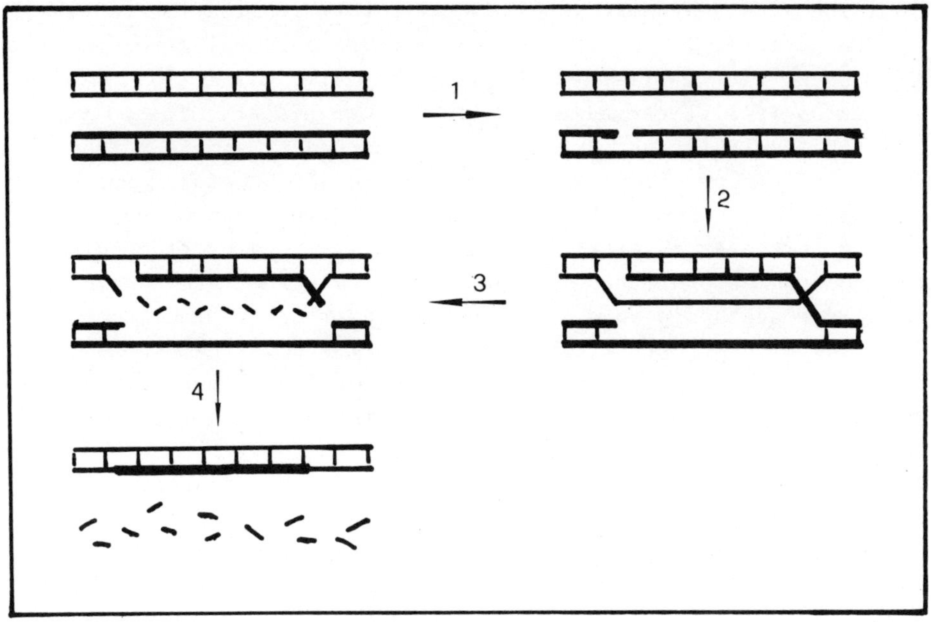

Fig. 5 - Hypothesis for participation of exo V in recombination proposed by Wiegand et al. (75).

3.1. The T7 gene 4 protein

A relatively simple in vitro system of DNA synthesis is the T7 DNA polymerase which works in combination with the protein coded by gene 4 of T7 phage. The T7 polymerase is a complex of a polymerase coded by gene 5 of T7 phage and the E.coli thioredoxin molecule (many and various are the ways of molecular parasitism!) The gene 4 protein turns out to be a DNA-dependent ATPase (80). Unlike the exonuclease V's and ATP-dependent restriction enzymes, but like all the other enzymes considered in sections 3.1.and 3.2., its ATPase activity is much more stimulated by single-stranded than by double-stranded DNA. Its molecular weight is in the region of 60,000, it is probably a single polypeptide chain, and may occur in two forms (81, 82). It has no nuclease, ligase or DNA polymerase activity (81) though it has another, surprising, enzymatic activity of which more below. The gene 4 product is essential for T7 replication in vivo.

The purified T7 polymerase can catalyse polymerisation of dNTP's using single-stranded DNA as a template, since this, by folding back on itself, can also provide its own primer. The polymerase by itself is inert when double-stranded or single-stranded circular DNA is provided as template. But addition of T7 gene 4 protein enables it to copy these templates. Copying of double-stranded template initiates at nicks (83) and is stimulated by ribonucleoside triphosphates, but these are not essential. They are however essential for synthesis on a single-stranded circular template (82) unless suitable primers are hybridised to the template. These effects of the gene 4 protein are specific for the T7 polymerase, and do not work for other polymerases. There may be a physical complex formed between the two proteins.

The gene 4 product seems to have two distinct functions - initiation in the absence of primer, and parent strand displacement.

The reason T7 polymerase cannot by itself replicate double-stranded DNA is probably that it cannot displace the strand ahead of it, for it can replicate to some extent when DNA-binding protein (DBP) is provided to stabilise the displaced strand. Presumably the gene 4 protein acts catalytically to displace this strand, but this is conjecture based on analogy to similar enzymes (see below),as a direct demonstration of the gene 4 protein's effect on DNA has not been published. It appears likely that for every shift of position along the template, the polymerase-ATPase complex hydrolyses one molecule of ATP: the ratio of triphosphate hydrolysis to deoxyribonucleotide incorporation is about 1:1.

In the absence of rNTP's synthesis,initiating at nicks in double stranded molecules uses the 3'-OH group there as primer, so that the newly synthesised DNA is covalently linked to old (83). The surprise finding is that in the presence of rNTP's the complex can synthesise a short oligoribonucleotide primer (82). Scherzinger et.al. even indicate that this may be an activity of the gene 4 protein rather than the polymerase (82). The primer can have any of a limited number of sequences such as pppApCpCpA, corresponding, it may be supposed, to a limited number of sites with the complementary sequence in the DNA strand where initiation may take place. If one supplies some of the oligoribo-primers T7 polymerase can work with the single-stranded circular template without needing gene 4 protein. In the light of this one can understand the essentiality of rNTP's with single-stranded circular template, and their non-essentiality for templates with primers (rNTP's are presumably not essential for strand displacement because the gene 4 ATPase also hydrolyses dNTP's).

Phage T4 has an enzyme (gene 44/62 protein) of activities and function similar to the T7 gene 4 protein (109). A DNA-dependent ATPase of no known function, but which appears to be a subunit of an enzyme complex, has been isolated from T4-infected cells (49).

3.2. rep and dnaB products

The replication of phage ØX174 DNA *in vivo* consists of three stages: stage I, conversion of the single (+) phage strand into the double-stranded circular replicative form (RF); stage II, multiplication of RF; and stage III, synthesis of the viral (+) strand coupled to encapsidation. By partial reconstitution experiments using proteins of stage II, the latter can be subdivided into two stages: (+) strand synthesis (stage II+) on RF template, and (-) strand synthesis (stage II-) using the products of stage II+ as template.

Stage II can be accomplished by the following proteins together: *E. coli* DNA polymerase III holoenzyme, DBP, and the proteins coded by the host *rep* gene and the phage *cisA* gene, together with dNTP's and rNTP's. The rep protein is a specialised endonuclease that can nick the (+) strand of the RF DNA at a specific site whose sequence is now known (118), remaining attached to the 5' end (84, 86). If the *rep* protein and ATP and DBP are added to this complex it is unwound, in the manner of Fig. 6 to form first a single stranded loop on the circle, and when the complex comes round to the starting point the single-stranded (+) linear molecule is liberated from

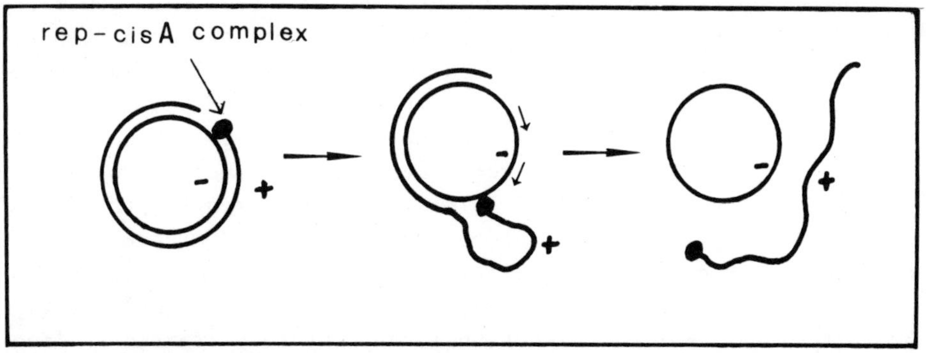

Fig. 6 - Probable mechanism of the cisA-rep complex in unwinding ØX DNA (85).

the (-) single-stranded circle (85). The cisA protein also has a ligase activity that enables it to rejoin the cisA-bound end to the free 3'-OH and circularise the linear (+) strand. The principle of strand separation appears to be very similar to that postulated by McKay and Linn for exonuclease V (see above). 1 molecule of ATP is hydrolysed for every basepair melted (85).

Polymerase III holoenzyme, like the T7 polymerase is unable to displace the single-strand ahead of it and achieve synthesis. In the model of Kornberg and coworkers (85) it closely follows the cisA-rep complex, polymerising, starting with the 3'-OH liberated by the cisA nick as primer. When the complex comes round again to the starting point it finds the original sequence on the (+) strand that the cisA endonuclease recognises, this is nicked and the displaced strand closed and released in circular form and the cycle can start again.

In uncertain relation with rep and with each other are two enzymes isolated by Abdel-Monem et.al. (88,89) and termed 'DNA unwinding enzymes' (unwindases) I and II, and an enzyme isolated by Richet and Kohiyama (90), all from E.coli. Any of these four may be the same, perhaps in different form. All are single-stranded DNA-dependent ATP ases, and unwindases I and II have been shown to denature both DNA-DNA and DNA-RNA partial duplexes in the presence of ATP, but they first have to attach to single-stranded tails (87,91). Unwindase II forms what is probably a not-very-stable 1:1 complex with single-stranded DNA and the investigators concluded that the denaturing action was distributive rather than processive, though this is hard to picture. A DNA molecule binds relatively firmly many molecules of unwindase I and a processive action in which these bound enzyme molecules co-operate has been attributed to it (92). The molecular weight of unwindase II, the Richet-Kohiyama enzyme and the rep protein are similar about 70,000, that of unwindase I is about 180,000, and the first two resemble each other in some other ways.

Another enzyme essential for cellular DNA replication as well as stages I and II- of ØX174 replication (see above) is that controlled by the dnaB gene of E.coli. This is another single-stranded DNA-dependent ATPase, though with some DNA-independent ATPase activity (93) and has recently been purified in milligram quantities (94). It is composed of subunits of 55,000 daltons and the active form is probably a tetramer. Its function is thought to be that of a 'mobile promoter' for synthesis of RNA primers for DNA synthesis (95). With the obligatory participation of some other proteins it forms a complex with the ØX(+) strand, and in some way in the presence of ATP enables transcription by the (dnaG-coded) RNA polymerase. dnaB protein is thought to travel round the ØX(+) strand in the 5'→3' direction, and in the presence of the dnaG-RNA polymerase but absence of DNA polymerisation this allows synthesis of primers at many points around the template, spaced about 200 nucleotides apart. Among the proteins required for formation of the dnaB-DNA complex is the dnaC gene product (96); this is an inhibitor of the ATPase activity of the dnaB protein (97) but a molecule of the complex, when formed, has the same activity of as a molecule of the dnaB protein. It is possible that the transfer of dnaB to DNA is from a dnaB-dnaC complex (97). The properties of the enzyme are thus suggestive of and consistent with the idea of it moving continuously in the 5'— 3' direction on the discontinuously copied parental strand of DNA just behind the replication fork, preparing it for transcription to Okazaki primers (95). The ATPase activity is presumably involved in 'propelling the molecule along the strand' (more strictly according to our previous argument in stopping it from moving back) and it will be interesting to see whether it is not involved also in putting the DNA into some special, thermodynamically unfavoured, conformation suitable for transcription at the sites of primer synthesis.)

There is still another DNA-dependent ATPase involved in ØX replication, and that is the 'Y factor' (112). This is distinct in its properties from rep and dnaB, and has the unusual property of being specific for one type of DNA - that of ØX itself.

3.3. DNA gyrases and replication

Yet further enzymes important in DNA replication with an ATP dependent action on DNA are the DNA gyrases, which convert relaxed double-stranded circular DNA into homogeneous supercoiled forms similar to the naturally occurring ones. These must be presumed to be ATPases, though this is not yet directly demonstrated, since supercoiled forms are thermodynamically less favoured than relaxed forms. Gellert et.al. comment 'DNA gyrase could either act as a swivel to counteract the positive superhelical turns introduced by replication or could pre-tension the DNA to produce a state of negative supercoiling strain' (98). One of the genes controlling a gyrase from E.coli is the cou locus, whose mutation confern resistance to coumeromycin and novobiocin (101). To change the winding number of a circular DNA molecule it is necessary to cut and reseal at least one strand. The component of gyrase that can do this appears to be controlled by a separate gene from cou, nalA (whose mutation confers resistance of DNA synthesis to nalidixic and oxolinic acids). The gyrase, by means of the nalA component can relax supercoiled DNA in the absence of ATP (but it requires high Mg^{2+}) (99,100). The mechanism may involve cutting of both strands; at any rate from an incubation of the enzyme with double-stranded relaxed or supercoiled DNA and the inhibitors oxolinic or nalidixic acids, linear DNA molecules to which the enzyme is probably still attached can be isolated (100). The nalA gene product can be purified separately from the gyrase; it is assayed by its capacity to confer naladixic acid sensitivity to the gyrase from a resistant strain (100). The picture that begins to emerge is of the nalA-coded product with nicking-closing activity, complexed to another protein with the ATPase activity, such that mechanistic coupling permits the energetically unfavoured supercoils to be introduced, rather than removed, in a circular duplex.

Host cell gyrase has been implicated in T7 replication (102) although T7 DNA is a linear molecule. A tentative explanation of this is in the anchoring of the T7 DNA to cellular structures such as membranes.

4. Termination of transcription

The rho factor which in catalytic quantities causes RNA polymerase to terminate transcription at certain sites is a RNA-dependent ATPase (103). When $\beta-\gamma$ -imino triphosphates which can serve as RNA polymerase but not ATPase substrates were incubated with polymerase and rho protein, transcription took place without termination at rho-dependent termination sites and it was concluded that the ATPase reaction is necessary for termination (104). The RNA cofactor specificity of the rho ATPase has been investigated (105); this is altered in a minority of mutant rhos (119). The structures of two rho-dependent transcription termination sites have recently been determined (106, 107). One of them has a potential stem-and-loop RNA structure just before the termination region, and the authors draw attention to a CAAUCAA sequence at or close to the (variable) point of transcription termination in both. Mutants with defective rho are disturbed in a remarkable variety of functions, they are uv-sensitive, deficient for recombination of a phage without its own recombination system, deficient in formation of stable lysogens and dead after a generation of growth at non-permissive temperature (108).

5. ATP-dependent restriction nucleases

Space limitations prevent any detailed discussion of these interesting and in many ways mysterious enzymes here, but they have been reviewed several times (9, 110-114). They are double-strand DNA-dependent ATPases that make a very limited number of double-stranded breaks of uncertain structure; two of the three known enzymes of this category hydrolyse a great number of ATP molecules, as many as 10^6 or more per break. Unlike most ('class II') restriction enzymes the breaks made by the known ATP-dependent ones are not at highly specific sites, and these nucleases also release a small quantity of acid-soluble oligonucleotides. As well as ATP, S-adenosyl methionine (SAM) is a necessary or stimulating cofactor for the nuclease action. These restriction molecules contain a subunit, which can however sometimes be purified separately from the nuclease activity, carrying a methylase activity which catalyses methylation of adenine residues in the DNA by SAM. Unlike the restriction sites, methylation sites are highly specific sequences (their structure has been determined (115, 116)).Their methylation prevents subsequent attack by the corresponding restriction activity. There is evidence in favour of an essentially processive mechanism of the nucleases. There is little evidence as yet of an effect on DNA such as unwinding.

Restriction enzymes by themselves produce only a limited number of breaks in a phage DNA molecule, but after this the broken molecules are rapidly degraded by other nucleases *in vivo*; exonuclease V may be the most important of these (117).

6. Conclusions

We have seen that DNA-dependent ATPases have important roles in vital processes in the molecular biology of the gene, such as recombination, replication, restriction, and termination of transcription, not to mention the nucleohydrolases involved in ribosomal translation (why not mention it? - Sheer ignorance! But not, we gather, only ours). They achieve separation of nucleic acid duplex partners and other energetically unfavourable changes of nucleic acid conformation such as supercoiling, and unidirectional processive actions. We have also seen that the ATPase is often part of the same molecule, or complexed with some molecule - the distinction is quite artificial - capable of some other action on DNA such as nuclease, nicking-closing, methylase or polymerase.

It is hardly speculation to predict that similar enzymes will be found acting in other vital processes. Two that obviously suggest themselves are other steps of transcription, and transport and packaging of nucleic acids. For if a processively acting nucleic acid-dependent ATPase is anchored to some structure, such as virus capsid or a membrane, it becomes a pump for transporting the nucleic acid from one location to another. Indeed, from this point of view the class of enzymes seems to lose its special character, and perhaps is just a member of the general class of transport systems involving an ATPase. We shall expect there to be a great deal of common structure in the family of nucleic acid-dependent enzymes (and perhaps with other transport systems), possibly some of the enzymes discussed will turn out to have common subunits (this has already been found in the case of the restriction nucleases).

A big blank likely to be filled in the next few years is knowledge of such enzymes in eukaryotes.

Acknowledgements

We thank Profs. C. Radding and S. Linn for providing us manuscripts prior to publication and Profs. G. Tocchini-Valentini and A. Falaschi for reading parts of the manuscript. This work was done under contract BIOI-160-76-I Commission of the European Communities -University of Rome. This is publication N°. 1516 of the Biology Department, Commission of the European Communities.

REFERENCES

1. Y. Tsuda and B.S. Strauss (1964), Biochemistry, 3, 1679-1684.
2. H.J. Hachmann and A.G. Lexins (1976) Eur. J. Biochem. 61, 325-330.
3. E. Paoletti, H. Rosemond-Hornbeak, and B. Moss (1974) J. Biol. Chem. 249, 3273-3280.
4. E. Paoletti and B. Moss, (1974), J. Biol. Chem. 249, 3281-3286.
5. Jovin T.J. (1976), Ann. Rev. Biochem. 45, 889-920.
6. A.J. Clark (1973), Ann. Rev. Genet. 67-86.
7. C.M. Radding (1978) Ann. Rev. Biochem. 47
8. Clark A.J. (1973) Ann. Rev. Genet. 17, 67-86.
9. Meselson M., Yuan R., Heywood J. (1972), Ann. Rev. Biochem. 41, 447-466.
10. Buttin G. and Wright M. (1968), Cold Spring Harbor Symp. Quant. Biol. 33, 259
11. S.D. Barbour and A. Clark (1970), Proc. Nat. Acad. Sci. USA, 65, 955-961.
12. M. Oishi (1969) Biochemistry 64, 1292-1299.
13. A.V. Chestukhin and M.F. Shemyakin and N.A. Kalinina and A.A. Prozorov (1972), FEBS Letters, 24, 121-125.
14. G.F. Vovis and Girard Buttin (1970), Biochim. Biophys. Acta 224, 42-54.
15. K.W.Wilcox and H.O. Smith (1975), J. Bacteriol. 122, 443-453.
16. B.Van Dorp, R. Renne and F. Palitti (1975), Biochim. Biophys. Acta 395, 446-454.
17. P.J. Goldmark and S. Linn (1972), J. Biol. Chem. 247, 1849-1860.
18. F.J.Ferdinand and R. Knippers (1975) Eur. J. Biochem. 52, 291-299.
19. S. Ohi and N. Sueoka (1973), J. Biol. Chem. 248, 7336-7341.
20. J. Doly and C. Anagnostopoulos (1976) Eur. J. Biochem. 71, 309-316.
21. G.F. Vovis and G. Buttin (1970). Biochim. Biophys. Acta, 224,29-41.
22. E.A. Friedman and H.O. Smith (1971) J. Biol. Chem. 247, 2846-2853.
23. M. Anai, T. Hirahashi, Y. Takagi (1970) J. Biol. Chem. 245, 767-774.
24. B. Van Dorp, M. Th.E. Guelen, H.P. Pouwels (1974) Biochem.Biophys. Acta, 340, 166-176.
25. A. Hout, R.A. Ooosterbaan, P.H. Pouwels, A.J.R. De Jonge (1970) Biochim. Biophys. Acta, 204, 632-635.
26. F.G. Winder, and M.F. Lavin (1971) , Biochim Biophys. Acta, 247 542-561.
27. J.D.C. Rosamond, M.R. Lunt (1977) Biochem. J. 163, 485-494.
28. M. Anai, T. Mihara, M. Yamanaka, T. Shibata, and Y. Takagi, J. Biochem. 78, 105-114.

29. M. Wright and G. Buttin (1971), J. Biol. Chem. 246, 6543-6555.
30. W.K. Wilcox, H.O. Smith (1976) J. Biol. Chem., 251, 6122-6126.
31. F.J. Ferdinand, R. Knippers (1975) J. Biochem. 52, 291-299.
32. B. Van Dorp, M.Th.E. Ceulen, H.L. Heijnekez, P.H. Pouwels (1973) Biochim. Biophys. Acta, 299, 65-81.
33. E.A. Friedman, H.O. Smith (1972) J. Biol. Chem. 247, 2859-2865.
34. T. Mukai, K. Matsubara, Y. Takagi (1973) Proc. Nat. Acad. Sci. USA, 70, 2884-2887.
35. Y. Takagi, K. Matsubara and M. Anai (1972) Biochim. Biophys. Acta 269, 347-353.
36. F.G. Nobrega, F.H. Rola, M. Pasetto-Nobrega, M. Oishi (1972) Proc. Nat. Acad. Sci. USA 69, 15-19.
37. S.R. Kushner (1974) J. Bacteriol. 120, 1218-1222.
38. M. Monk, J. Kinross (1972), J. Bacteriol. 109, 971-978.
39. R. Benziger, L.W. Enquist, Ahlzalka (1975) J. Virology 15,861-871.
40. M.F. Shemyakin, A.V. Chestukhin, N.A. Kalinina and A.A. Prozorot (1973), FEBS Letters 31, 31-34.
41. F.G. Winder, P.A. Sastry (1971), FEBS Letters, 17, 27-30.
42. H. Gutfreund (1972) Enzymes, Physical Principles publ. Wiley Interscience 157-160.
43. E.A. Karu, V. MacKay, P.J. Goldmark, S. Linn (1973) J. Biol. Chem. 248, 4874-4884.
44. V. MacKay, S. Linn (1974) J. Biol. Chem. 249, 4286-4294.
45. K.W. Wilcox, H.O. Smith (1976) J. Biol. Chem. 251, 6127-6134.
46. E.A. Friedman and H.O. Smith (1973) Nature New Biology 241, 54-58.
47. F. Palitti, G. Cerio-Ventura, C. Salerno, A. Vellante, P.M. Fasella and E.P. Whitehead. Unpublished work.
48. D.C. Eichler, I.R. Lehman (1977), J. Biol. Chem. 252, 499-503.
49. R.M. Purkey and K. Ebisuzaki (1977) Eur. J. Biochem. 75, 303-310.
50. E.C.B. Banks, C.A. Vernon (1970) J. Theor. Biol. 29, 301-306.
51. F. Winder (1972) Nature new biol. 236, 75-76.
52. M. Orlosky, O.H. Smith (1976), J. Biol. Chem. 251, 6117-6121.
53. A.E. Karu, S. Linn (1972) Proc. Nat. Acad. Sci. USA 69, 2855-2859.
54. R.P. Liebermann and M. Oishi (1973) Nature New Biology 243, 75-77.
55. R.P. Liebermann and M. Oishi (1974), Proc. Nat. Acad. Sci USA 71, 4816-4820.
56. R.W. Hendler, M. Pereira, R. Scharff (1975) Proc. Nat. Acad. Sci. USA, 72, 2099-2103.
57. U. Hermanns, W. Wackernagel (1977) Eur. J. Biochem. 76, 425-432.
58. N. Sarkar and H. Paulus (1972), Proc. Natl. Acad. Sci. USA 69, 3570-3574.

59. A.E. Karu, Y. Sakaki, S. Linn (1975) J. Biol. Chem., 250, 7377-7387.
60. W. Wackernagel, U. Hermanns (1974) Biochem. Biophys. Res. Comm. 60, 521-527.
61. Y. Yamazaki (1971) Biochim. Biophys. Acta, 247, 535-541.
62. M. Oishi, S.D. Cosloy (1972), Biochem. Biophys. Res. Comm. 49, 1568-1572.
63. W. Wackernagel (1973) Biochem. Biophys. Res. Comm. 51, 306-311.
64. L. Grossman, A. Braun, R. Feldberg, I. Mahler (1975) Ann. Rev. Biochem. 44, 19-43.
65. R.B. Settlow, J.K. Settlow (1972) Ann. Rev. Biophys. Bioeng. 2, 293-346.
66. P.R. Lehmann, B.A. Bridges (1977) Essays in Biochemistry, 13, 71-119.
67. V.N. Iyer, W.D. Rupp (1971), Biochim. Biophys. Acta, 228, 117-126.
68. A.K. Ganesan (1974) J. Mol. Biol. 87, 103-119.
69. K.C. Smith and D.H.C. Muen (1970) J. Mol. Biol. 61, 459-472.
70. W.D. Rupp, C.E. Wilde III, D.L. Reno, H.P. Flanders (1971), J. Mol. Biol. 61, 25-44.
71. P.K. Cooper, P.G. Hanawalt (1972), Proc. Nat. Acad. Sci. USA 69, 1156-1160.
72. D.S. Kapp, K.C. Smith (1970), J. Bacteriol. 103, 49-54.
73. R.S. Cole, O. Levitan, R.R. Sinden (1976), J. Mol. Biol. 103, 39-59.
74. R.S. Cole (1973), Proc. Nat. Acad. Sci. USA, 70, 1064-1068.
75. R.C. Wiegand, K.L. Beattie, W.K. Hallman, C.M. Radding (1977), J. Mol. Biol., 116, 805-825.
76. T.R. Broker and I. R. Lehman (1971), J. Mol. Biol., 60, 131-149.
77. S. Ohi, D. Bastia, N. Sueoka (1974), Nature, 248, 586-588.
78. O.H. Smith, A.E. Friedman (1972), J. Biol. Chem. 247, 2854-2858.
79. M. Anai, T. Hirahashi, M. Yamanaka, Y. Takagi (1970), J. Biol. Chem. 245, 775-780.
80. R. Kolodner and C.C. Richardson (1977) Proc. Natl. Acad. Sci. USA 74, 1525-1529.
81. R. Kolodner, Y. Masanume, J.E. Le Cleve and C.C. Richardson (1978), J. Biol. Chem. 253, 566-573.
82. E.Scherzinger, G. Morelli, D. Seiffert and A. Yuki (1977), Eur. J. Biochem. 72, 543-558.
83. R. Kalodner and C.C. Richardson (1977) J. Biol. Chem. 253, 574-584.

84. S. Eisenberg, J.F. Scott and A. Kornberg (1976), Proc. Natl. Acad. Sci. USA 73, 3153-3155.
85. J.F. Scott, S. Eisenberg, L.L. Bertach and A.Kornberg (1977), Proc. Natl. Acad. Sci. USA 74, 193-197.
86. S.E. Eisenberg, J. Griffith and A.Kornberg (1977), Proc. Natl.Acad Sci. USA 74, 3198-3202.
87. M. Abdel-Monem, H. Lauppe, J. Kartenbeck, H. Durwald and H. Hoffmann-Berling (1977), J. Mol. Biol. 110, 667-686.
88. M. Abdel-Monem and H. Hoffmann-Berling (1976), Eur. J. Biochem. 65 431-440.
89. M. Abdel-Monem, M-C. Chanal and H. Hoffmann-Berling (1977), Eur. J. Biochem. 79, 33-38.
90. E. Richet and M. Kohiyama (1976) J.Biol. Chem. 251, 802-812.
91. M. Abdel-Monem, H. Durwald and H. Hoffmann-Berling (1977), Eur. J. Biochem. 79, 39-45.
92. M. Abdel-Monem H. Durwald and H. Hoffmann-Berling (1976), Eur. J. Biochem. 65, 441-449.
93. M. Wright, S. Wickner and J. Hurwitz (1973), Proc. Natl. Acad. Sci. USA 70, 3120-3124.
94. K. Ueda, R. McMacken and A. Kornberg (1978), J. Biol. Chem. 253, 261-269.
95. R. McMacken, K. Ueda and A. Kornberg (1977), Proc. Natl. Acad. Sci. USA, 74, 4190-4194.
96. J.H. Weiner, R. McMacken and A. Kornberg (1976), Proc. Natl. Acad. Sci. USA, 73, 752-756.
97. S. Wickner and J. Hurwitz (1975), Proc. Natl.Acad. Sci. USA, 72, 921-925.
98. M. Gellert, K. Mizuuchi, M.H. O'Dea, and H.A. Nash (1976) Proc. Natl. Acad. Sci. USA 73, 3872-3876.
99. M. Gellert, K. Mizuuchi, M.H. O'Dea, T. Itoh, J-I. Tomizawa (1977) Proc. Natl. Acad. Sci. USA 74, 4772-4776.
100. A. Sugino, C.L. Peebles, K.N. Kreutzer and N.R. Cozzarelli (1977) Proc. Natl. Acad. Sci. USA 74, 4767-4771.
101. M. Gellert, M.H. O'Dea, T. Itoh and J.-I Tomizawa (1977) Proc. Natl. Acad. Sci, USA 73, 4474-4478.
102. T. Itoh and J-I. Tomizawa (1977) Nature 270, 78-80.
103. C. Lowery-Goldhammer and J.P. Richardson (1974), Proc. Natl. Acad Sci. USA 71, 2003-2007.
104. B.H. Howard and B. de Crombrugge (1976), J. Biol. Chem. 251, 2520-2524.
105. C. Lowery and J.P. Richardson (1977), J. Biol. Chem. 252, 1381-1385.

106. M. Rosenberg, D. Court, H. Shimatake, C. Brady and D.L. Wulff (1978), Nature, 272, 414-423.
107. H. Küpper, T. Sekiya, M. Rosenberg, J. Egan and A. Landy (1978) Nature, 272, 423-428.
108. A. Das, D. Court and S. Adhya (1976), Proc. Natl. Acad. Sci.USA 73, 1959-63.
109. B. Alberts, and R. Sternglanz (1977) Nature, 269, 655-661.
110. R.J. Roberts (1976) Crit. Rev. Biochem. 4, 123-164.
111. S. Linn, J.A. Lautenberger, B. Eskin and D. Lackey (1974), Fed. Proc. 33, 1128-1134.
112. H.W. Boyer (1974) Fed. Proc. 33, 1125-1127.
113. H.W. Boyer (1971. Ann. Rev. Microbiol. 25, 153.
114. W. Arber (1974), Prog. Nucleic Acid Res. Mol. Biol. 14, 1.
115. B.J.P. Brockes, P.R. Brown and K. Murray (1974), J. Mol. Biol. 88, 437-443.
116. J.A. Lautenberger, N.C. Kan, D. Lackey, S. Linn, M.H. Edgell and C. Hutchinson (1978). Proc. Natl. Acad. Sci. USA, in press.
117. V.F. Simmon and S. Lederberg (1972) J. Bact., 112, 161-169.
118. S.A. Langeveld, A.P.M. van Mansfeld, P.D. Baas, H.S. Jansz, G.A.van Arkel, P.J. Weisbeck (1978), Nature 271, 417-420.

BIOSYNTHESIS OF IRON-SULFUR STRUCTURES IN IRON-SULFUR PROTEINS

P. Cerletti, F. Bonomi and S. Pagani

Department of General Biochemistry, University of Milan

I 20133, Milano, Italy

Iron sulfur proteins are comparatively newcomers on the biochemical scene since active interest in them has developed in the last twenty years and definite information on the structure of representative types has been gained only in this decade. The development of recent research in the field is reflected in the volumes edited by Lovenberg (1).

These proteins are widespread in microorganisms, plants and animals and they all contain as prosthetic group iron coordinated by four sulfur atoms, either cysteine sulfur or "labile sulfide" this latter being liberated as inorganic sulfide upon degradation. The amount of iron and of sulfur and their coordination geometry vary in different clusters. One iron structures do not contain labile sulfide. Iron and sulfide are in equivalent amounts in 2Fe-2S and 4Fe-4S complexes: in the first ones each iron atom is bound in tetrahedral coordination by 2 cysteine sulfurs and 2 inorganic sulfide atoms. In 4Fe-4S clusters iron and sulfide are organized in cubic coordination. All these species display typical EPR signals. Schematic models of the cluster and the g value of their EPR signal are shown in fig. 1.

Iron sulfur proteins are involved in quite different metabolic pathways such as anaerobic fermentations, nitrogen fixation, photosynthesis, hydroxylation reactions, mitochondrial electron transport. Most of them have redox capacities, in general with very low redox potential. Some typical examples of iron-sulfur proteins, with values of molecular weight and of redox potential are listed in table 1.

Conversion of Apoferredoxin to Holoferredoxin

The biosynthesis of the protein part of ferredoxins has been

TABLE 1

TYPES OF IRON SULFUR PROTEINS

	PROTEIN	other prosthetic group	M.W.		E'_o (mV)
I.	Proteins containing iron but not sulfide				
	A. Anaerobic organisms				
	1. Clostridial rubredoxin		6,000		-57
	B. Aerobic organisms				
	1. *Pseudomonas oleovorans* rubredoxin		20,000		
II.	Fe_2S_2 proteins.				
	A. Photosynthetic organisms: chloroplast ferredoxin		11,500		-420
	B. Nonphotosynthetic organisms:				
	1. Adrenodoxin		13,000		-325
	2. Putidaredoxin		12,000		-240
	3. *A. vinelandii* protein 1		21,000		-350
	4. *A. vinelandii* protein 2		24,000		-350
	5. Mitochondrial b-c_1 Rieske protein		26,000		+220
III.	Fe_4S_4 proteins: high-potential iron protein (HiPIP)		10,000		+350
IV.	2 Fe_4S_4 proteins: bacterial ferredoxins				
	A. Photosynthetic organisms:				
	1. *Chromatium* ferredoxin		10,000		-490
	2. *Chlorobium* ferredoxin		6,000		
	B. Nonphotosynthetic organism:				
	1. Clostridial ferredoxin		6,000		-390:-420
	2. Azotobacter ferredoxin		15,000		-420
V.	Complex iron-sulfur proteins				
	1. Nitrogenase, component 1	Mo	220,000		
	2. Aldehyde oxidase	Mo,FAD	280,000		
	3. Xanthine oxidase	Mo,FAD	290,000		
	4. Dihydroorotate dehydrogenase	FAD,FMN	115,000		
	5. NADH dehydrogenase	FMN	78,000		
	6. Succinate dehydrogenase subunit 70	FAD	69,000	S_1 : Fe_2S_2	-5
				S_2 : Fe_2S_2	-260:-400
	subunit 30		27,000	S_3 : HiPIP	+60

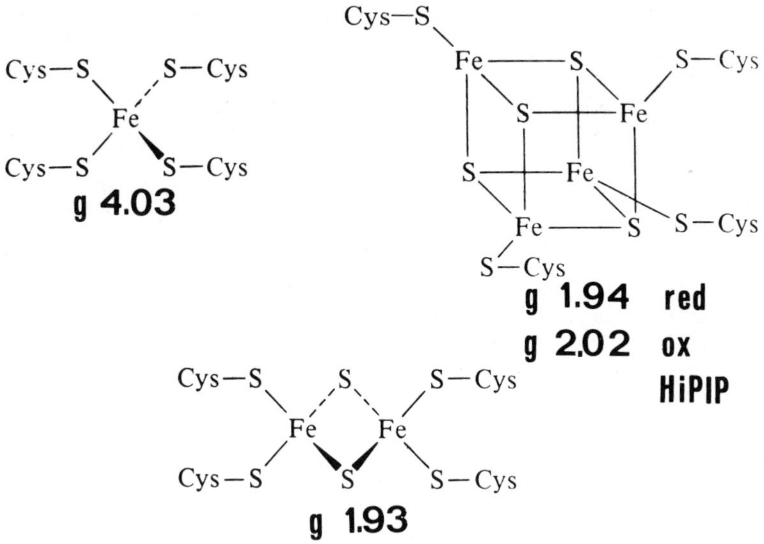

Fig. 1. Structure and EPR signals of active sites of iron sulfur proteins.

extensively studied; it occurs on cytoplasmic ribosomes: clostridial ferredoxin, which has only 53 aminoacids, has been obtained with isolated polysomes (2, 3); rubredoxin contains N-formyl-methionine at its N-terminal (4). Synthesis is regulated: light stimulates ferredoxin production in plants, algae and photosynthetic bacteria (5-7), iron salts induce it in bacteria (8).

On the other hand very little information is available on the biological formation of the iron sulfur component. Experimental interest has gone particularly to structures containing sulfide beside iron. It has been found that ferredoxins can be restored by treating with Fe^{II} or Fe^{III} and sulfide in the presence of a suitable reductant, either mercaptoethanol or dithiothreitol, the apoprotein produced by mercurial treatment of the native protein. Reconstitution has been found to be an all or none process and only fully restored molecules with two 4Fe-4S structures are produced from decayed clostridial ferredoxin or molecules with two 2Fe-2S clusters from spinach ferredoxin (9, 10). However with more complex iron sulfur proteins, e.g., succinate dehydrogenase, after chemical reactivation the molecule retains much contaminant iron and sulfide (11) and procedures suggested to remove the latter do not yield unequivocal results (12).

These artificial reconstitution systems provide little information on the biological formation of iron sulfur structures. Indeed iron and sulfide can be chemically inserted also in proteins which do not have any iron sulfur centers in their native state such as bovine serum albumin. Treated BSA has 4 to 7 moles sulfide and 5 to 8 moles iron per mole and shows absorbtion spectra and Cotton effects in the CD spectrum resembling those found in bacterial ferredoxins. No sulfide is incorporated unless iron is present and the reaction requires excess mercaptoethanol to generate free sulfhydryls in the protein by reducing part of the disulfides (13). It is unlikely that FeS, Fe_2S_3 or other iron salts are utilized directly for ferredoxin formation *in vivo* , since these compounds are extremely insoluble.

In vivo exchange of ^{35}S has been observed between labelled sulfide in the medium and clostridial ferredoxin, in the presence of an extract of the microorganism (14). Adrenodoxin has been reconstituted from the apoprotein, mercaptopyruvate and a crude preparation of mercaptopyruvate sulfurtransferase (15), but no information is available on the role of the enzyme. The Rome group had shown that rhodanese (thiosulfate: acceptor sulfurtransferase, EC 2.8.1.1, Rd) and thiosulfate may substitute for sulfide in the system for chemical reconstitution of spinach ferredoxin (16, 17). The reaction mechanism of rhodanese does formally not allow for reduction of sulfur from the sulfane oxidation state in thiosulfate to sulfide. The enzyme indeed functions by double-displacement mechanism between a sulfane containing anion, such as thiosulfate, and a thiophilic anion, the intermediate being a persulfide formed on the active site of the transferase. Nonetheless when the acceptor is a dithiol, the immediate product, a persulfide, spontaneously decomposes to sulfide and the intramolecular disulfide of the dithiol, the overall reaction yielding reduction to sulfide of the outer sulfur of thiosulfate (18). The reaction has been observed to occur also with dithiothreitol (19) which was a component of the medium in the reconstitution experiments mentioned: therefore it might have been understood that the enzyme and its substrate restored the system for chemical reconstitution.

Action of Rhodanese on Ferredoxins

In order to make sure whether the observations mentioned really represented a mechanism of biological formation of iron sulfur structures we aimed to establish whether reconstitution with rhodanese occurred in the absence of external reducing agents and what was the role of the transferase in the reaction.

The system for the overall reaction contained unlabelled rhodanese and thiosulfate labelled in the outer sulfur or, when the half reaction from transferase to acceptor was studied, thiosulfate was omitted and sulfane sulfur for incoporation was labelled persulfide sulfur at the active site of the transferase. As acceptor we used either spinach or clostridial ferredoxins having respectively 2Fe-2S or 4Fe-4S structures, and a complex iron-sulfur flavoprotein, succinate dehydrogenase (succinate: acceptor oxidoreductase, EC 1.3.99.1, SDH) containing one HiPIP and two 2Fe-2S clusters. Except where otherwise stated, experimental conditions and the materials used were as described in ref. 20 and 21.

Spinach and clostridial ferredoxin incubated with rhodanese were protected from loss of sulfide and in proper conditions the sulfide content was even increased, showing that protection was the balance between insertion by rhodanese and decay. Sulfur was incorporated into the ferredoxin molecule and thiosulfate, which supplies new sulfur to the transferase, enhanced the binding. No transfer occurred when rhodanese was omitted (20).

Labelled labile sulfide was formed but when the incorporation was high part of the sulfur transferred was not reduced; it was probably bound as trisulfide similar to the endogenous sulfane which, as shown by Petering et al. (22) and confirmed in our data, derives from decayed iron sulfur clusters.

Some of the sulfide formed did not originate from external sulfur supplied as rhodanese or as thiosulfate: the increase of total sulfide in clostridial ferredoxin exceeded many times the measured bound (labelled)sulfur derived from the trnasferase or from thiosulfate. Also, in experiments where labelled rhodanese was the only external sulfur donor, the measured increase in total sulfide was larger thant the amount of sulfur that the enzyme carried into the reaction mixture (20). The other possible source of sulfur in the reaction was ferredoxin-bound sulfane sulfur resulting from decayed iron sulfur structures which was found to saturate practically all available cysteines.

This unlabelled sulfur if used diluted the ^{35}S rhodanese: the data in table 2 may be interpreted in this sense, though decreased label on rhodanese might as well be related to increased transfer to ferredoxin.

Ferredoxin bound unlabelled sulfane also competed with ^{35}S from thiosulfate in forming the sulfur-substituted rhodanese intermediate: indeed (unlabelled) rhodanese reacted with spinach ferredoxin appeared not to incorporate as much sulfur from ^{35}S thiosulfate as it could (up to 1.35 mol/mol, dr. Cannella personal communication) and as it did for example with succinate dehydrogenase (21), and it was less labelled with increasing turnover (table 2). These results indicate that, *via* the transferase, ferredoxin bound sulfane sulfur may serve as a source for sulfide formation.

It is interesting that the dilution mentioned did not occur with the clostridial protein where the impact of ferredoxin-bound sulfane sulfur was most evident and where the (labelled) sulfur of rhodanese was diluted in the sulfide on the acceptor (20).

Clostridial ferredoxin was a preparation of commercial origin which has lost much of its native labile sulfide. The residual la-

TABLE 2

DILUTION OF LABEL ON RHODANESE IN SULPHUR EXCHANGE WITH FERREDOXIN

^{35}S donor	^{35}S in Rd, $mol(mole)^{-1}$		^{35}S in isolated Fd $mol(mole)^{-1}$
	before reaction	after reaction and isolation	
^{35}S Rd	0.56	0.20	0.003
	0.62	0.10	0.010
	1.28	0.19	0.033
Unlabelled Rd + 1 mM $^{35}SSO_3^{-2}$	0	0.85	0.023
	0	0.39	0.230
	0	0.36	0.450

Spinach ferredoxin (Fd) was incubated at 0°C in 0.15 M tris buffer pH 7.3 with rhodanese (Rd) added in a molar ratio 1:3, without or with thiosulfate. After 60 min the mixture was fractionated by gel filtration chromatography on Sephadex G 25 (2 x 20 cm) followed by diethyl aminoethyl cellulose DE 52 (0.8 x 3 cm). Rhodanese is not retained on the ion exchanger and ferredoxin was eluted with 0.15 M tris buffer pH 7.3 containing 1 M NaCl. Fractions of 1.8 ml were collected. Proteins, sulfide content, enzymic activity and radioactivity were measured in the reaction mixture and on the separated reactants. The radioactivity measured was normalized for the specific activity of the ^{35}S.

bile sulfide did not decay significantly in the incubation conditions (20). Also spinach ferredoxin preparations with lower initial labile sulfide were more stable and rhodanese had less effect on them. Formed or forming structures were more labile in more native preparations and in experiments in which rhodanese had more effect: in either conditions more sulfide was lost as inorganic sulfide in separating the ferredoxin from the reaction mixture (table 3).

Spectral changes in the processes studied occurred with either ferredoxin at wavelengths typical for iron sulfur chromophores and rhodanese counteracted the decrease in absorbancy due to aging (figures 2 and 3).

In the interaction with both ferredoxins rhodanese was inactivated. The decay was logarithmic with time (fig. 4). It appeared related to the extent of rhodanese effect and varied inversely to the presence of bound sulfane on the ferredoxin: indeed it was smaller with the clostridial protein, a heavily decayed preparation

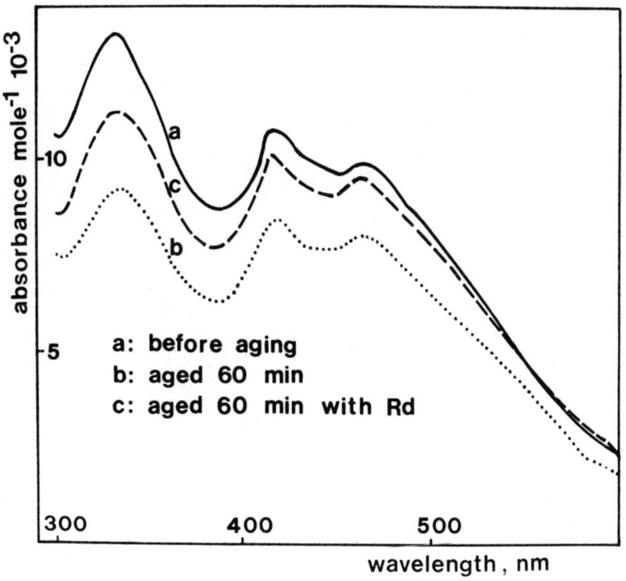

Fig. 2. Optical spectra of spinach ferredoxin incubated without and with rhodanese. Ferredoxin was incubated at 0°C in 0.15 M tris buffer pH 7.3 without and with rhodanese added in a molar ratio 1:3. Spectra were recorded at 12°C before reaction and after 60 min incubation.

TABLE 3

THE EFFECTS OF RHODANESE ON SPINACH FERREDOXIN AND THE INACTIVATION ON THE SULPHURTRANSFERASE

^{35}S sulfane donor	Sulfide, mol(mole Fd)$^{-1}$ before reaction	in the reaction mixture - Rd a	+ Rd b	b-a	lost during chromatography	^{35}S transferred to Fd mol(mole)$^{-1}$	Residual transferasic activity %
	1.884	1.045	1.765	0.720	1.219	0.033	76
$^{35}S.Rd$	1.788	1.164	1.692	0.528	0.624	0.010	87
	1.138	0.951	1.136	0.185	0.398	0.003	100
$^{35}SSO_3^{-2}$	1.728	1.188	1.968	0.780	0.468	0.230	100
	1.360	0.664	1.001	0.337	0.131	0.023	100

Spinach ferredoxin (Fd) was incubated with and without rhodanese (Rd) the experimental conditions were those of table 2.

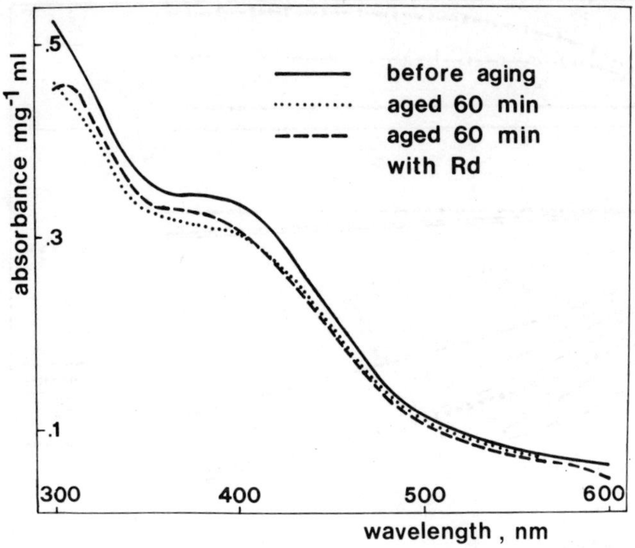

Fig. 3. Optical spectra of clostridial ferredoxin incubated without and with rhodanese. Experimental conditions were those of fig. 2, except for the buffer having pH 8.1.

(see above), and with preparations of ferredoxin with less initial sulfide (table 3). It also did not occur or was much less in the presence of thiosulfate, which strongly confirms that supply of sulfane was important for stability of the transferase. Loss of activity become particularly evident after separating the transferase from the other reactants (fig. 4).

Action of Rhodanese on Succinate Dehydrogenase

Succinate dehydrogenase contains two 2Fe-2S centers of the spinach ferredoxin type, centers S-1 and S-2, likely located in the flavin containing subunit and one 4Fe-4S center of the HiPIP type, center S-3, probably residing in the smaller iron-protein subunit. In the soluble enzyme the midpoint potentials of centers S-1 and S-2 are -5 ± 15 mV and -400 ± 15 mV respectively and the one of center S-2 is, in particulate preparations, $+60 \pm 15$ mV (23, 24). Center S-2 therefore is not reducible by succinate.

The flavoprotein was assayed in the oxidized state and after succinate reduction. The results were very similar to those obtained with the ferredoxins: the dehydrogenase bound as labile sulfide sulfane sulfur from rhodanese and, *via* rhodanese, from thiosulfate. Sulfide binding was associated with a) either increase or smaller decay of typical iron sulfur absorbancies of the flavoprotein during

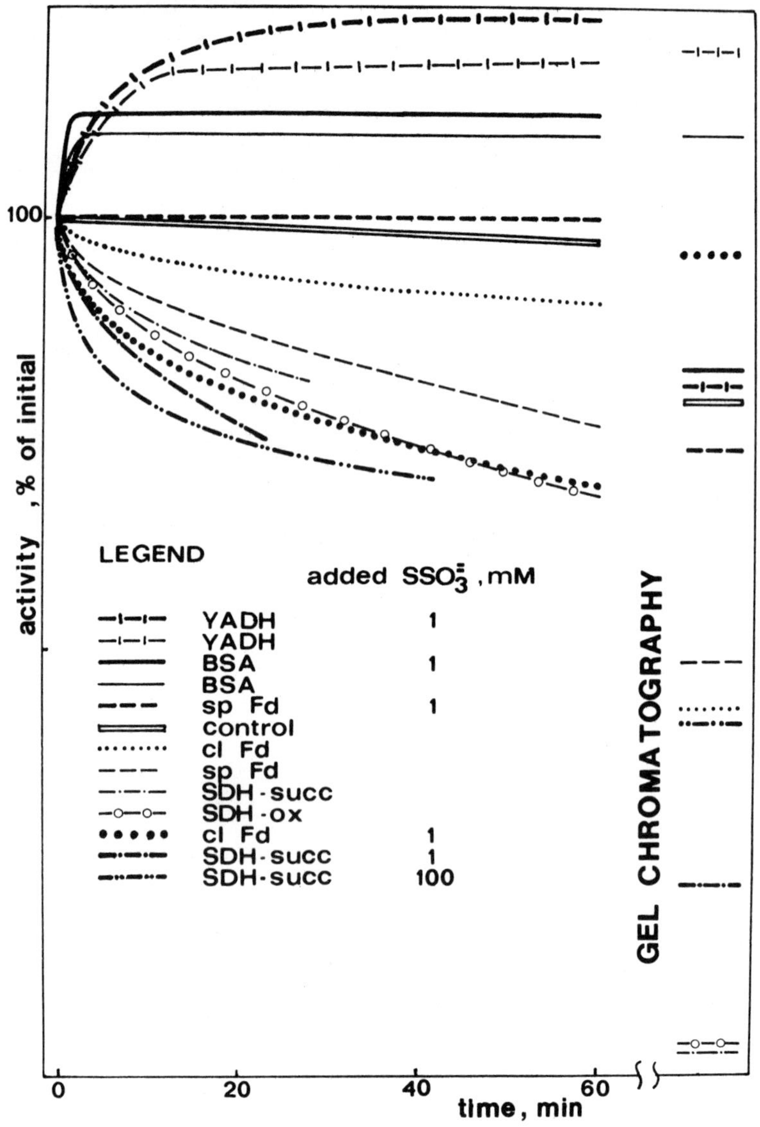

Fig. 4. Deactivation of rhodanese in the interaction with protein acceptors. Rhodanese was incubated in a molar ratio 1:3 with the proteins indicated. Temperature was 8°C and buffer 50 mM phosphate, 20 mM succinate pH 7.6 in the control, for yeast alcohol dehydrogenase (YADH), bovine serum albumin (BSA) and succinate dehydrogenase (SDH succ). Succinate dehydrogenase was also incubated without succinate (SDH ox). For spinach (spFd) and clostridial (clFd) ferredoxins conditions were those of fig. 2 and 3. At the time indicated the samples in phosphate were chromatographed on a column of Ultrogel AcA 54 (2x20 cm) and ferredoxins were treated as specified in table 2.

aging, b) protection of its reconstitutive capacity, which depends on the state of center S-3 in the flavoprotein, and c) inactivation of rhodanese (21, 25, 26).

Other proteins which in their native state do not possess an iron sulfur component behaved quite distinctly. Namely with yeast alcohol dehydrogenase, which contains per gm approximately as many sulfhydryl groups as succinate dehydrogenase, sulfur was discharged from rhodanese and from thiosulfate and was found in the reaction medium and in part also on the protein, but at the sulfane oxidation level. With bovine serum albumin, which has a large number of intramolecular disulfide groups but only one, scarcely reactive, sulfhydryl, no sulfur was released (21). In either case rhodanese was not inactivated: on the contrary its activity was enhanced (fig. 4). It was already known that bovine serum albumin stabilizes rhodanese: indeed the protein is added to this aim to the assay medium of the transferase (27).

The dehydrogenase isolated after reaction appeared less labelled when reacted in oxidized than in a reduced condition also if thiosulfate supplied a large amount of exchangeable sulfur. Nonetheless the label lost by the transferase and not bound to the flavoprotein was in the medium as sulfide: practically none was found in the absence of succinate dehydrogenase (21, 26). Limited incorporation therefore appears due to rapid decay of formed or forming structures in the flavoprotein and heavy labelling of initially cold rhodanese indicated active turnover of available thiosulfate to restore them. The two processes are interrelated and are enhanced in the absence of stabilizing succinate.

With succinate present labelled protein bound sulfide accounted for all sulfur incorporated, whereas in the oxidized enzyme only part of bound sulfur was reduced. Uncomplete reduction may be related to lively turnover: indeed also in the reduced flavoprotein only part of incorporated sulfur was sulfide in conditions of active turnover (high binding on the dehydrogenase and heavy labelling of rhodanese), such as with 100 mM thiosulfate (26). This suggests presence of sulfane sulfur on the acceptor protein preliminary or alternative to sulfide formation. These possibilities are discussed in a further section in connection with the mechanism of the reactions studied.

The kinetics of rhodanese action on oxidized succinate dehydrogenase are showin in fig. 5. All the spectrum from 300 to 600 nm was recorded but absorbancy changes at 370, 410, 450 and 540 nm are given since at these wavelengths the spectra were modified most and they correspond to typical absorptions of oxidized 2Fe-2S chromophores (410 and 450 nm) and of HiPIP ones (370 and 450 nm). All chromophores had a rapid increase which then vanished.

Direct evidence for an effect of rhodanese on the iron sulfur clusters came from EPR spectra. Both the HiPIP type signal in the

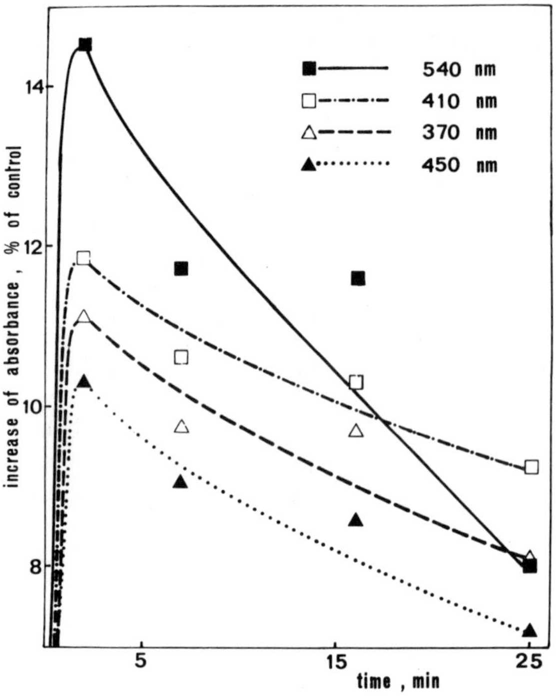

Fig. 5. Kinetics of absorbancy changes in succinate dehydrogenase during rhodanese action. The flavoprotein was incubated (2.58 mg ml^{-1}) at 12°C in 50 mM phosphate buffer pH 7.5 in a nitrogen atmosphere without and with rhodanese added in a molar ratio 1:3. The spectra were recorded at the intervals indicated. The differences in absorbancy at the given wavelengths between samples with and without rhodanese are reported. At the end of incubation the reconstitutive capacities in samples incubated without and with rhodanese were respectively 25.8% and 47.0%, and sulfide contents 31.74 and 43.73 nmol (mg protein)$^{-1}$.

oxidized flavoprotein and the g = 1.93 signal after succinate addition were affected. They varied in parallel with labile sulfide content, reconstitutive capacity and optical spectra being protected from decay during aging. The signal from the flavin semiquinone at g = 2.00, decreased by aging, diminished further: data in the literature induce to interpret this behaviour as indicating that rhodanese restored, at least in part, the molecular integrity of succinate dehydrogenase (26).

The interaction with succinate dehydrogenase modified rhodanese more heavily than the one with ferredoxins. Loss of activity was rapid and increased where turnover was more lively, e.g. in the presence of high thiosulfate (fig. 4).

The most severe inactivation appeared after separating the transferase from the other reactants and a preceding protective effect of thiosulfate was manifest: it had probably taken place during the reaction since the compound is removed by the separative treatment. Proteins which released the sulfur of rhodanese but did not induce its reduction like yeast alcohol dehydrogenase, or which did not interact like bovine serum albumin, did not inactivate the transferase (fig. 4).

Something about Mechanisms

Transfer by rhodanese of sulfane sulfur to a protein acceptor containing free SH groups had so far not been reported but it does not appear irrealistic since persulfides are formed by the transferase and they are present in some proteins. Indeed the results with yeast alcohol dehydrogenase and with bovine serum albumin indicate that transfer occurs provided the acceptor has free thiols. The very point to explain is reduction of transferred sulfur in the absence of external reducing agents.

We believe that reducing equivalents came from cysteines in the proteins participating to the reaction, in a similar process to what described for small molecular weight dithiols (18, 19), namely formation of a disulfide concomitant to reduction of a persulfide sulfur.

The measured decrease of thiols in rhodanese reacted with succinate dehydrogenase candidates the transferase as being at least one of the donors of reducing equivalents in the mechanism suggested. If the disulfide formed involves the SH group at the active site of the enzyme, the transferase is inactivated. When thiosulfate is present and turns over with rhodanese it produces sulfite which may regenerate by sulphitolysis of the disulfide one free SH group: if this is the catalytic thiol, the transferase is reactivated: presence of one sulfite residue on rhodonase does not inactivate it (Dr. Cannella, personal communication). Thiosulfate is much less effective if it is added to the inactivated transferase after incubation with the protein: it is likely that the compound, when present during the reaction with the acceptor binds to the transferase molecule and prevents irreversible conformational changes connected with disulfide formation.

Sulfide formation in the hypothesis outlined would occur on the rhodanese molecule as an independent process from sulfane transfer to the acceptor protein. As a possible alternative sulfide might be formed on the acceptor protein: in this case the first step would be sulfane transfer from rhodanese with formation of persulfide followed by sulfur reduction and intramolecular disulfide formation, or alse reduction might involve directly the trisulfide sulfur in the acceptor protein, derived from decayed iron sulfur structures. However little free cysteine is available in ferredoxin preparations decayed in air; in conditions of anaerobiotic aging oxygen is consistently decreased but not totally excluded. Rhodanese, particularly with exogenous sulfane readily supplied by thiosulfate, may effectively compete with oxidation of cysteines of decaying clusters; nonetheless as long as thiols in the acceptor protein are the only source of reducing equivalents, it may limit loss of existing structures but cannot account for a net increase in the total content of labile sulfide.

In synthetic analogues of 2Fe-2S centers, the cysteine ligands to each iron atom are at a distance allowing disulfide bond formation (28). These distances should be valid also in spinach ferredoxin and in succinate dehydrogenase and it is conceivable that unless major conformational changes occur they are preserved after decay of the native clusters. Thiol ligands involved in 4Fe-4S clusters are too far apart to give a disulfide but the number of cysteines in clostridial ferredoxin and the small molecular size make possible interactions between and possibly also within molecules.

The varying rhodanese inactivation, the data on sulfur incorporation and sulfide content suggest that one or the other of the mechanisms proposed may operate depending on the type of substrate and on availability of sulfane sulfur and of reducing equivalents. A substancial conformational mobility of rhodanese in response to substrates has been documented (29-31) and may allow for different characteristics of catalysis.

With succinate dehydrogenase oxidation of rhodanese has been directly evidenced and is witnessed by the very high inactivation of the enzyme. The stoichiometry of sulfide incorporation indicates that rhodanese alone is sufficient to provide the reducing power needed, except in experiments with 100 mM thiosulfate. In this case presence of sulfane sulfur on the flavoprotein might be indicative of persulfide formation prior to local reduction to sulfide.

The size of the dehydrogenase makes intermolecular disulfide bond formation unlikely. If disulfides were formed between cysteine ligands to 2Fe-2S centers these clusters cannot be restored and too little is known about the relative location of the clusters on the molecule to predict if the formed sulfide may be used for a nearby structure. Little can be said about the residual 8-9 thiols not involved in the iron sulfur structures (32) since it is not known where they are located. In conclusion if reduction were intramolecular, full restoral of the iron sulfur structures appears doubtful and the resulting flavoprotein should in part be modified.

With ferredoxins the limited deactivation of the transferase, the stoichiometry of transfer and of reducing equivalents indicate that the acceptor protein contributes sulfur and reducing power. The used sulfane may first be bound to rhodanese as the dilution of label in the interaction with spinach ferredoxin indicates, or else it is reduced and transferred directly, which may be the case with the clostridial protein probably due to its molecular qualities. Reduction of sulfane nevertheless occurs only in the presence of rhodanese.

Concluding remarks

The results reported and the mechanisms outlined offer a plausible pathway for *in vivo* formation of labile sulfide. However from a metabolic standpoint they elucidate only half reaction namely origin and reduction of sulfane. Information is still needed on restoral of reducing equivalents used and on an alternative source of reducing power, not involving protein thiols. Indeed these groups must necessarily be affected in our experimental conditions but appear an unlikely reductant in the cell; free cysteines available in newly synthetized apoferredoxins are required as ligands for clusters.

A further problem is how clusters are put together and iron is inserted. Direct evidence on this point has still to come but some hypothesis, based or available information, can be presented. It is known that in iron sulfur proteins iron from decayed iron sulfur structures remains at least in part bound to the protein; in succinate dehydrogenase it is not liberated at all (G. Palmer, personal communication). It is therefore available for cluster formation. With synthetic analogues and also with plant and bacterial ferredoxins extrusion and replacement of the iron sulfide core of 2Fe-2S and 4Fe-4S clusters has been produced. Interconversion of dimeric and tetrameric structures has been observed in synthetic analogues (28). Therefore we may think that once iron, sulfide and the proper apoprotein ligand are present, formation of the cluster may occur spontaneously.

SUMMARY

Little is known on the biosynthesis of the iron sulfur prosthetic groups in iron sulfur proteins. Successful attempts of chemical reconstitution have been done with ferredoxins, but with more complex proteins results were less satisfactory. Informations on *in vivo* processes are scanty.

We found that rhodanese inserts as sulfide sulfane sulfur from thiosulfate and the protein bound trisulfide sulfur from decayed iron sulfur clusters into spinach ferredoxin, which has two 2Fe-2S clusters, clostridial ferredoxin, a protein having two 4Fe-4S groups, and succinate dehydrogenase, which contains two 2Fe-2S clusters and one 4Fe-4S center of the HiPIP type. The physical, chemical and catalytic properties of the acceptors are modified and indicate a more native state of the molecule.

Transferred sulfur is reduced only if the acceptor is an iron sulfur protein. Presence of free cysteines alone induces transfer but not reduction.

Rhodanese is inactivated in the interaction: two of its four thiol group are oxidized and probably this is related to reduction of the transferred sulfane to sulfide. The extent of inactivation varies with different acceptors and alternative mechanisms for sulfane reduction are proposed.

The implications of the data obtained for the biosynthesis of iron sulfur structures are discussed.

ACKNOWLEDGMENTS

This investigation is part of a joint program with Dr. C. Cannella of the Department of Biological Chemistry of the University of Rome. It was supported in part by a grant of the Italian National Research Council (C.N.R.).

REFERENCES

1. W.Lovenberg ed. Iron-sulfur proteins, vol.I Biological Properties, 1973, vol. II Molecular Properties, 1973, vol. III Structure and Metabolic Mechanisms, 1977. Academic Press.
2. Nepokroeff, C., and Aronson, A.I. (1970) Biochemistry 9, 2074-2081.
3. Brodrick, J.W. (1974) Ph.D.Dissertation, University of California, Berkeley.
4. McCarthy, K.F., and Lovenberg, W. (1970) Biochem.Biophys.Res. Commun. 40, 1053-1057.
5. Haslett, B.G., Cammack, R., and Whatley, F.R. (1973) Biochem.J. 136, 697-703.
6. Armstrong, J.J., Surzycki, S.J., Moll, B., and Levine, R.P. (1971) Biochemistry 10, 692-701.
7. Shanmugam, K.T., Buchanan, B.B., and Aron, D.I. (1972) Biochim. Biophys. Acta 256, 477-486.
8. Knight, E., Jr., and Hardy, R.W.F. (1966) J.Biol.Chem. 241, 2752-2756.
9. Hong, J.S., and Rabinowitz, J.C. (1970) J.Biol.Chem. 245, 6574-6581.
10. Fee, J.A., and Palmer, G. (1971) Biochim.Biophys.Acta 245, 175-195.
11. Baginsky, M.L., and Hatefi, Y. (1969) J.Biol.Chem. 244, 5313-5319.

12. King, T.E., Winter, D., and Steele, W. (1972) in Structure and Function of Oxidation Reduction Enzymes (Akesson, B., Ehrenberg A., eds.), vol. 18, Pergamon Press, 519-532.

13. McCarthy, K., and Lovenberg, W. (1969) J.Biol.Chem. 243, 6436-6441.

14. Jeng, D., and Mortenson, L.E. (1968) Biochem.Biophys.Res.Commun. 32, 984-991.

15. Taniguchi, T., and Kimura, T. (1974) Biochim.Biophys.Acta 364, 284-295.

16. Finazzi-Agrò, A., Cannella, C., Graziani, M.T., and Cavallini, D. (1971) FEBS Letters 19, 172-174.

17. Tomati, U., Matarese, R., and Federici, G. (1974) Phytochemistry 13, 1703-1706.

18. Westley, J. (1973) Adv.Enzymol. 39, 327-368.

19. Pecci, L., Pensa, D., Costa, M., Cignini, P.L., and Cannella, C. (1976) Biochim.Biophys.Acta 455, 104-111.

20. Bonomi, F., Pagani, S., and Cerletti, P. (1977) FEBS Letters 84, 149-152.

21. Bonomi, F., Pagani, S., Cerletti, P., and Cannella, C. (1977) Eur.J.Biochem. 72, 17-24.

22. Petering, D., Fee, J.A., and Palmer, G. (1971) J.Biol.Chem. 246, 643-653.

23. Ohnishi, T., Salerno, J.C., Winter, D.B., Lim, J., Yu, C.A., and King, T.E. (1976) J.Biol.Chem. 251, 2094-2104.

24. Ohnishi, T., Lim, J., Winter, D.B., and King, T.E. (1976) J. Biol.Chem. 251, 2105-2109.

25. Bonomi, F., Pagani, S., and Cerletti, P. (1977) in Flavins and Flavoproteins, Physicochemical Properties and Functions (Ostrowski, W., ed.), Polish Scientific Publishers, Warsaw, 151-164.

26. Bonomi, F., Pagani, S., and Cerletti, P. (in press) Flavins and Flavoproteins (Yagi, K., and Yamano, T., eds.), Scientific Societies Press, Tokyo, Japan.

27. Sörbo, B.H. (1953) Acta Chem.Scand. 7, 1129-1137.

28. Holm, R.M., and Ibers, J.A. (1977) in Iron-Sulfur Proteins, vol. III (Lovenberg, A., ed.), Academic Press, Inc. 205-281.

29. Wang, S.F., and Volini, M. (1973) J.Biol.Chem. 248, 7376-7384.

30. Volini, M., and Wang, S.F. (1973) J.Biol.Chem. 248, 7386-7391.

31. Volini, M., and Wang, S.F. (1973) J.Biol.Chem. 248, 7392-7395.

32. Pagani, S., Bonomi, F., and Cerletti, P. (1974) FEBS Letters 39, 139-143.

SOME PROPERTIES OF TWO PROTEINS INVOLVED IN MEMBRANE TRANSPORT

G.L. Sottocasa, E. Panfili, G. Sandri, G.F. Liut,
C. Tiribelli*, M. Luciani and G.C. Lunazzi

Istituto di Chimica Biologica
Istituto di Patologia Medica*
University of Trieste, Italy

INTRODUCTION

One of the most intensively investigated fields of biological research nowadays is membrane structure and function.The reason for this great interest resides in the great number of biological functions connected with these microscopic structures. It is not appropriate to recall here all the concepts which have developed in this area since some decades (see for a recent review 1). The most generally accepted structure today is that illustrated by Singer (2) and referred to as fluid-mosaic model. In the model phospholipids are arranged in a bilayer system,with the hydrophobic tails in close association to each other,whereas the polar heads are facing the water phase on both sides of the membrane. Associated to the phospholipid bilayer are the protein molecules which are to be classified in peripheral and integral ones. The type of interaction of the two classes of proteins with phospholipids are totally different. In the former case protein are bound to the surface <u>via</u> polar groups. Changes in ionic strength and/or chelation of divalent cations is often sufficient to cause the peripheral proteins to be released. On the contrary,integral proteins are firmly bound to the lipid bilayer and hydrophobic interactions contribute greatly to the stabilization of the structure. Obviously depending on the physical properties of the surface of the protein three possibilities are open: a) a protein molecule is completely embedded in the lipid bilayer; b) a protein molecule is only partially embedded in hydrophobic core of the membrane and faces one of the two surfaces with a hydrophilic tail and c) a protein molecule may be so arranged that only the intermediate portion of it shows a hydrophobic outer surface and the protein may be visualized as a transmembrane component. The acceptance of such a model has had important implications concerning the possible transport mechanisms known to occur at the membrane level. On thermodynamic grounds,two models have been considered unlike,namely the mobile carrier model and the revolving door one. The former suggests that a hydrophilic protein binds a ligand,gets associated to the lipid bilayer,moves across the membrane and releases the ligand on the opposite side. The latter postulates the existence of an integral-type protein which binds the ligand on one side and,changing its orientation,may release it on the opposite one. Always on the basis of thermodynamic considerations,a model which assumes the existence of specialized pores across the membrane has been favoured. Pores are visualized as consisting of more than a protein subunit,an arrangement which ensures both the existence of specificity for a given ligand and the possibility of conformational changes favouring a unidirectional movement.

The purpose of this presentation is to discuss whether a general type of mechanism is to be accepted in all cases of transport across membranes. The discussion will be based on some experimental data derived from the study of two independent proteins extracted and purified from two different biological membranes in our laboratory. We shall first of all summarized some properties of the two isolated components in the two following paragraphs and,subsequentely,we shall try to draw some conclusions from a comparative analysis of the data in order to discuss the general question asked above.

PROPERTIES OF THE CALCIUM-BINDING GLYCOPROTEIN ISOLATED FROM MITOCHONDRIA

Table I summarizes some of the main features of the mitochondrial calcium-binding glycoprotein isolated in our laboratory (3-5). Chemically the protein does not show any peculiarity with regard to aminoacid composition,characterized only by a relatively high content of acidic residues such as glutamic and aspartic (6). As for neutral sugars,it contains glucose,mannose,galactose and xylose,in addition to the aminosugars N-acetylgalactosamine,N-acetylglucosamine and sialic acid. Interestingly the K_D of the calcium-glycoprotein complex is compatible with the K_M for calcium transport. Lanthanides and ruthenium red,known inhibitors of mitochondrial calcium-movements efficiently block calcium-binding by the glycoprotein. Mitochondria genetically devoid of calcium-transport activity lack the glycoprotein which is replaced by other glycoproteins,usually of higher molecular weight,totally devoid of calcium-binding activity (7). The properties summarized speak in favour of an involvement of the glycoprotein in calcium-transport. More convincing evidence comes,however,from two other lines of experimentation which we shall discuss in more detail. Figure 1 shows an experiment in which calcium-movements in respiring mitochondria have been followed spectrophotometrically using murexide as a calcium indicator in the medium. When 500 nmoles of calcium

TABLE I

Some properties of the Ca^{2+}-binding glycoprotein from mitochondria

Molecular weight	30,000 daltons
Sugar moiety	~ 10 %
Calcium bound	present
Magnesium bound	present
Phospholipid associated	variable
K_D	$\sim 10^{-7}$ M
Inhibitors of Ca^{2+}-binding	La^{+3}, ruthenium red
Distribution	ubiquitous in calcium-transporting mitochondria, absent in calcium-non-transporting mitochondria

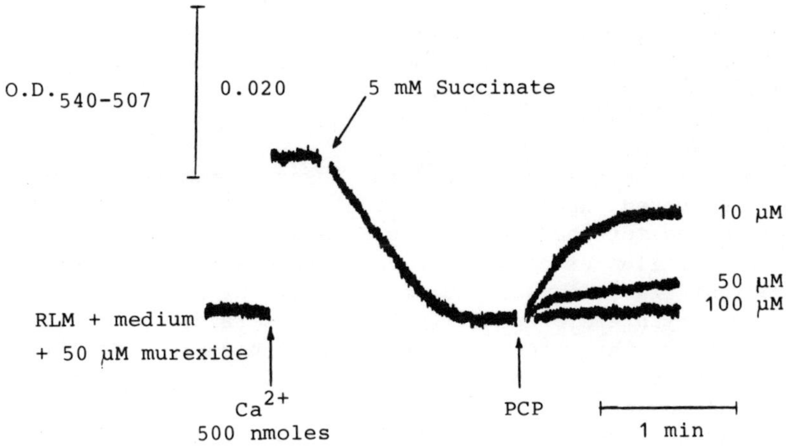

Figure 1. Influence of pentachlorophenol concentration on the passive efflux of calcium from rat-liver mitochondria. Experimental conditions: 3.5 mg rat-liver mitochondria were added to a medium containing 0.1 M mannitol, 50 mM KCl, 25 mM TRIS-HCl buffer at pH 7.4, 8 mM $MgSO_4$, 15 μM rotenone, 5 mM acetate buffer at pH 7.4 and 5 mM phosphate at pH 7.4, other additons as indicated in the Figure. Dual Wavelength Phoenix Recording Spectrophotometer. The reaction was run at room temperature. (From Sandri et al. (8), permission of the Editors requested).

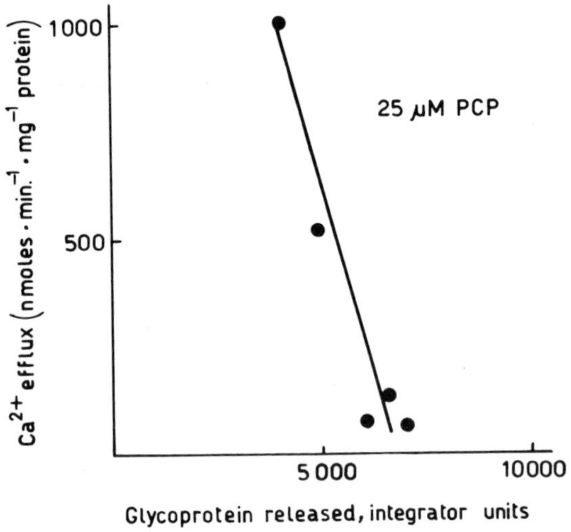

Figure 2. Correlation between decrease of rate of calcium-efflux and glycoprotein release. Experimental conditions:the rate of calcium-efflux has been measured as described in Figure 1. Two min after addition of pentachlorophenol the samples were centrifuged with an Eppendorf mod.3200 centrifuge and the supernatant subjected to electrophoresis , toluidine blue staining and densitometric scanning.(From Sandri et al.(8)).

are added to an aerobic mitochondrial suspension,a change in absorbance is evoked. Upon addition of a respiratory substrate calcium is taken up and retained inside the organel. Addition of an uncoupler promotes the efflux of calcium,which is released down the concentration gradient. The somewhat surprising finding is that,if the concentration of pentachlorophenol(PCP)added is raised well above its uncoupling value,the rate of calcium efflux becomes virtually zero, despite of the high concentration gradient across the membrane and in the total absence of energy available to maintain it. The only possible explanation for this finding is that the mitochondrial inner membrane,under the conditions described,had become totally impermeable to calcium. When in the medium the concentration of the glycoprotein has been measured,we found that,simultaneously to the uncoupling,pentachlorophenol had promoted a glycoprotein release. As shown in Figure 2,a linear negative correlation exists between the rate of calcium efflux and the amount of glycoprotein dissolved. This finding suggests that the rate-limiting step in passive calcium-efflux is the amount of glycoprotein present in mitochondria.

Even more direct evidence comes from antibody studies. Affinity-

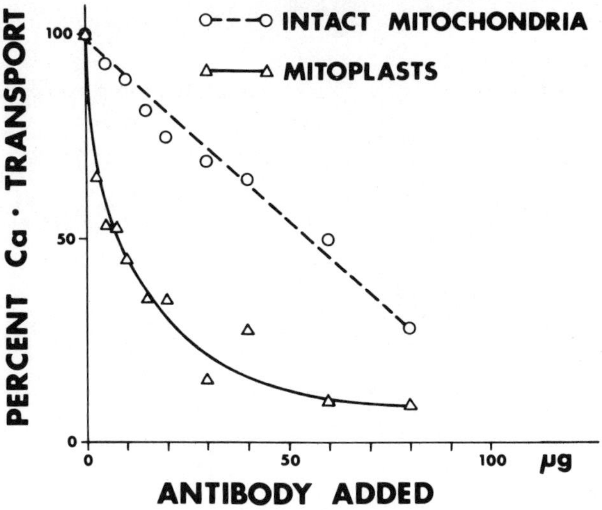

Figure 3. Effect of antibodies against the glycoprotein on calcium-transport in rat-liver mitochondria and mitoplasts. Experimental conditions:mitochondria and mitoplasts were prepared according to Schnaitman and Greenawalt(19) ;calcium transport was followed as described in previous Figures. The reaction mixture(3 ml)consisted of 5 mg mitochondria or mitoplasts protein,100 µM murexide,70 mM sucrose,220 mM mannitol,2 mM HEPES pH 7.4, 5 mM acetate pH 7.4, 1 mM phosphate pH 7.4, 15 µM rotenone,165 µM calcium, 5 mM succinate. Mitochondria and mitoplasts were pre-incubated with the antibody preparation for 1 hour at 0°C in a final volume of 60 µl. (From Panfili et al. (9),permission of the Editors requested).

chromatography purified antibodies may efficiently impair active calcium transport,as shown in Figure 3: it may be noted that with mitoplasts some 6 µg of antibody in a system containing 5 mg of mitochondrial protein is sufficient to elicit a 50 % inhibition of the rate of calcium uptake. This finding is indicative of a very specific mechanism of action and, also, of a high degree of purity of the antibody preparation. With mitochondria the relationship is linear, as expected in view of the presence of a diffusion barrier represented by the outer mitochondrial membrane. It is worth mentioning that other mitochondrial functions are virtually unaffected by the antibody treatment, as shown in Table II. Of great interest is also the observation that not only the rate of influx, but also that of efflux, induced by uncouplers, is blocked by antibodies, as shown in Figure 4. The limits of this presentation do not allow to discuss a number of other experiments in which other types of calcium movements have been tested for their sensitivity to the inhibitory action of the antibodies. It is enough to recall here

TABLE II

Effect of specific antibodies against the calcium-binding glycoprotein on calcium-transport, electron transport and respiratory control ratio

Antibodies added (μg)	% Inhibition Calcium-transport	Electron-transport	R.C.R.
5	45	0	0
20	70	17	10
40	84	37	19

A Clark oxygen electrode was used at 30°C (20). The substrate succinate and reaction mixture were as described before. R.C.R.=respiratory control ratio . (From Panfili et al. (9)).

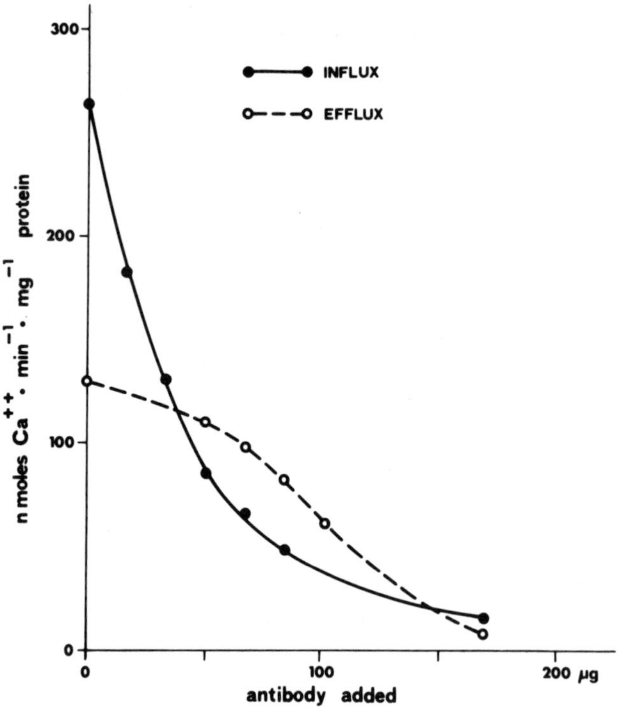

Figure 4. Influence of antibodies on the rate of calcium-influx and efflux in intact mitochondria. Experimental conditions:as in Figure 3. Calcium-efflux has been evoked by the addition of 10 μM pentachlorophenol.

that, so far, all calcium movements tested could be efficiently inhibited by the antibody preparation. The evidence summarized convincingly speaks in favour of a necessary involvement of the calcium-binding glycoprotein in the mechanism of mitochondrial calcium transport. Some features have been found, which may suggest a closer insight to the mechanism of operation of the glycoprotein: the glycoprotein undergoes conformational changes upon binding of calcium and/or magnesium: the phenomenon is documented both by physical-chemical studies (7) and by the finding that in the presence of the divalent cations the glycoprotein may be extracted in decane (10). A second point of interest concerns the intramitochondrial location of the compound, which was found originally free in the intermembrane space, but could be detected also associated to the inner membrane. The glycoprotein has never been found in the matrix space. The degree of association to the inner mitochondrial membrane was shown to be dependent on the presence of calcium in the medium, as well as on calcium transport (8). The association phenomenon, in addition, was shown to be freely reversible. The effect is specific for the inner membrane and does not occur with the outer mitochondrial membrane (11). On the basis of these findings, we proposed a mechanism by which the glycoprotein could promote both the influx and the efflux of calcium in mitochondria. The mechanism is schematically represented in Figure 5, which indicates that calcium penetrates freely across the outer mitochondrial membrane, and becomes associated to the glycoprotein. As a consequence, the glycoprotein acquires a more hydrophobic conformation which facilitates its association to the hydrophobic core of the inner mitochondrial membrane. Here the energy-tranducing mechanism of the mitochondria may provide the energy necessary for the conformational

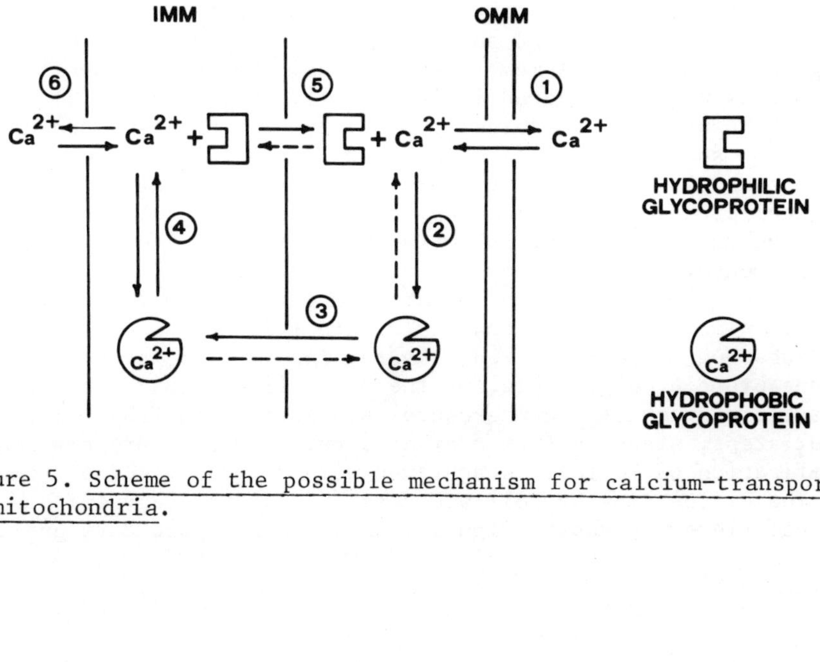

Figure 5. Scheme of the possible mechanism for calcium-transport in mitochondria.

TABLE III

Purification of BSP-binding protein from liver plasma membrane

Fraction	Total protein (mg)	BSP-binding(⊠) /mg protein	BSP-binding(⊠) total
Crude membranes	975.6	1.01	985.3
Extract	66.0	10.80	712.8
Sephadex-G 100	7.0	53.25	372.7
AG 1x8	4.1	90.00	369.0

(⊠) binding measured at a constant BSP concentration of 10 μM

change leading to the hydrophilic dissociated form of the glycoprotein. As a result, calcium is released towards the matrix space, and the free glycoprotein in the opposite direction. A necessary assumption is an asymmetrical arrangement of the inner mitochondrial membrane. Such an assumption may be easily accepted on the basis of the well known asymmetry of this membrane structure, documented both at the morphological and chemical level. A more comprehensive review of these experimental data has recently appeared (12).

PROPERTIES OF BILITRANSLOCASE, A LIVER PLASMA MEMBRANE PROTEIN INVOLVED IN BILIRUBIN UPTAKE

The process of bilirubin uptake by liver has been regarded for a while as a simple diffusion mechanism across the plasma membrane. This assumption was based on the observation that bilirubin is a lipid-soluble substance and should not find difficulties in diffusing across biological membranes. On the other hand, specific binding proteins were discovered inside the liver cytoplasm (13). On the basis of the study of the binding capacity of rat plasma and hepatic cytosol for sulfobromophtalein (BSP), we suggested that an active transport should occur in liver at least for this diagnostic dye (14). This conclusion was reinforced by the finding that isolated rat liver plasma membrane could bind BSP with high affinity. The dissociation constant measured for the complex of the membrane with this dye was found to be 4 μM, a value one order of magnitude lower than that found for plasma proteins, and two orders lower than for cytoplasm (15). Subsequentely we devised a system for the purification of the membrane protein responsible for the high affinity BSP-binding (16). The recovery sheet of the procedure is presented in Table III. Clearly in four steps, starting from a crude plasma membrane preparation, a purification of 90 fold is achieved with a total yield of some 37 %. The process may be followed also by sodium dodecylsulphate (SDS) gel electrophoresis. Figure 6 shows such a procedure applied to the material coming from the three last steps presented in Table III. Clearly the final preparation (gel n.3)consists virtually of a single band in SDS, whose molecular weight in the presence of detergents is around 170,000 daltons, as measured by gel filtration. Its binding characteristics with regard to BSP are illustrated in Figure 7, which shows an apparent hyperbolic dependence on substrate concentration, with a half saturation of 4 μM. As a matter of fact, on a more expanded scale, the curve is not hyperbolic, but rather sigmoidal, with a Hill coefficient of 2, indicating the cooperation of at least two molecules in the binding.

Before proceeding further in the study of this protein molecule, we considered necessary to prove inequivocally its physiological involvement in liver bilirubin uptake. This has been achieved by a technique similar to that utilized with the previous protein, namely by the use of affinity chromatography purified antibodies. Figure 8 shows the effect of the purified antibody on BSP-binding in vitro. It may be noted that 50 % inhibition is obtained when equal amounts of protein and antibody are present in the mixture. This finding, in consideration that γ-globulins and the proteins have approximately the same molecular weight, speaks in favour of the high degree of purity of the antibody preparation. When 3 mg of it were injected into the portal vein of a rat, a remarkable increase in bilirubin blood level ensues. This phenomenon is illustrated in Figure 9. Clearly suitable controls were without any ef-

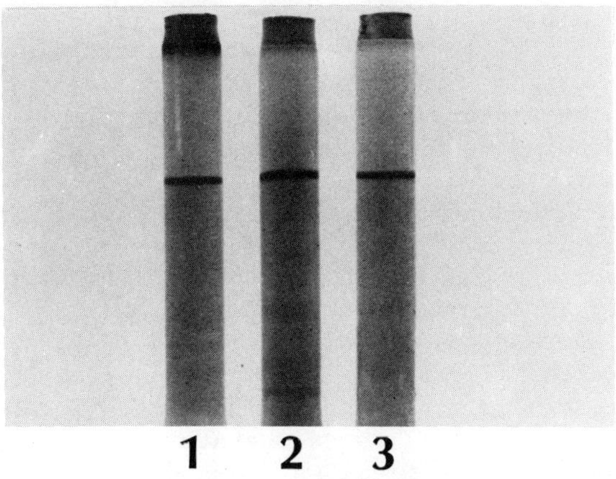

Figure 6. Sodium dodecylsulphate polyacrylamide gel electrophoresis of the different purification steps. 1:extract, 2:pooled fractions from Sephadex G-100, 3:after ion exchange chromatography on AG 1x8. Experimental conditions: 150 μg protein/gel; gel electrophoresis according to Weber and Osborne(21). (From Tiribelli et al.(16), permission of the Editors requested).

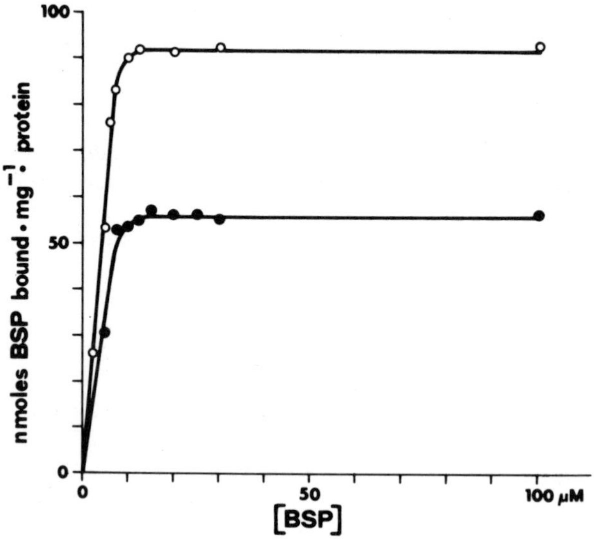

Figure 7. Binding of BSP by the isolated protein. Experimental conditions: 300 ug of the fraction has been tested for BSP binding as described by Tiribelli et al. (16). Free BSP in the buffer varied from 0 to 100 μM. O---O last stage of purification, ●---● eluate from Sephadex G-100 (from Tiribelli et al.(16)).

fect on bilirubin levels. Such a dramatic result was terribly encouraging, but was not considered conclusive proof of the physiological involvement of the protein in bilirubin transport. It is well known, in fact, that binding to plasma membranes of multivalent ligands such as antibodies may lead to profound changes in permeability properties. We decided therefore to obtain monovalent antibodies by papain digestion, as suggested by Porter (17). Such a preparation was found to be still capable of inhibiting BSP binding *in vitro*, though had lost the ability to form immunoprecipitates with the specific antigen. When such a preparation was injected into a rat, again jaundice developed. As indicated in Figure 10, such a situation of hyperbilirubinemia persisted, and 36 hours after the injection, the bilirubin plasma level was still twice the starting value. The effect, using the monovalent antibodies, was less dramatic than in the case of the divalent preparation, a finding which may be attributed either to a lower affinity of the second preparation as compared to the first, or to a combination of specific with aspecific effect in the case of divalent antibodies. The data summarized seem to provide very strong evidence that the protein isolated is a necessary ingredient in bilirubin translocation from plasma to liver. For this reason we suggested for this membrane compound the name of bilitranslocase (18).

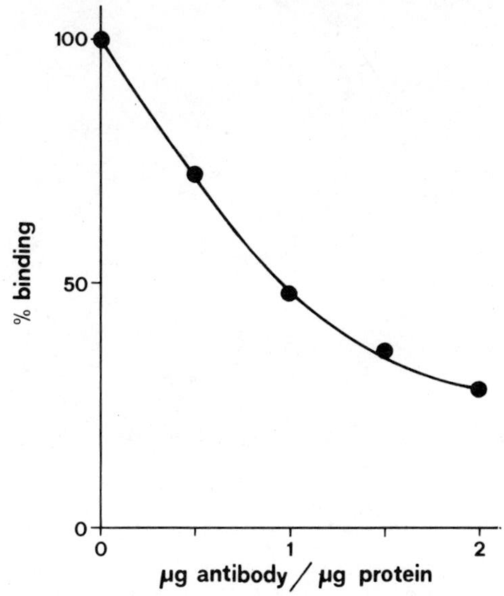

Figure 8. In vitro antibody inhibition of BSP binding to the isolated protein. Experimental conditions: the sample(148 µg) was incubated in ice for 1 hour in the presence of antibodies. The reaction mixtures contained in addition 72 µg, 145 µg, 217 µg and 290 µg purified antibodies in a final volume of 0.2 ml. BSP binding has been measured as in Figure 7.

DISCUSSION

Data presented in the two previous paragraphs clearly indicate that two proteins isolated by us are necessary components in two transport phenomena. Stricking differences exhist however, between the two. The calcium binding glycoprotein when isolated is water soluble and does not seem to be firmly bound to the membrane. In this sense it should be considered a peripheral protein. Perhaps the definition of peripheral applied to it, is not completely correct, because the protein has the tendency to acquire a hydrophobic nature when calcium bound. In this form (decane extractable) the protein may be rather deeply embedded in the lipophylic core of the membrane, a situation inconsistent with the definition of peripheral. The evidence so far collected indicates that the spontaneous conformational change induced by calcium binding would per se promote the conversion of the glycoprotein into an integral component of the inner mitochondrial membrane, thus making possible the process of calcium transport. This situation is entirely different

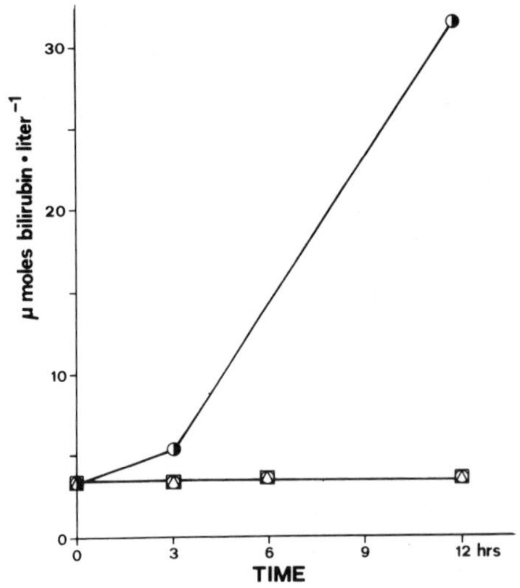

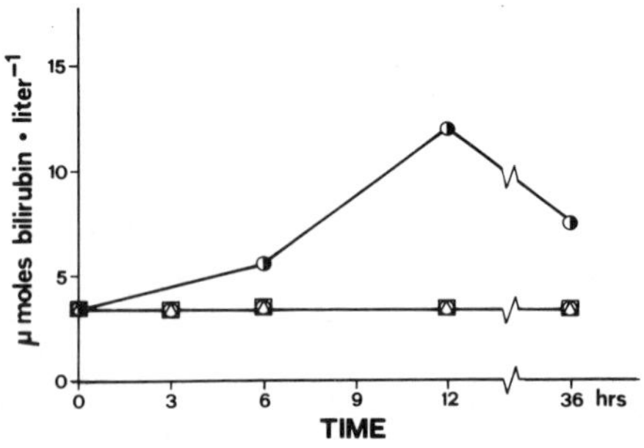

Figure 9. Changes in plasma bilirubin levels induced by affinity-chromatography purified antibodies directed to the isolated protein. Experimental conditions: the animals were operated under light ether anesthesia; the bilirubin concentration in blood was measured by the technique of Jendrassik and Grof (22). ◑---◑ specific antibodies, O---O aspecific γ-globulins, □---□ saline solution, △---△ sham operation.

Figure 10. Changes in plasma bilirubin levels induced by Fab-fragments of anti-bilitranslocase. Experimental conditions: as described in Figure 9. ◑---◑ specific antibodies, O---O aspecific γ-globulins, □---□ saline solution, △---△ sham operation.

from the pre-existence of a protein pore or channel transversing the membrane. On the contrary, in the case of bilitranslocase, we are dealing with a very hydrophobic protein, devoid af any sugar moiety, extractable from plasma membrane only under rather drastic conditions such as acetone treatment. In addition, the protein tends to aggregate in the absence of detergents, and shows a sigmoidal dependence of binding on substrate concentration. This finding suggests a facilitation of binding by multimeric structures. On these grounds, it is quite feasible that bilitranslocase is arranged in the membrane so as to constitute a specialized channel for bilirubin and other anions. If our conclusions are correct, it follows that two entirely different mechanisms are operating at the mitochondrial and plasma membrane level for calcium and bilirubin respectively. In other words, in two different districts of the cell transport occurs differently. It is perhaps acceptable so far that where a single membrane is present, a channel or pore is the structural ingredient necessary for transport, but where a second membrane is present, such in the case of mitochondria, alternative mechanisms may be operating. The function of outer mitochondrial membrane could well be to confine in a limited domain proteins capable of undergoing reversible association with the inner mitochondrial membrane.

REFERENCES

1. " Cell Membranes:Biochemistry, Cell Biology and Pathology", In G.Weismann and R.Klaiborne, HP Publishing Co., Inc.N.Y. (1975).
2. Singer,S.J." Architecture and Topography of Biologic Membranes", Ibidem, pp.35-44.
3. Sottocasa,G.L., G.Sandri, E.Panfili and B.de Bernard. A glycoprotein located in the intermembrane space of rat liver mitochondria. FEBS Lett. (1971) 17 100-105.
4. Sottocasa,G.L., G.Sandri, E.Panfili and B.de Bernard. "Glycoprotein in the mitochondrial compartments of rat liver", In G.F.Azzone, E.Carafoli, A.L.Lehninger, E.Quagliariello and N.Siliprandi, Biochemistry and Biophysics of Mitochondrial Membranes, Academic Press N.Y. (1972) pp.431-443.
5. Sottocasa,G.L., G.Sandri, E.Panfili, B.de Bernard, P.Gazzotti, F.D.Vasington and E.Carafoli. Isolation of a soluble calcium binding glycoprotein from ox liver mitochondria. Biochem.Biophys. Res.Commun. (1972) 47 808-813.
6. Carafoli,E., P.Gazzotti, C.Saltini, C.S.Rossi, G.L.Sottocasa, G.Sandri, E.Panfili and B.de Bernard. "Further studies on the mitochondrial calcium -binding glycoprotein", In G.F.Azzone, L.Ernster, S.Papa, E.Quagliariello and N.Siliprandi, Mechanisms in Bioenergetics, Academic Press, N.Y. (1973) pp.293-307.
7. Panfili,E., G.Sandri and G.L.Sottocasa. Some properties of an isolated glycoprotein possibly related to calcium transport in mitochondria. Acta Vitamin. et Enzymol. (1974) 28 323-330.
8. Sandri,G., E.Panfili and G.L.Sottocasa. The calcium binding glycoprotein and mitochondrial calcium movements. Biochem. Biophys.Res.Commun. (1976) 68 1272-1279.
9. Panfili,E., G.Sandri, G.L.Sottocasa, G.Lunazzi, G.Liut and G. Graziosi. Specific inhibition of mitochondrial calcium transport by antibodies directed to the calcium binding glycoprotein. (1976) 264 185-186.
10. Prestipino,G., D.Ceccarelli, F.Conti and E.Carafoli. Interactions of a mitochondrial calcium binding glycoprotein with lipid bilayer membranes. FEBS Lett. (1974) 45 99-103.
11. Sandri,G., E.Panfili and G.L.Sottocasa. Specific association of the calcium binding glycoprotein to inner mitochondrial membrane. Bull.Mol.Biol.Med. (1978) submitted for publication.

12. Sottocasa,G.L., E.Panfili and G.Sandri. The problem of mitochondrial calcium transport. Bull.Mol.Biol.Med. (1977) 2 1-28.
13. Levy,A.J., Z.Gatmaitan and I.M.Arias. Two hepatic cytoplasmic protein fractions,Y and Z, and their possible role in hepatic uptake of bilirubin, sulfobromophtalein, and other anions. J.Clin.Invest. (1969) 48 2156-2167.
14. Frezza,M., C.Tiribelli, E.Panfili and G.Sandri. Evidence for the existence of a carrier for bromosulphonphthalein in the liver cell plasma membrane. FEBS Lett. (1974) 38 125-128.
15. Tiribelli,C., E.Panfili, G.Sandri, M.Frezza and G.L.Sottocasa. "Liver bromosulphonphthalein transport as a carrier mediated process", In Deseases of the Liver and Biliary Tract. C.M.Leevy. S.Karger A.G., Basel (1976) pp. 55-59.
16. Tiribelli,C., G.C.Lunazzi, M.Luciani, E.Panfili, B.Gazzin, G.F. Liut, G.Sandri and G.L.Sottocasa. Isolation of a sulphobromophthalein binding protein from hepatocyte plasma membrane. Biochem. Biophys. Acta (1978) 532 105-112.
17. Porter,R.R. The hydrolysis of rabbit γ-globulin and antibodies with crystallin papain. Biochem. J. (1959) 73 119-126.
18. Tiribelli,C., G.C.Lunazzi, M.Luciani, B.Gazzin and G.L.Sottocasa. Bilitranslocase: a protein from rat liver plasma membrane involved in hepatic bilirubin uptake. J. Clin. Invest. (1978) submitted for pubblication.
19. Schnaitman,C. and J.W.Greenawalt. Enzymatic properties of the inner and outer membranes of rat liver mitochondria. J. Cell. Biol. (1968) 38 158-175.
20. Estabrook,R.W. "Mitochondrial respiratory control and the polarographic measurement of ADP:O ratios". In R.W.Estabrook and M.E.Pullman, Methods in Enzymology, Academic Press, (1967) 10 pp. 41-47.
21. Weber,K. and N.Osborne. The reliability of molecular weight determinations by dodecylsulphate-polyacrylamide gel electrophoresis. J. Biol. Chem. (1969) 244 4406-4412.
22. Jendrassik,L. and P.Grof. Vereinfachte photometrische Methode zur Bestimmung des Blutbilirubins. Biochem. Zeit. (1938) 297 81-89.

ENZYME INDUCTION AND DEINDUCTION

IN ANIMAL CELL ASSOCIATIONS

R.I. Salganik

Institute of Cytology and Genetics
Siberian Branch of the U.S.S.R. Academy of Sciences
Novosibirsk 630090 U.S.S.R.

INTRODUCTION

Enzyme induction is widely used in unicellular organisms, especially in procaryotes, where chemical change in the environment prompts the single cell adaptation by shifting the enzymatic programmes.

The aim of this paper is to discuss the role of enzyme induction in multicellular organisms, particularly in animals, where the distribution of the functions between cells and cell specialization provides a specific mode of adaptation.

THE ROLE OF ENZYME INDUCTION IN MULTICELLULAR ORGANISMS

The various regulatory mechanisms in an animal organism may be evaluated in terms of their energy costs. Most of the physiological functions depend, as a rule, on definite enzymic reactions and normal physiological performance is supported by mechanisms regulating these enzymes.

Changes in enzyme activities may depend on: 1) changes in enzyme specific activities or 2) changes in the absolute amounts of enzymes. Let us consider these two aspects.

1) Changes in enzyme specific activities are achieved by:
   a. allosteric control (through feedback regulation)
   b. chemical modification of enzymes (phosphorylation, adenylation, ADP ribosylation and other modifications); limited proteolysis also belongs to this class.

2) Enzyme content is increased by an activation of their synthesis from the level of transcription or translation. Decreased enzyme content is brought about by their proteolytic degradation.

Allosteric regulation is the cheapest means of controlling enzyme activites. Chemical modification of enzymes requiring the expenditures of ATP or its energy equivalents are more costly. The energy costs of enzyme synthesis rise from the translation to the transcription levels. The degradation of enzymes is just as costly, because it wastes the energy used for enzyme synthesis.

To illustrate, an average-sized enzyme protein, composed of 500 amino acid residues, contains 1-3 residues of serine, which may be phosphorylated, and 1-3 ATP molecules correspondingly expended for the process. To synthesize a protein of this kind from ready-made amino acids and using an available RNA template, a cell consumes not less than 1500 ATP molecules. Were this preceded by synthesis of mRNA coding for this protein, energy expenditures would by covered by, minimally 4500 ATP molecules.

In view of these considerations, the questions posed are: What is the advantage of this expensive transcription-translation control and how far do its implications extend? A fixed set of ready-made enzymes are under allosteric control and/or subjected to chemical modification. Eventually, this set is either activated or inactivated. By contrast, genetic induction culminates in a new set of enzymes and functional programmes, it ensures the passage of a cell to a new state of activity. By means of genetic induction a single polyfunctional cell can replace, perhaps, a set of monofunctional ones. Hepatocytes may serve as an example. The conversion of a whole variety of exogeneous compounds is within the competence of liver cells. In these cells cortisol induces glycogenic enzymes and insulin induces glycolytic enzymes; amino acids initiate the activities of enzymes involved in their metabolism and galactose stimulates the enzymes which transform this sugar into glucose (Table I). The concentrations of the diverse substances entering the liver constantly vary. Perhaps it would be more economical to keep a single polyfunctional cell in a site requiring functional shifts brought about by enzymic adaptation than many monofunctional cells with constant sets of various enzymes. Similar reasoning, although in the opposite direction, may be applied to a tissue or organ performing a constant external function, where monofunctional cells controlled by chemical modification are more advantageous. For instance, lypocytes serve as suppliers of fatty acids, this being their sole external function supported by the activation of lipase as a result of its phosphorylation by 3',5'-AMP dependent protein kinase. A number of hormones (the luteonizing, adenocorticotrophic and thyroid-stimulating hormones, secretin, glucagon) are capable of activating this system through adenylate cyclase and, hence, of stimulating lypolysis (Rodbell, 1972).

Table I. Induction of Enzymes and Other Functional Proteins in Liver Cells

Inducers	Enzymes (and other functional proteins) which are induced	References
glucocorticoids	gluconeogenic enzymes	Weber et al.,1965
insulin	glycolytic enzymes	Salas et al.,1967
galactose	galactose transforming enzyme	Salganik et al., 1973
aminoacids	urea cycle enzymes	Schimke, 1961
hepatectomy	enzymes of DNA synthesis	Gottlieb et al., 1964
xenobiotics	mixed function oxygenases	Parke,1975 (review)
Hypoproteinemia	albumin	Marsch et al., 1966
estrogens	vitellogenin (in egg-laying animals)	Tata, 1976

The logical inference is that each mode of enzyme activity regulation has its own place and functional role.

Feedback control seems to be the fundamental way in which the enzymes of the cell household are regulated. This control provides the internal functions of the cell, and the intermediate metabolites act as allosteric effectors (see reviews by Moyed and Umbarger, 1962; Dagley and Nicholson, 1970). For instance, glycolytic and glycogenic enzymes (Underwood and Newsholm, 1965; Weber et al., 1968), enzymes of the tricarbonic cycle (Krebs, 1970), and those involved in the synthesis of fatty acids (Numa et al., 1969) are under feedback control of corresponding intermediate metabolites in animal cells.

As compared with allosteric regulation, chemical modification is more energy-consuming, but guarantees the stability of enzymes in an active or inactive state. By means of chemical modification, an enzymic process in transferred from one regime to another. Chemical modification mainly serves the external functions of animal cells; it is aimed at satisfying the needs of the whole organism.

External factors are regulators of the chemical modifications of cell enzymes. Enzymes of monofunctional cells, performing one of the external functions and possessing corresponding constant (although renewing) enzyme sets, depend on chemical modification. Many hormones exert their control over such monofunctional animal cells (or one of the functions of a polyfunctional cell) by phosphorylation of enzymes or other cell proteins (see reviews by Sutherland et al., 1968; Robison et al., 1971; Pastan, 1972; Rodbell, 1972).

Control of genetic induction of enzymes seems to extend to polyfunctional animal cells that accomplish more than one essential external function by setting up a new programme under the effect of external inducers.

All these considerations are summarized in Table II.

Besides reprogramming polyfunctional cells, genetic induction is also directly concerned with the sequential replacement of functional programmes during cell differentiation and development (Davidson, 1969).

THE INTERACTION OF ENZYME INDUCERS IN ASSOCIATIONS OF POLYFUNCTIONAL CELLS

The total number of inducers for polyfunctional units such as liver cells may be as high as 15-20. For this reason, the liver may be considered as an example of how associations of polyfunctional cells may respond to inducers.

Under physiological conditions, the blood stream simultaneously supplies the liver with most inducers at varying concentrations. The question is whether each cell is capable of responding simultaneously to all these substances by synthesizing all the sets of inducible enzymes. This seems very unlikely, taking into account the high energy costs of protein biosynthesis, as well as the limited number of initial monomers and cell space. A more realistic situation is as follows. Functions are distributed among liver cells in such a way that, at a given time point, each hepatocyte is activated by a single or very few inducers. There is evidence lending credibility to this kind of cell function distribution. According to immunofluorescent data, albumin is detectable only in 10% of liver cells and fibrinogen is identifiable in 1% of these cells. These are the cells that seem to intensively synthesize, store and secrete such proteins. Their synthesis appears to be controlled by different inducers. Relevant is the fact that these proteins are predominantly detected in various cells and that only 0.2% of cell population contain albumin and fibrinogen (Hamashima et al., 1964). By means of fluorescent antibodies against prothrombin, the presence of this protein has been demonstrated only in

Table II. Regulation of Enzyme Activities in Multicellular Organisms

Means	Energetic costs (per average enzyme molecule of 500 aminoacids)	REGULATORS	Subjects of regulation	Aim of regulation	Specific features
1. Changes in enzyme specific activities					
a) allosteric control	—	Internal intracellular metabolites, acting as ligands	all cells, household enzymes	to keep cell household	operative changes of activity of pre-existing enzymes
b) chemical modification of enzymes (Phosphorylation, adenylation, etc.)	1-3 ATP	External intercellular signals-hormones, mediators, triggering enzyme modification	monofunctional cells (e.g.lypocytes) external function enzymes	to provide the needs of the whole organism	relatively stable changes of activity of pre-existing enzymes
2. Changes in absolute amounts of enzymes		External intercellular signals-hormones metabolites, inducing transcription or/ and translation	polyfunctional cells (e.g.hepatocytes) external function enzymes	to provide the needs of the whole organism	changes of functional enzymatic programmes
a) synthesis through transcription - translation	4500 ATP				
b) proteolytic degradation	— " —				

a portion of liver cells (Barnhart and Anderson, 1962). Liver cells are also heterogenous with respect to the histochemically identified enzymes: glucose-6-phosphate, glycogen synthetase, and glycogen phosphorylase (Sasse et al., 1975).

The versatile external functions of liver cells, controlled by incoming signals, seem to be distributed among the various hepatocytes. Does this mean that the functions of such cells are predetermined and that, despite their morphological similarity, they are designated to accomplish different functions? A stable distribution of functions among cells would hardly be an optimal strategy for maintaining adaptation to fluctuating conditions of the cell environment. The best way out would be inconstancy of the number of cells fit to be engaged in different enzyme programmes. Their number would change proportionately to shifts in the concentration of inducers. In this situation, hepatocytes would differ only in sensitivity to inducers, their sensitivity being supposedly determined by the different number of cell receptors for each inducer.

Let us assume that the number of inducers (I) interacting with a liver cell is n, and that the set of inducers consists of I_1, I_2, $I_3 \ldots I_n$. Let us further assume that each cell is capable of responding, although differently, to the majority of inducers. The differential response makes possible the characterization of each cell according to its sensitivity to the whole set of inducers.

For example, a given cell is most sensitive to the inducer I_1 and less sensitive to I_2 and still less to I_3 and so on. Under these assumptions, all cells which are most sensitive to I_1 would respond to the threshold concentration of the inducer I_1.

When all the inducers are supplied at threshold concentrations $[I_1]_1$, $[I_2]_1$, $[I_3]_1 \ldots [I_n]_1$, the cell population is subdivided into subpopulations, each consisting of cells most sensitive to a corresponding inducer; each subpopulation is engaged in the synthesis of a specific set of enzymes providing functions that are controlled by a particular inducer (Fig. 1). However, when the concentration of the inducer $[I_1]_1$ rises from $[I_1]_1$ to $[I_1]_2$, the less sensitive cells become capable of responding to the inducer I_1, i.e. the inducer starts to act on the lower ranking cells.

The expected consequence is a redistribution of cells among the subpopulations - a definite number of cells sensitive to a higher concentration of inducer $[I_1]_2$ would assume the function of the subpopulation programmed by the given inducer and would stop performing their previous function.

Any inducer acting on these cells may be treated within the framework of the model suggested. Its validity has been confirmed

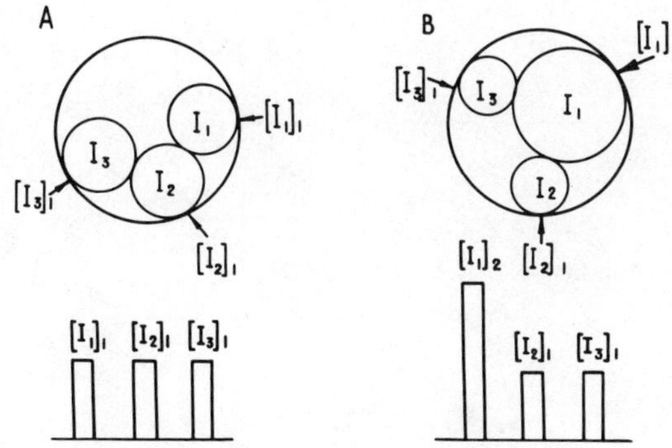

Fig. 1. Distribution of functions in a population of polyfunctional cells under the effect of various inducers. (A) Concentrations of the inducers I_1, I_2 and I_3 are equivalent. (B) Increase in the concentration of the inducer I_1: $[I_1]_2 > [I_1]_1$.

by experimental data (Salganik et al., 1977).

As shown in Fig. 2, rat liver cells are heterogeneous with respect to the content of cortisol-induced tyrosine aminotransferase (TAT). This enzyme has been isolated and purified in this laboratory (Mertvetsov et al., 1976) and identified in liver cells with the help of fluorescent antibodies. The activity of rat liver TAT rose with increasing doses of cortisol (Fig. 3). We have attempted to answer the questions: How is this dose-dependent effect accomplished? Is enzyme synthesis enhanced in those cells that were earlier involved in the process or does the hormone mobilize additional cells?

Rat liver cells were dissociated with sodium tetraphenylborate (Rapoport and House, 1966), the suspension of hepatocytes was layered on Ficoll density gradients (Drochmans et al., 1975), the number of cells in each fraction was counted and TAT activity was assayed (Diamondstone, 1966). It was found that 5 hours after cortisol injection to rats, increased TAT activity was paralleled by a sharp rise in the cell number of the light fraction (d = 1.04) and a decline in that of the heavy fraction (d = 1.14) (Table III). Cells showing cortisol induced TAT activity were concentrated in the light fraction. The decreased buoyant density of cells which responded to the administered cortisol by an induction of TAT is the possible result of an augmentation in the amounts of granular endoplasmic reticulum caused by induction (Khristolyubova et al., 1973). The experimental results indicate that cortisol treatment is associated with an increase in the number of cells involved in the synthesis of TAT and also, presumably, with a higher level of TAT synthesis in the induced cells.

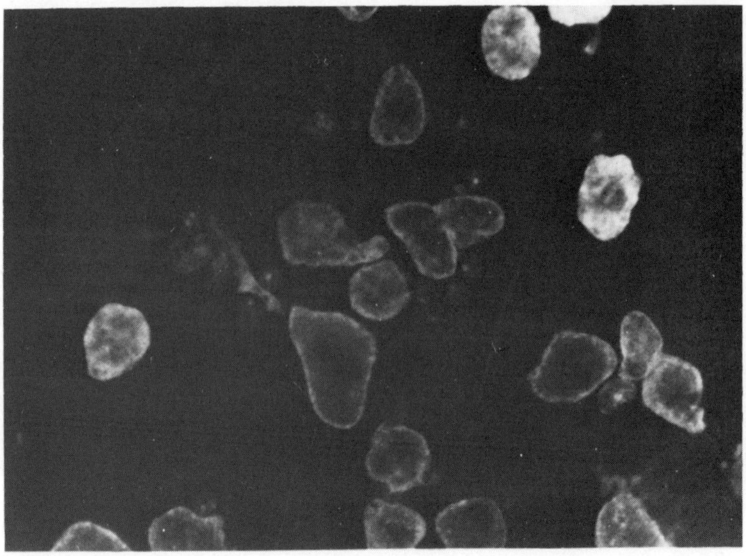

Fig. 2. Heterogeneity of rat liver cells. Cortisol induced TAT (light cells) was demonstrated by the method of fluorescent antibodies.

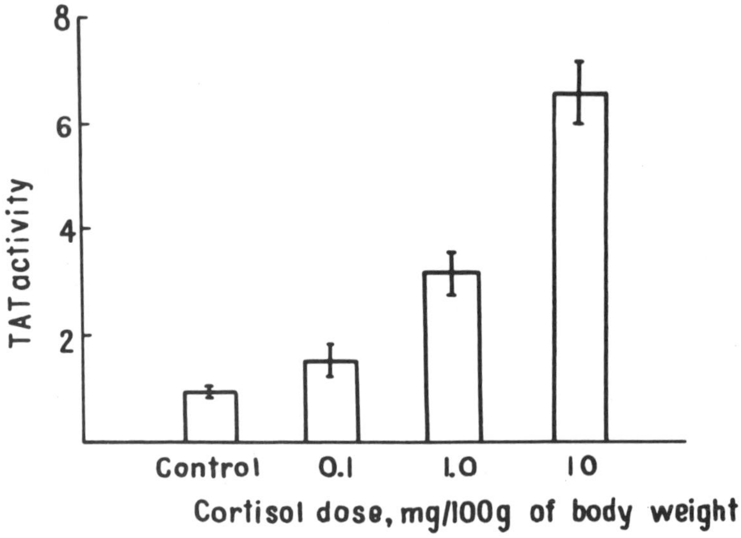

Fig. 3. Enhanced activity of rat liver TAT under the effect of increased doses of cortisol. TAT activity is expressed as μM of p-hydroxyphenylpyruvate formed per mg of protein for 10 min. The enzyme activity was determined 5 hrs after cortisol administration.

Table III. Redistribution of the Hepatocytes Between Fractions After Cortisol Treatment (in %)

Control (Adrenalectomized rats)		5 hours after cortisol administration to rats (5 mg per 100 g of body weight)	
"Light" cells	"Heavy" cells	"Light" cells	"Heavy" cells
58.8 ± 4.3	41.2 ± 4.1	83.2 ± 2.0	16.8 ± 1.7

It seems plausible that an elevated concentration of blood glucocorticoids results in an induction of the less sensitive cells which were previously irresponsive at a relatively low (physiological) concentration of the hormones (the experiments were performed on adrenalectomized rats). As the dose of the inducer increases, so does the number of responding cells and, inevitably, the number of cells previously responding to other inducers diminishes. This appears to be a flexible and economical mode of controlling the multitude of functions of an organ composed of polyfunctional cells. With this control, simultaneous or sequential administration of large doses of two (or more) inducers would give rise to their competitive interaction and, consequently, lower the efficiency of either or both.

The following experimental series confirm the existence of these competitive relationships of inducers. The interaction of cortisol, the effect of which was estimated on the basis of increasing TAT activity (Diamondstone, 1966), and phenobarbital, which induces mixed-function oxygenases in the liver, was investigated. The effect of phenobarbital was measured on the basis of changes in the activity of aryl hydrocarbon hydroxylase (AHH) activity (Nebert, Gelboin, 1968). It is known that maxium TAT activity in rat livers is attained 5 hours after cortisol administration. As to AHH, its activity rises not earlier than 24 hours after phenobarbital administration; a substantial rise takes place after daily treatment for 4-5 days (Nedelkina et al., 1972).

Under the experimental conditions used cortisol induced TAT in rat liver, while phenobarbital did not affect TAT activity. The reverse was observed for AHH (Salganik et al., 1977). Rats were injected with phenobarbital for 4 days. After this treatment, the activity of AHH augmented approximately three-fold. In phenobarbital-pretreated rats, the cortisol-induced rise in TAT activity was 30-40% lower than in controls. Cortisol administration did not affect the response of the cells to the continuous phenobarbital treatment by AHH induction. When rats received both cortisol and phenobarbital for 5 days, AHH activity did not increase at all or even fell to a level below the control (in control rats that were injected with phenobarbital alone, there occured a threefold increase in the activity of the enzyme) (Fig. 4B); TAT activity was induced, although much weaker than after a single injection of the hormone (Fig. 4A). The data obtained may be expained by the competition of the inducers for the cells. A number of observations made by other researchers may also be explained by the interference of inducers. Thus, insulin inhibits the induction of key glycogenic enzymes induced by cortisol; on the contrary, cortisol suppresses the inductive effect exerted by insulin on glycolytic enzymes (Weber et al., 1966; Mertvetsov, 1969).

The induction of ornithine transaminase and threonine dehydrogenase, along with that of other enzymes of amino acid metabolism, has been observed in rats fed a diet containing casein hydrolysate (Schimke, 1961; Pitot and Cho, 1961). When casein hydrolysate is supplemented with glucose, the induction of ornithine transaminase and threonine dehydrogenase does not occur (Pitot and Peraino, 1963). This inhibition may be the result of the inductive effect of glucose on glucokinase and other glycolytic enzymes (Gunn and Taylor, 1973; Drevich and Salganik, 1975). The addition of estradiol to cultures of Xenopus liver cells induces vitellogenin synthesis which is accompanied by a deinduction of albumin synthesis (Tata, 1976). In all likelihood, under these experimental conditions there arises a competition of inducers for cells.

Competitive interactions of inducers may have far reaching practical implications. Treatment of patients with high doses of cortisol can suppress the activity of mixed-function oxygenases and, hence, interfere with the elimination of foreign compounds; this would make toxic even routine therapeutic doses of drugs. It now may seem that the incompatibility of some nutrients is the result of their competitive inductive interactions.

The suggested system may be sufficiently flexible to enable a cell population to cope with large fluctuations in environmental conditions. At a normal environmental input of inducers, the entire cell population is subdivided so as to guarantee an optimum level for each process controlled by the given inducer. Some limitations must be imposed on the system to prevent an inducer, even a very strong one, from directing all the population of polyfunctional cells to the performance of a single function. The distribution of cells according to the sensitivity to various inducers cannot be random and is, probably, under genetic control. Under the conditions described, genetic induction may operate as a means of regulating not independent single cells, but their associations acting as a polyfunctional unit.

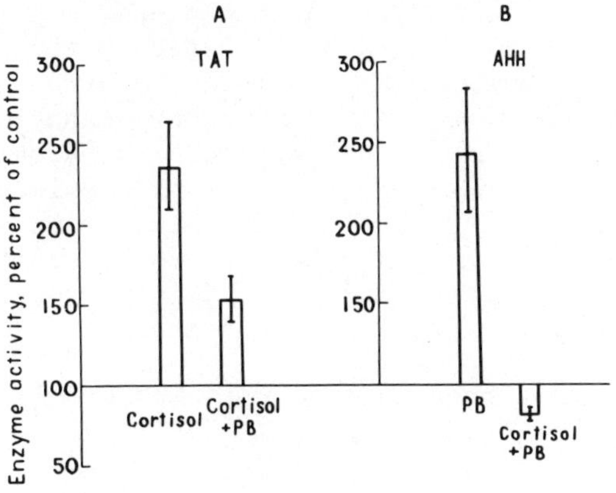

Fig. 4. Tyrosine aminotransferase (TAT) and aryl hydrocarbon hydroxylase (AHH) activities in rat liver under combined cortisol and phenobarbital treatment. (A) "Cortisol"-TAT activity 5 hrs after cortisol administration to untreated rats; "cortisol + PB"-TAT activity 5 hrs after cortisol administration to rats treated daily with cortisol phenobarbital for 5 days. (B) "PB-AHH" activity in rats treated daily with phenobarbital for 5 days; "cortisol + PB"-AHH activity in rats treated daily with cortisol and phenobarbital for 5 days (doses as indicated in the legend to Fig. 7).

DEINDUCTION AND ITS MECHANISMS

Genetic induction in animal cells is associated with not only the synthesis of m-RNA molecules and adaptive enzymes, but also with the de novo formation of the whole protein-synthesizing machinery: t-RNA's, r-RNA and ribosomal proteins, as well as membranes of the endoplasmic reticulum (Orrhenius et al., 1965; Salganik et al., 1967; Khristolyubova et al., 1973). When the inducer supply dimishes, the activity of the induced enzymes and the amount of the protein-synthesizing structures fall to the initial level. Thus, 5 hours after cortisol administration to rats, the activity of TAT increases severalfold (Fig. 5). This rise in enzyme activity is the result of the synthesis of new molecules of TAT, not of the activation of the preexisting molecules of the enzyme (Kenney, 1967). However, 12-15 hours after a single injection of cortisol, when the effect of the injected hormone is abolished, the activity of TAT returns to basal values. After cortisol administration, the surface density of the endoplasmic reticulum membranes sharply increases; after abolition of hormonal action, the surface density of the endoplasmic reticulum membranes also falls to the initial level (Khristolyubova et al., 1973).

Obviously, the retention of inducible enzymes, newly synthesized ribosomes and endoplasmic reticulum membranes without any further exploitation would encourage the indolence of structures and futile storing. For polyfunctional cells, which are capable of enzymic reprogramming under the effect of new inducers, clearing of cell space for the complete take-over of newly synthesized structures and enzyme is, evidently, a vital act. Deinduction may be defined as the elimination of induced enzymes and cell structures after cessation of the action of an inducer. Deinduction appears to play a role that is not less significant than that of induction.

Some attempts have been made to gain insight into the mechanisms of deinduction in this laboratory.

Examination of the curve for TAT induction (Fig. 5) raised the question: How can a cell that is no longer induced return the activity of the previously induced enzyme precisely to the initial values?

This question was partly answered by the following experiments. Gel electrophoresis of TAT enzymes yielded double-banded patterns,

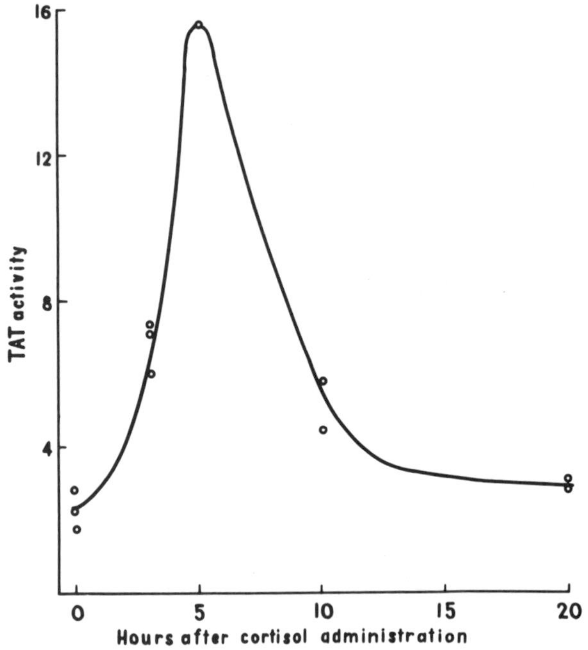

Fig. 5. Changes in tyrosine aminotransferase (TAT) activity in rat liver after a single cortisol administration. The inducible isoenzyme activity was determined according to Lin et al. (1958) in agar gels. (From Mertvetsov et al., 1974.)

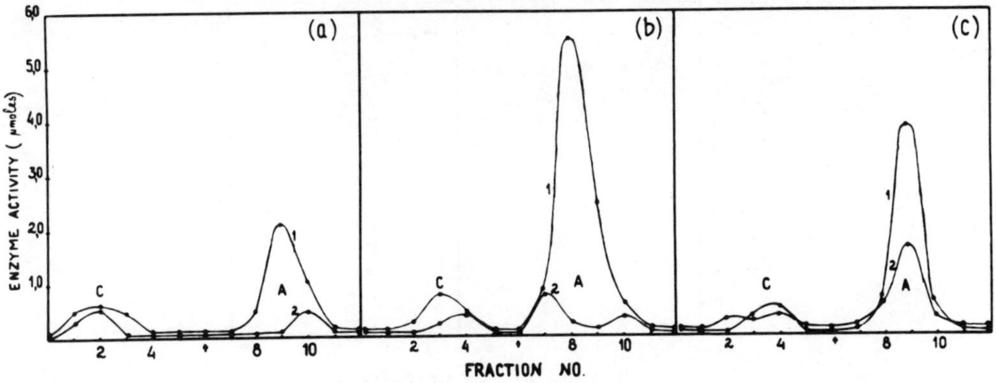

Fig. 6. The effect of proteases on the activity of non-inducible (C) and inducible (A) isoenzymes of rat liver tyrosine aminotransferase (TAT). Gel electrophoretic patterns of: (a) untreated rat liver: 1-control, 2 - gel incubated with pronase; (b) rat liver 5 hrs after cortisol administration: 1-control, 2 - gel incubated with pronase; (c) rat liver 5 hrs after cortisol administration: 1 - cortisol, 2 - liver supernatant incubated with trypsin. (From Mertvetsov et al., 1974.)

one migrating to the cathode and the other to the anode. 5 hours after cortisol administration, the activities of only the enzymes which move anodally increased, whereas those of the enzymes moving towards the cathode remained unchanged. 14-16 hours after cortisol treatment, when the inductive effect of the hormone was abolished, the activities of the more anodally situated TAT isoenzymes returned to the basal levels, the activities of those closer to the cathode being unaltered.

A comparative analysis of the properties of the inducible and non-inducible isoenzymes of TAT demonstrated that the former are very sensitive to proteases (Fig. 6). Incubation with trypsin decreased the activities of the inducible TAT isoenzymes by 2.5 times and those of their non-inducible counterparts by only 15%. These differences were also observed after incubation with other proteolytic enzymes such as pronase, chymotrypsin, lysosomal proteases (Mertvetsov et al., 1974). The inducible and non-inducible isoenzymes of TAT have been obtained in a highly purified form. These isoenzymes also differed in their sensitivities to proteases (Fig. 7). The apoenzyme of the inducible protease-sensitive enzyme was loosely bound to the coenzyme; by contrast, its non-inducible counterpart was firmly bound to pyridoxal 5-phosphate. The non-inducible variant can utilize not only α-ketoglutarate, but also oxalacetate as acceptor of NH_2-groups; its affinity for

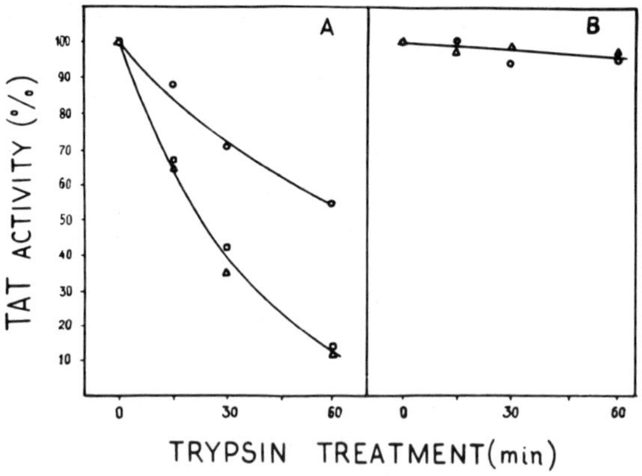

Fig. 7. Effect of trypsin on the activity of isolated inducible (A) and non-inducible (B) isoenzymes of rat liver tyrosine aminotransferase (TAT). (A) -o- inducible TAT isoenzyme in a reaction medium containing pyridoxal 5'-phosphate; -Δ- inducible isoenzyme in a reaction medium not containing pyridoxal 5'-phosphate; -□- apoenzyme of the inducible isoenzyme in a reaction medium not containing pyridoxal 5'-phosphate. (B) -o- non-inducible TAT isoenzyme in a reaction medium containing pyridoxal 5'-phosphate; -Δ- non-inducible isoenzyme in a reaction medium not containing pyridoxal 5'-phosphate. (From Mertvetsov et al., 1976.)

α-ketoglutarate is higher than that of the non-inducible isozyme (the apparent Km's for α-ketoglutarate are $1.5 \cdot 10^{-5}$ and $35 \cdot 10^{-5}$, respectively).

From these data, it may be concluded that cortisol induces TAT exclusively at the expense of the inducible isozyme and that the high protease-sensitivity of the inducible variant makes possible its preferential elimination.

It is known that insulin induces hexokinase (HK) in animal liver. Of the four HK isoenzymes of rat liver, only HK II proved to be inducible. It was the inducible isozyme of HK distinguished by its high sensitivity to proteases (Fig. 8) (Chesnokov et al., 1974).

It seems likely that, in a number of situations, inducers may involve in transcription the genes which code for special inducible isozymes. Were these isozymes very sensitive to proteases, this would be an efficient mechanism of deinduction.

Accepting that high sensitivity of inducible isozymes to proteases provides deinduction, one is still baffled by the restoration of the amount of endoplasmic reticulum membranes (and ribosomes) during this period.

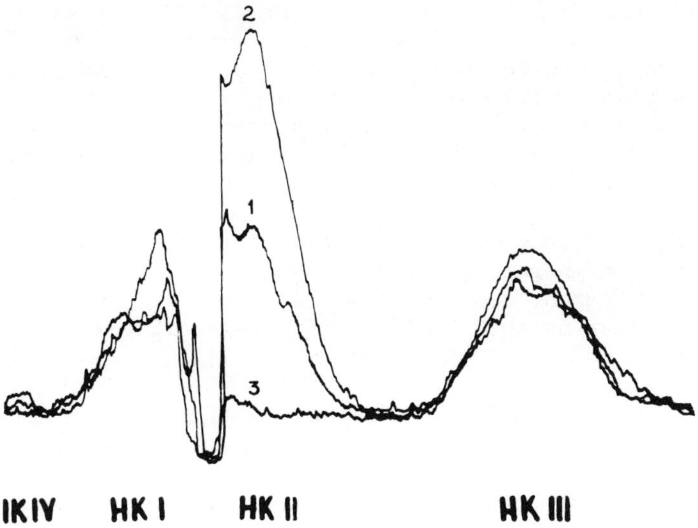

Fig. 8. Effect of trypsin treatment on the activity of insulin inducible and non-inducible isoenzymes of rat liver hexokinase. Densitometry of agar gel electropherograms: 1 - untreated rats, 2 - 12 hrs after the last insulin administration (insulin injected for 10 days), 3 - before electrophoresis; the supernatant (18,000 x g) of rat liver homogenate from experiment 2 was incubated with trypsin (5 μg per mg of protein). (From Chesnokov et al., 1974.)

It may be conjectured that the newly induced cell components are synthesized and then located in some isolated cell compartments. These compartments may be formed by endoplasmic reticulum membranes adhering to segments of the cell nucleus where the chromosomes responsible for the newly synthesized enzymes and protein synthesizing structures are located. The effect of hydrolase (possibly of lysosomal origin) would then be restricted to the cell compartments containing the inducible cell structures and enzymes. High ordering of the biochemical processes in the inner cell space is, presumably, an indispensable condition for cell function.

Deinduction may also be based on changes in the activities of the hydrolytic enzymes of lysosomes and other specific hydrolases: proteases hydrolyzing the apoenzymes of pyridoxal enzymes or NAD-dependent enzymes, for example (Katunuma et al., 1971 a,b).

Much work on the biochemistry of deinduction remains to be done.

SUMMARY

Data are presented suggesting that the major function of the enzyme induction in the multicellular organism is to provide changes in the enzymatic programmes of polyfunctional cells. The cells are located mainly in those sites of the organism where the chemical milieu is constantly changing.

The association of polyfunctional cells is assumed to be distributed in temporary subgroups engaged in various functions provided by different sets of inducible enzymes. The associations are flexible systems able to adapt to changes in the chemical environment.

The interference of the inducers, based on the competition of inducers for the polyfunctional cells is demonstrated.

Deinduction is considered to be an important means ensuring changes in the enzymatic programmes.

Highly increased sensitivity of the inducible isoenzymes to proteases, as compared with the non-inducible ones, serves as a possible route for the selective degradation and elimination of inducible isoenzymes, when the action of the inducer ceases.

REFERENCES

Barnhart, M.I., and Anderson, C.F. (1962) Biochem. Pharmacol. 9, 23-27.

Chesnokov, V.N., Mertvetsov, N.P., and Salganik, R.I. (1974). Biokhimia 39, 294-299.

Dagley, S., and Nicholson, D.E. (1970) "An Introduction to Metabolic Pathways". Backwell Scientific Publication, Oxford and Edinburgh.

Davidson, E.H. (1969). "Gene Activity in Early Development", Academic Press, New York, London.

Diamondstone, T. (1969). Anal. Biochem. 16, 395-398.

Drevich, V.F., and Salganik, R.I. (1975). Vopr.Med.Khimii 21, 503-506.

Drochmans, P., Wanson, G.C., and Mosselmans, R. (1975). J. Cell.Biol. 66, 1-22.

Gottlieb, L.I., Fausta, N., and Van Lancker, J.L. (1964). J. Biol. Chem. 239, 555-559.

Gunn, J.M., and Taylor, C.B. (1973). Biolchem. J. 136, 455-464.

Hamashima, G., Harter, J.G., and Coons, A.H. (1964). J. Cell.Biol. 20, 271-279.

Katunuma, N., Kito, K., and Kominami, E. (1971 a). Biochem.Biophys. Rer.Commun. 45, 76-81.

Katunuma, N., Kominami, E., and Kominami, S. (1971 b). Biochem. Biophys.Res.Commun. 45, 70-75.
Krebs, H.A. (1970). Adv.Enz.Regul. 8, 325-332.
Khristolyubova, N.B., Shilov, A.G., Kiseleva, E.B., and Salganik, R.I. (1973). Tsitologiya 15, 254-259.
Marsch, J.B., Drabkin, D.L., Braun, G.A., and Parkes, I.S. (1966). J. Biol.Chem. 241, 4168-4172.
Mertvetsov, N.P., Saprykina, V.A., Chesnokov, V.N. and Salganik, R.I. (1974). Biokhimia 39, 3-8.
Mertvetsov, N.P., Chesnokov, V.N., Sakhno, L.V., and Salganik, R.I. (1976). Biokhimia 41, 1352-1366.
Moyed, H.S., and Umbarger, H.E. (1962). Physiol.Rev. 42, 444-466.
Numa, S., Borts, W.M., and Lynen, F. (1965). Adv.Enz.Regul. 3, 407-424.
Orrenius, S., Ericsson, L.E., and Ernster, L. (1965). J. Cell.Biol. 25, 627-639.
Pastan, I.H. (1972). In "Current Topics in Biochemistry" (C.B. Anfinsen, R.F. Goldberger, A.N. Schechter, Eds.), pp.65-100, Academic Press, New York.
Pitot, H.C., and Cho, Y.S. (1961). Cold Spring Harbor Symp.Quant. Biol. 26, 371-377.
Rappoport, O., and Howse, G. (1966). Proc.Soc.Exptl.Biol.Med. 121, 1010-1021.
Rodbell, M. (1972). In "Current Topics in Biochemistry" (C.B. Anfinsen, R.F. Goldberger, A.N. Schechter, Eds.), pp. 187-218, Academic Press, New York.
Salganik, R.I., Khristolyubova, N.B., Kiknadze, I.I., Morozova, T.M., Gryaznova, I.M., and Valeeva, F.S. (1967). In "Structure and Function of Cell Nucleus" (I.B. Zbarsky, Ed.), str. 20-23, Nauka, Moscow.
Salganik, R.I., Solovyeva, N.A., Drevich, V.F. (1973). Dokl.AN SSSR 209, 489-502.
Salganik, R.I., Drevich, V.F., Mertvetsov, N.P., Deribas, V.I. (1978), in preparation.
Sasse, D., Katz, M., and Jungman, K. (1975). FEBS Lett. 57, 83-86.
Schimke, R.T. (1961). Cold Spring Harbor Symp.Quant.Biol. 26, 363-366.
Sutherland, E.W., Robison, G.A., Butcher, R.W. (1968). Circulation 37, 279-307.
Tata, J.R. (1976). Cell 9, 1-14.
Turkington, R.W. (1970). Biochim.Biophys.Acta 213, 478-483.
Underwood, A.H., and Newsholm, E.A. (1965). Biochem. J. 95, 868-875.
Weber, G., Singhal, R.L., and Srivastava, S.K. (1965b). Adv.Enz. Regul. 3, 43-75.
Weber, G., Lea, M.A., Stamm, N.B. (1968). Adv.Enz.Regul. 6, 101-123.

PART III

MACROMOLECULE CHANGES IN EMBRYONIC DEVELOPMENT AND CELL DIFFERENTIATION

ENZYMATIC MODIFICATIONS OF NUCLEAR DNA IN THE EARLY EMBRYONIC DEVELOPMENT OF THE SEA URCHIN

L. Tosi and E. Scarano

Stazione Zoologica and
Laboratory of Molecular Embryology, CNR
Naples, Italy

The encoding in DNA of the structure of a protein and the basic control mechanisms of protein synthesis in prokaryotes by repressor and activator proteins are well understood. In principle we can understand how, by spontaneous assemblage of components of the lower level, viruses, subcellular organelles, bacterial cells and even the cells of monocellular eukaryotes arise.

On the other hand, the problem of the encoding of the form of higher organisms and therefore of ontogenesis is still unsolved. This problem can be approached through the methods of molecular biology and biochemistry using theoretical models. At the present stage of knowledge of experimental embryology, of molecular genetics and of molecular biology, a deductive hypothetical logic can play a very important role in embryological research. In order to apply this logic to molecular embryology, four basic phenomena must be taken into consideration.

The first is expressed in the central dogma of molecular biology that states that the flow of genetic information is unidirectional. That is, genetic information is transmitted from nucleic acids to proteins; never the other way around.

The second phenomenon concerns cell lineage. In the ontogenesis of the higher metazoa there are two classes of cell duplication: normal cell duplication by which a cell gives rise to two identical daughter cells that are also identical to the mother cell, and a second class by which new cells are formed.

The third phenomenon refers to the fact that the control of

gene expression in the new cell lines occurs at the transcriptional level; not only different proteins but also different messenger RNAs occur at the phenotypic level in differentiated cells.

The fourth basic phenomenon is cell determination. Experimental embryology has shown that in the ontogenesis of the higher metazoa, cells arise that, although morphologically and by current knowledge also biochemically identical to other embryonic cells, are determined to forming specific structures of the embryo. The findings by Speman on self-wise differentiation in contrast to neighbour-wise differentiation, and the imaginal disc of drosophila are classic examples of cell determination.

The synchron model (1,2,3) proposed by Scarano more than 10 years ago, is a molecular-genetic model of determination and of ontogenesis of the higher metazoa.

The synchron model attempts to answer two questions. Firstly, how can we explain the encoding of the structure of the higher organisms in DNA, namely, what is the mechanism of action of morphogenes? Secondly, what is the biological meaning of enzymatic modifications of DNA in higher organisms?

The synchron model is based essentially on three hypotheses. The first is that new basic control mechanisms of structural gene expression arose during the course of evolution of the higher metazoa, namely mechanisms which could explain cell lineage and the action of morphogenes. The second hypothesis is that the morphogenic determinants are genetic determinants. The third hypothesis postulates that cell determination is caused by enzymatic nuclear DNA modifications, namely cell determination occurs at the genotype level (Fig.1). Specific DNA modifying enzymes would cause molecular changes in specific control segments of DNA so that the dependent structural genes become transcribable or potentially transcribable at the time a new determined cell line arises in embryonic development. Sequential synthesis of specific DNA-modifying enzymes during ontogenesis would determine the orderly spatial-temporal rise of new cell lines, which by the recognition of specific cell surface macromolecules (probably glycoproteins) would form the embryo structures.

An intermediate synchron is depicted in Fig.2. The true developmental mutants, as for instance the homeotic mutants of *Drosophila* and the hereditary phocomelia in man, would be mutants in the loci of the DNA modifying enzymes or in their recognition loci.

We emphasize that it is necessary to distinguish at least two

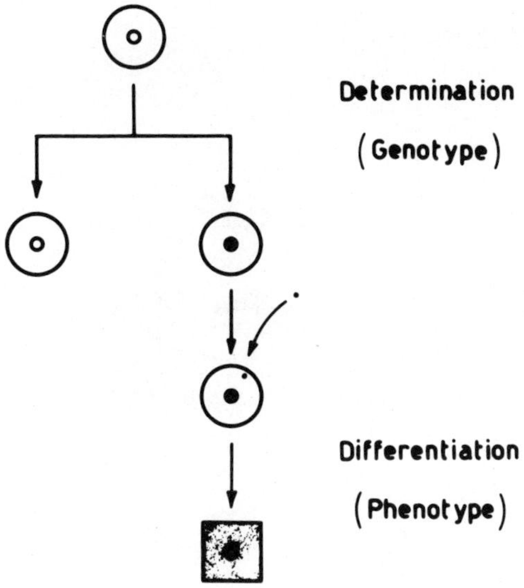

Figure 1. Hypothetical occurrence of the molecular events of cell determination at the genome level. A signal from the cell environment triggers the realization of the specific differentiated phenotype.

classes of cell differentiation. For instance, in plants no cell determination initiates cell differentiation. In animals, cell determination preceeds in general cell differentiation. We assume that different molecular mechanisms underlie cell determination and cell differentiation. Only in those cases in which cell differentiation is initiated by cell determination, heritable DNA modifications would occur.

Given the present knowledge about cell determination, the data obtained with chimeric mice (4) and the findings on nuclear transplantation (5), we think it is misleading to discuss "cell differentiation" in general and to attempt to explain cell differentiation with one mechanism. The findings on nuclear transplantation and on gametogenesis also suggest that, in animals, some cell differentiations not requiring a preceeding determination step occur. Concerning the ontogenesis of gametes, an abundant litera-

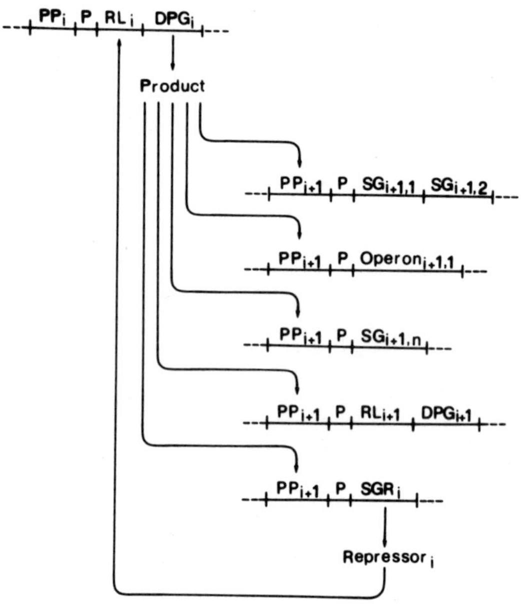

Figure 2. Diagram of a synchron.

ture demonstrates that complex cell interactions in gonads lead to the formation of eggs and spermatozoa, i.e. in the germ cell line no heritable DNA modifications would occur and differentiation to gametes would be directed by specific cell interactions.

At the VI International Congress of Embryology in Paris in 1969 Scarano (1) stated "The DNA enzymes can be grouped in three classes: enzymes that synthesize DNA; enzymes that degrade DNA; and enzymes that modify DNA. One might envisage mechanisms by which each type of DNA enzyme could be involved in cell differentiation. For instance, specific DNA nucleases might be involved in the disappearance of DNA during the maturation of erythroid cells. But in the present report only enzymes which modify DNA will be considered. Several types of DNA-modifying enzymes have been found, but many more will certainly be discovered; the enzymology of chromosome physiology is a very promising field."

With each group of DNA-modifying enzyme, a model may be constructed. With specific DNA endonucleases, for instance of the "restriction endonuclease" type, one could build a model based on the so-called "transponsons or jumping genes" of bacteria. An insertion segment could change its position or orientation and make an adjacent structural gene transcribable or potentially transcribable.

Other classes of DNA-modifying enzymes might be considered: DNA methylases, DNA-aminohydrolases, namely enzymes that might carry out transitions or transversion of bases. Enzymes that can change the frame of reference of transcription, namely enzymes that might cause changes of the type occurring in frameshift mutations.

In our work we have assumed that at least two and very likely several different mechanisms based on enzymatic DNA modifications play a role in the embryogenesis of the higher metazoa, i.e. of organisms in which determination events exist (1,2,3,6,7,8,9). We have investigated in the developing sea urchin embryo the base transitions AT to GC and GC to AT by culturing the embryos in the presence of labeled pyrimidine or purine precursors. We have never found in DNA hydrolyzates hypoxanthine and xanthine, which are intermediates from A to G, or uracil, which is a possible intermediate of the transition GC to AT.

In this review we will discuss experiments performed with nuclei isolated from embryos at different stages of the early development of the sea urchin and incubated in vitro in the presence of S-adenosyl methionine (SAM). The methylation of nuclear DNA and the possible formation of DNA minor thymine were investigated by using this in vitro system.

It has long been known that in eukaryotes, about 4% of the cytosines in DNA are methylated in the 5 position (10,11). No biological function can as yet be assigned to this phenomenon. In prokaryotes, a fraction of DNA methylation, of adenine, as well as of cytosine, is known to be involved in the modification-restriction mechanism (12). Moreover, a role in the development of the phage has been assigned to the single 5 methylcytosine (5-CH_3C) in the DNA of ØX 174 (13).

Riggs (14), and Holliday and Pugh (15) have suggested that DNA methylation in eukaryotes might also control cell differentiation. Sager and Kitchin (16) have proposed that modification and restriction of DNA regulate various developmental processes in eukaryotes.

Only methylation of specific nuclear DNA cytosines to 5-CH_3C occurs in the developing sea urchin embryo (17). The transfer of the methyl group into DNA takes place at the polynucleotide level (17). The methyl donor is S-adenosyl methyonine (8-17). The methylation of DNA cytosine in the developing sea urchin embryo is non random. In fact, 60% of the synthesized 5-CH_3C is present in the monopyrimidine fraction, although only 25% of the pyrimidines occur in this fraction (17). Moreover, in the dinucleotides of a pancreatic DNAse digest, more than 90% of 5-CH_3C is found in the dinucleotide CpG and the remaining in CpT and CpC (17). These data demonstrate that 5-CH_3C is distributed non randomly with respect to the primary sequence of DNA in the developing sea urchin embryos.

Razin and Cedar have demonstrated that 5-CH_3C is distributed non randomly (18) with respect to the chromosome structure in Drosophila DNA. By digestion of chromatin with micrococcal nuclease, they find 50% of the DNA to be nuclease-resistant, while over 75% of the 5-CH_3C is protected from nuclease digestion by chromatin proteins. However, a fractionation based on the nucleosome structure that is made up of units of about 200 base pairs, might not give the best separation between the control segments of DNA, which should be rich in 5-CH_3C as predicted by the synchron model (1,2,3), and the dependent structural genes.

Nuclei isolated from developing sea urchin embryos by centrifugation in sucrose gradients and incubated in the presence of methyl-labeled SAM, methylate their own DNA (8,9). A striking activation of DNA methylation occurs when small amounts of trypsin are added to the incubation mixtures (9). The activity determined in the absence of trypsin is defined basal DNA methylase activity and that measured in the presence of optimal concentration of trypsin, maximal DNA methylase activity. The DNA methylase activity of nuclei isolated from embryos at different stages of early embryonic development is typical of the stage in the two species, Sphaerechinus granularis and Paracentrotus lividus, which we have investigated (19).

Figures 3,4 and 5 show the DNA methylase activity, as a function of trypsin concentration, of nuclei isolated from three stages of the early embryonic development of the sea urchin Sphaerechinus granularis. Cultures of embryos and preparation of nuclei were performed as previously described (8). The DNA methylase activity was determined by measuring the incorporation of ^3H from $[C^3H_3]$-SAM in DNA 5-CH_3C. The probable error per cent of a measurement is 10. By averaging the data of three sets of experiments using Sphaerechinus granularis the basal DNA methylase

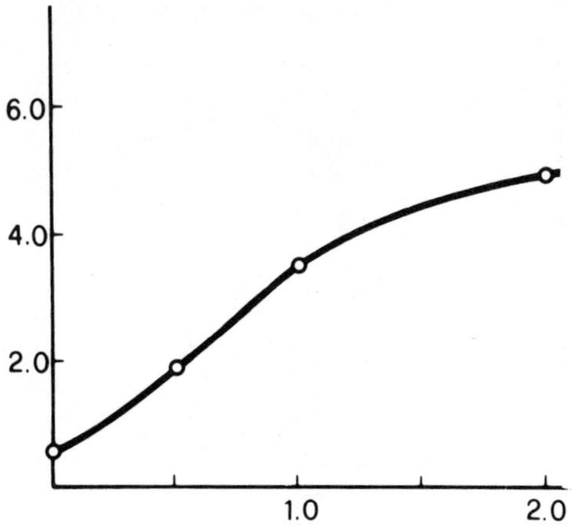

Figure 3. DNA methylase activity as a function of trypsin concentration, of nuclei isolated from Sphaerechinus granularis embryos at the hatched blastula stage. The assay mixtures, final volume 250 µl, pH 8.2, contained 25 µ moles Tris, 80 µ moles glucose, 0.2 µ moles Mg^{2+}, 100 µg of Salmon sperm DNA, nuclei containing about 5 mg protein, trypsin as indicated, 6 µ moles $[C^3H_3]$-SAM 6 µCi·nmole^{-1}. After 10 min. of pre-incubation at 26°C the reaction was started by addition of $[C^3H_3]$-SAM. The incubation was for 30 min. at 26°C. Preparation of DNA, hydrolysis of DNA to free bases, isolation of 5-CH_3C by paper chromatography and determination of radioactivity were performed as previously described (9).

$$\text{Abscissa} = \frac{\mu\text{g trypsin}}{250\ \mu\text{l}}\ ; \quad \text{Ordinate} = \frac{\text{dpm}}{\text{mg DNA}} \cdot 10^{-5}$$

activity of nuclei isolated from hatched blastulae, that from gastrulae with primary mesenchyme and that from gastrulae with secondary mesenchyme are as 10 : 2 : 1. For the maximal DNA methylase activity the corresponding figures are 60 : 7 : 1.

The rates of DNA synthesis measured in vivo at the same stages by pulses of labeled deoxythimidine are relatively unchanged. Other parameters directly proportional to the rate of DNA synthesis such as nuclear DNA accumulation, increase in the number of cells per embryo and frequency of mitosis per embryo all

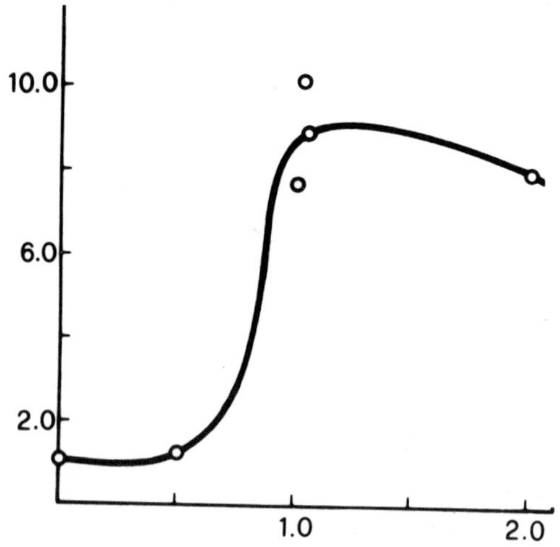

Figure 4. DNA methylase activity as a function of trypsin concentration, of nuclei isolated from Sphaerechinus granularis embryos at the gastrula with primary mesenchyme stage. Experimental conditions as in the legend to Fig.3.

$$\text{Abscissa} = \frac{\mu\text{g trypsin}}{250\ \mu\text{l}}\ ; \quad \text{Ordinate} = \frac{\text{dpm}}{\text{mg DNA}} \cdot 10^{-4}$$

Three identical samples containing 1 μg of trypsin per 250 μl of incubation mixture were used.

vary within factors of about four at the hatched blastula stage, at the gastrula with primary mesenchyme stage and at the gastrula with secondary mesenchyme stage.

These findings indicate that the nuclear DNA methylase activity is not correlated to the rate of DNA biosynthesis in the early embryonic development of the sea urchin. At the hatched blastula stage, a fraction of the DNA methylase activity might have a function other than that of just replicating the pattern of DNA methylation.

The work of S. Hörstadius has demonstrated that important determination events take place at the blastula stage in the sea urchin embryo (20). Since a strikingly high DNA methylase activity is found at the blastula stage (when important determination events occur) we propose that it is at this stage that specific single-strand DNA methylases exist. These specific DNA methylases would recognize the regulatory loci of the synchrons.

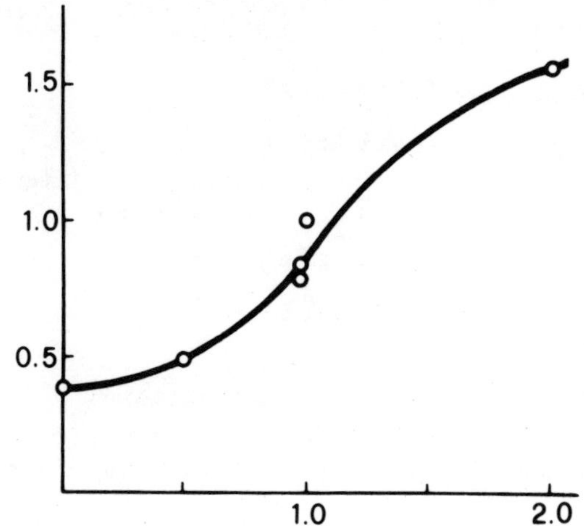

Figure 5. DNA methylase activity as a function of trypsin concentration, of nuclei isolated from Sphaerechinus granularis embryos at the gastrula with secondary mesenchyme stage. Experimental conditions as in the legend to Fig.3.

$$\text{Abscissa} = \frac{\mu\text{g trypsin}}{250\ \mu\text{l}}\ ; \qquad \text{Ordinate} = \frac{\text{dpm}}{\text{mg DNA}} \cdot 10^{-4}$$

Three identical samples containing 1 μg of trypsin per 250 μl of incubation mixture were used.

This particular mechanism of DNA enzymatic modification and how a modified DNA arises in one chromosome during cell lineage is outlined in Fig.6.

A prediction on the mechanism of mitosis of any model of cell lineage based on DNA enzymatic modifications is outlined in Fig.7. Because of diploidy it is clear that at mitosis to get all the modified chromosomes in one of the two daughter cells, the spindle fibers of one pole must recognize the plus strand of the chromosomal DNA, and the spindle fibers of the other pole must recognize the minus strand. The DNA modifications based on specific single strand DNA methylases and one unspecific DNA replicative methylase might be reversible as illustrated in Fig.8.

The second mechanism of enzymatic DNA modifications that we have been investigating since 1964 (21) involves GC to AT transitions by the intermediate methylation of specific DNA cytosines (1,2,3,6,7). At the time a new cell line is determined, a cell

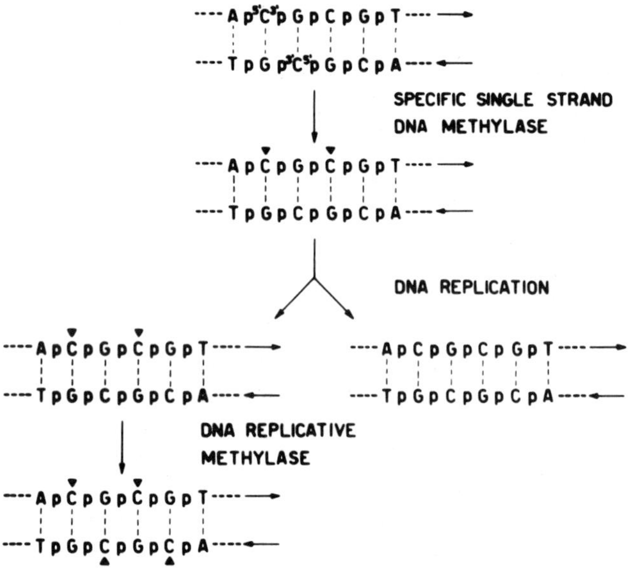

Figure 6. Hypothetical mechanism of a class of DNA enzymatic modifications: specific single-strand DNA methylases and one unspecific DNA replicative methylase would explain cell lineage.

lineage specific DNA 5-methylcytosine aminohydrolase will catalize the formation of the DNA minor thymine (DMT) in the regulatory loci of the synchrons. The properties of DMT are obvious from this definition. First, it will be detectable only in a cell undergoing determination. Second, since it originates from DNA 5-CH_3C it must have the intact methyl group of methionine and it must be non-randomly distributed with respect to both DNA sequence and chromosome structure (6,7). Because of this operational definition we challenged (7) the finding of DMT in Novikoff hepatoma cells cultured in vitro (22) and we did not question the subsequent report by Sneider (23).

The DNA modifications involving GC to AT transitions are irreversible and consequently cell lineage based on them would be irreversible also.

The finding of the non-random distribution of (C^2H_3) thymine (7) in DNA from developing sea urchin embryos cultured in the

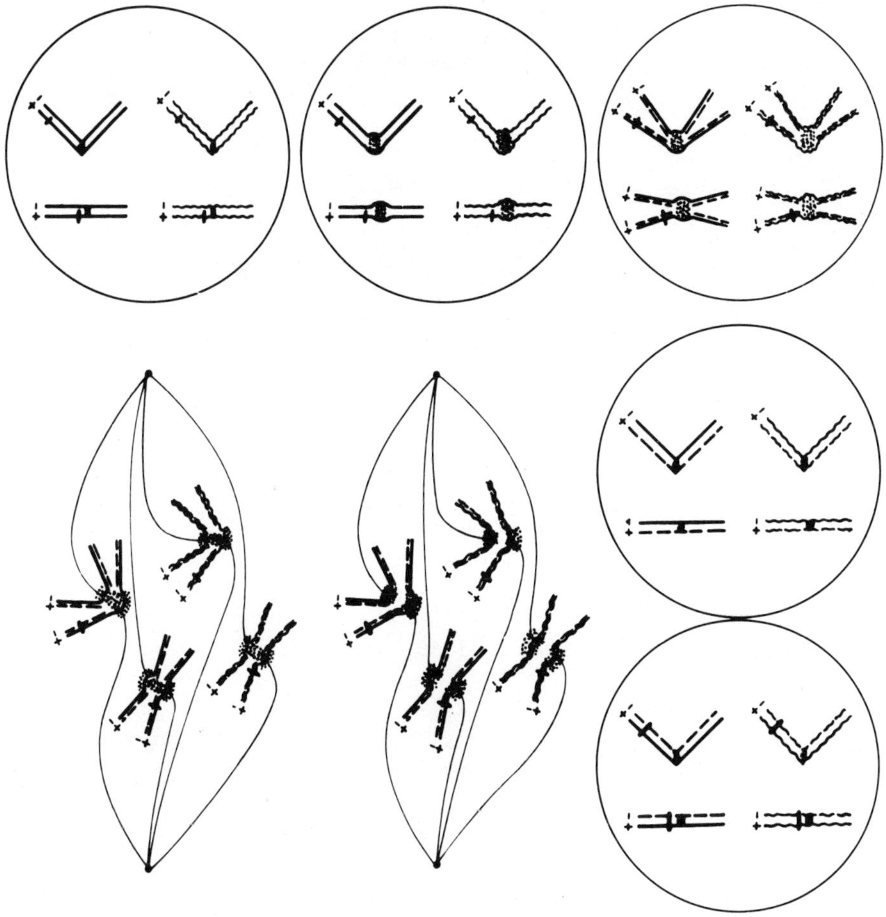

Figure 7. Hypothetical mechanism of a specific mitosis that would occur when a new cell line originates in embryonic development. A diploid nucleus with two sets of chromosomes is depicted during the successive stages of cell division. Each chromosome contains one single DNA molecule. The DNA of one of the two homologous chromosomes is indicated by straight lines and that of the other, by wavy lines. The newly synthesized DNA strand is depicted by discontinuous straight or discontinuous wavy segments. The components of the centrosome structure are indicated by dots. The DNA of the centrosome structure is the last to be replicated.

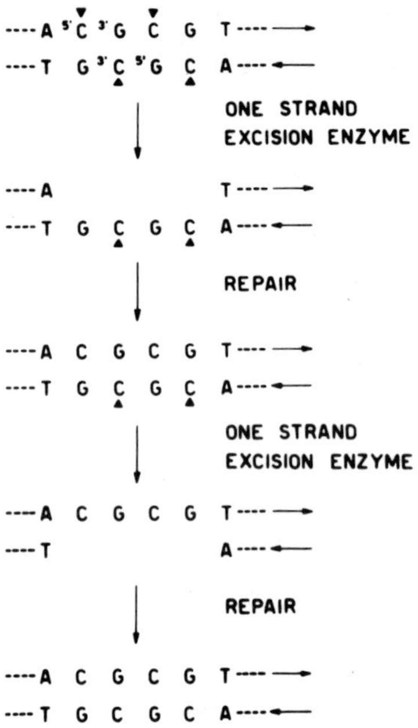

Figure 8. Reversibility of DNA modifications based on specific single strand - DNA methylases and an unspecific DNA replicative methylase.

presence of (C^2H_3) methionine gives an example of our _in vivo_ experimental approach.

We have investigated _in vitro_ the formation of DMT in experiments with isolated nuclei such as those reported in Fig.3. The only DNA bases which are labeled in experiments with isolated nuclei from developing sea urchin embryos incubated in the presence of isotopic methyl SAM are 5-CH_3C and thymine. But the distribution of the label between the two bases varies widely from experiment to experiment (9). The distribution might depend on the species of the sea urchin because of differences in time of developmental events, on the stage of the embryos and on the method of the preparation of the nuclei. In the instances in which more than 10% of the label is found in thymine it seems unlikely that the labeled thymine originates because of some technical problem in the isolation and purification procedures of the base. However, at the present stage of the experiments both _in_ vitro and in vivo, no mechanism for the origin of DMT can be inferred with certainty. It is noteworthy that the addition of trypsin appears to increase mainly the methylation of DNA cytosine. On the basis of the hypothesis of DNA-modification complexes, i.e. the DNA-cytosine methylase apparatus and the DNA-5-methylcytosine aminohydrolase apparatus (8) it is possible that the methods to detect one complex might not be those optimal for the detection of the other.

Concerning our in vitro approach on DMT we notice that we have never reported that dCMP-aminohydrolase deaminates DNA-5-methylcytosine to DNA thymine. For instance we wrote "...the most important physiological function of dCMP-aminohydrolase could be the regulation of the deoxynucleotide pool."(24).

In summary, the findings with isolated nuclei of the developing sea urchin embryo demonstrate that at the blastula stage the DNA methylase activity is not correlated to the replication of the pattern of DNA methylation. Though further experimental data are necessary to substantiate our contention that specific DNA-cytosine methylases and specific DNA-5-methylcytosine aminohydrolases play the primary role in cell lineage, both the simple theoretical base of the synchron model and the experimental data warrant continuation of the work.

REFERENCES

1. Scarano, E. Enzymatic modifications of DNA and embryonic differentiation. Ann. Embryol. Morphogen. (1969) Suppl. 1 51-61.
2. Scarano, E. The control of gene function in cell differentiation and in embryogenesis. Adv. Cytopharmacol. (1971) 1 13-24.
3. Scarano, E., L. Tosi and A. Granieri. Enzymatic modifications of DNA: A model for the molecular basis of cell differentiation. In F. Salvatore, E. Borek, V. Zappia, H.G. Williams-Ashman and F. Schlenk, The Biochemistry of Adenosylmethionine, Columbia University Press, New York (1977) pp. 369-382.
4. Mintz, B. Gene control of mammalian differentiation. Ann. Rev. Genet. (1974) 8 411-470.
5. Gurdon, J.B. Control of gene expression in animal development. Harvard University Press (1974).
6. Scarano, E., M. Iaccarino, P. Grippo and E. Parisi. The heterogeneity of thymine methyl group origin in DNA pyrimidine isostichs of developing sea urchin embryos. Proc. Natl. Acad. Sci. USA (1967) 57 1394-1400.
7. Grippo, P., E. Parisi, C. Carestia and E. Scarano. A novel origin of some deoxyribonucleic acid thymine and its non-random distribution. Biochemistry (1970) 9 2605-2609.
8. Tosi, L., A. Granieri and E. Scarano. Enzymatic DNA modifications in isolated nuclei from developing sea urchin embryos. Exptl. cell Res. (1972) 72 257-264.
9. Tosi, L. and E. Scarano. Effect of trypsin on DNA methylation in isolated nuclei from developing sea urchin embryos. Biochem. Biophys. Res. Commun. (1973) 55 470-476.
10. Wyatt, G.R. The purine and pyrimidine composition of deoxypentose nucleic acids. Biochem. J. (1951) 48 584-590.
11. Vanyushin, B.F., S.G. Tkacheva and A.N. Belozersky. Rare bases in animal DNA. Nature (1970) 225 948-949.
12. Arber, W. DNA modification and restriction. Progr. Nucleic Acids Res. & Mol. Biol. (1974) 14 1-37.
13. Razin, A., D. Goren and J. Friedman. Studies on the biological role of DNA methylation: Inhibition of methylation and maturation of the bacteriophage ØX 174 by nicotinamide. Nucleic Acids Res. (1975) 2 1967-1974.
14. Riggs, A.D. X inactivation, differentiation and DNA methylation. Cytogenet. Cell Genet. (1975) 14 9-25.

15. Holliday, R. and J.E. Pugh. DNA modification mechanisms and gene activity during development. Science (1975) 187 226-232.
16. Sager, R. and R. Kitchin. Selective silencing of eukaryotic DNA. Science (1975) 189 426-433.
17. Grippo, P., M. Iaccarino, E. Parisi and E. Scarano. Methylation of DNA in developing sea urchin embryos. J. Mol. Biol. (1968) 36 195-208.
18. Razin, A. and H. Cedar. Distribution of 5-methylcytosine in chromatin. Proc. Natl. Acad. Sci. USA (1977) 74 2725-2728.
19. Scarano, E. and L. Tosi. Differential DNA methylation by isolated nuclei of developing sea urchin embryos. 10th International Congress of Biochemistry. Hamburg (1976). Abstract 11-8-384, p. 533.
20. Hörstadius, S. Experimental embryology of echinoderms. Clarendon, Oxf. (1973).
21. Scarano, E., M. Iaccarino, P. Grippo, and D. Winckelmans. On methylation of DNA during development of the sea urchin embryo. J. Mol. Biol. (1965) 14 603-607.
22. Sneider, T.W. and Van R. Potter. Methylation of mammalian DNA: Studies on Novikoff hepatoma cells on tissue culture. J. Mol. Biol. (1969) 42 271-284.
23. Sneider, T.W. On the source of "Mynor Thymine" in DNA from a Novikoff rat hepatoma cell line. J. Mol. Biol. (1973) 79 731-734.
24. Rossi, M., G. Geraci and E. Scarano. Deoxycytidylate aminohydrolase. III. Modifications of the substrate sites caused by allosteric effectors. Biochemistry (1967) 6 3640-3645.

REGULATION OF MACROMOLECULAR SYNTHESIS DURING SEA URCHIN DEVELOPMENT

A. Arzone, V. Matranga, G. Giudice, V. Mutolo, H. Noll, A.M. Rinaldi, I. Salcher, and M.L. Vittorelli

The University of Palermo
Institute of Comparative Anatomy
Palermo, Italy

Immediately following fertilization the sea urchin egg enters a period of very rapid cell division that cleaves the egg cell into about one thousand proportionately smaller cells, which form the swimming blastula, i.e. a larval form that is less vulnerable to environmental injuries since it is capable of actively swimming away from them.

In order to form these one thousand cells rapidly, by a process of cell doubling, the first cell divisions have to be very frequent, and actually follow each other at a bacterial pace, about every thirty minutes (1). Because of this, the egg has stored in the cytoplasm many macromolecules whose synthesis is ordinarily coupled with that of nuclear DNA, such as ribosomes (2-6), histones (7), and tubulin (8). Preformed mitochondria are also stored in the cytoplasm of the unfertilized egg (9). The egg, therefore, represents a suitable system for studying the mechanisms of the coupling and uncoupling of nuclear DNA synthesis with other macromolecular syntheses. Among those, the relationship between nuclear DNA synthesis and mitochondrial RNA and DNA synthesis appears of special interest for two reasons. First, the relationships between these two genomic sets within the same cell are still far from being clear. Second, the egg can be easily divided, by centrifuging it through two sucrose layers, into a nucleated half and a non-nucleated half containing almost all the mitochondria (1,10).

By using this technique Rinaldi *et al.* (11) have been able to demonstrate that labeled RNA precursors are incorporated into mitochondrial RNA at a very low rate by unfertilized eggs. The rate is somewhat increased if the egg is fertilized or parthenogenetically

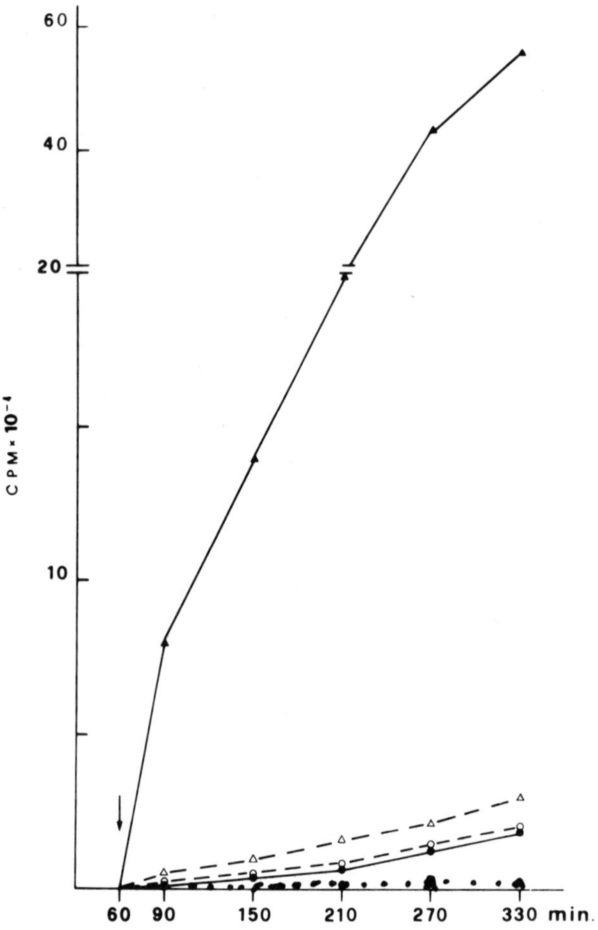

Fig. 1. P^{32} incorporation into mitochondrial RNA. P^{32} was administered to unfertilized (•••••••), fertilized (○ —— ○), parthenogenetically activated (●——●) eggs, or to fertilized (Δ— —Δ), parthenogenetically activated (▲——▲) non-nucleated halves.

activated; but if the nucleus has been removed by centrifugation, and the non-nucleated half parthenogenetically activated, then the incorporation into mitochondrial RNA is increased at least ten fold (Fig. 1).

This increase has to be attributed to increased mitochondrial RNA synthesis since permeability to the labeled precursors does not change under these conditions. These results strongly suggest that it is the cell nucleus that exerts a negative control over mitochondrial RNA synthesis, which becomes liberated if the nucleus is removed and the non-nucleated half parthenogenetically activated. As a further proof of this, if a nucleus is introduced into the non-nucleated half by fertilizing it instead of parthenogenetically activating it, then the synthesis of RNA remains at the same low level as that of the entire fertilized egg.

The kind of RNA synthesized by the mitochondria of the non-nucleated half is mostly mitochondrial type ribosomal RNA. This suggested that mitochondria once freed from the nuclear control might begin replication.

In order to investigate this possibility, Rinaldi et al. (12) administered H^3-labeled thymidine to the eggs and then measured its incorporation into the mitochondrial DNA purified by centrifugation in cesium chloride-ethidium bromide gradients. The results show that very little or no thymidine incorporation occurs into the mitochondrial DNA of parthenogenetically activated entire eggs, in which, however, nuclear DNA, as a control, shows a very active incorporation. On the other hand, the mitochondrial DNA of the non-nucleated activated halves shows a lively thymidine incorporation. Here again, if a nucleus is introduced into the non-nucleated egg by fertilizing it, no thymidine incorporation into mitochondrial DNA occurs. In order to prove that this incorporation is indicative of DNA replication and not merely of repair, the authors have looked for mitochondrial DNA replicative forms with the electron microscope. Out of 156 molecules observed, 24 were found in a replicative state (Fig. 2) when mitochondrial DNA was extracted from non-nucleated activated halves. None were found out of 94 molecules studied when mitochondrial DNA was found out of 94 molecules studied when mitochondrial DNA was extracted from the entire activated egg.

These results clearly show that it is the nucleus which by means of some unknown factor negatively controls mitochondrial DNA and RNA synthesis during the sea urchin embryogenesis.

It was demonstrated years ago (13-17) that sea urchin embryos can be dissociated into isolated cells which, under suitable conditions, can be reaggregated into cell clumps which are able to

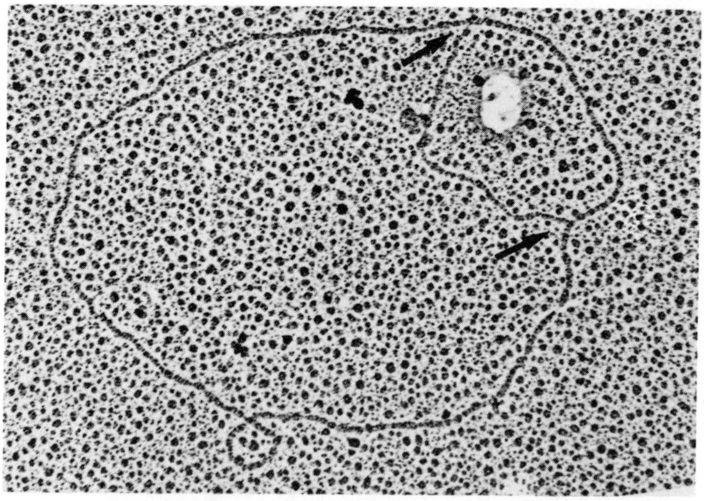

Fig. 2. Electron micrograph of a DNA molecule extracted from mitochondria of non-nucleated, parthenogenetically activated halves. The arrows point at the replication forks (x50,000).

differentiate into larva-like structures (Fig. 3). The relevance of macromolecular synthesis in the process of sea urchin cell reaggregation has been studied in the past (18).

A close correlation between intercellular contact and DNA synthesis was, for example, shown in (19), and a surface glycoprotein which may be involved in the control partially characterized in (20). In order to investigate the role of surface macromolecules in the process of cell interactions, reaggregation, and differentiation, Noll and Vittorelli (21) have tried to extract some macromolecular factor from the cell surface by treating the dissociated cells with diluted butanol. Following such treatment the cells are no longer able to reaggregate. If, on the other hand, the dyalized butanol extract is added to the butanol treated cells, they reacquire the ability of reaggregating and differentiating into embryo-like structures.

The same effect is shown by the butanol extract of isolated cell membranes. A preliminary electrophoretic investigation of the butanol extract shows a complex pattern of protein migration.

This offers a new approach for studying the role of cell surface macromolecules in cell recognition and embryonic differentiation.

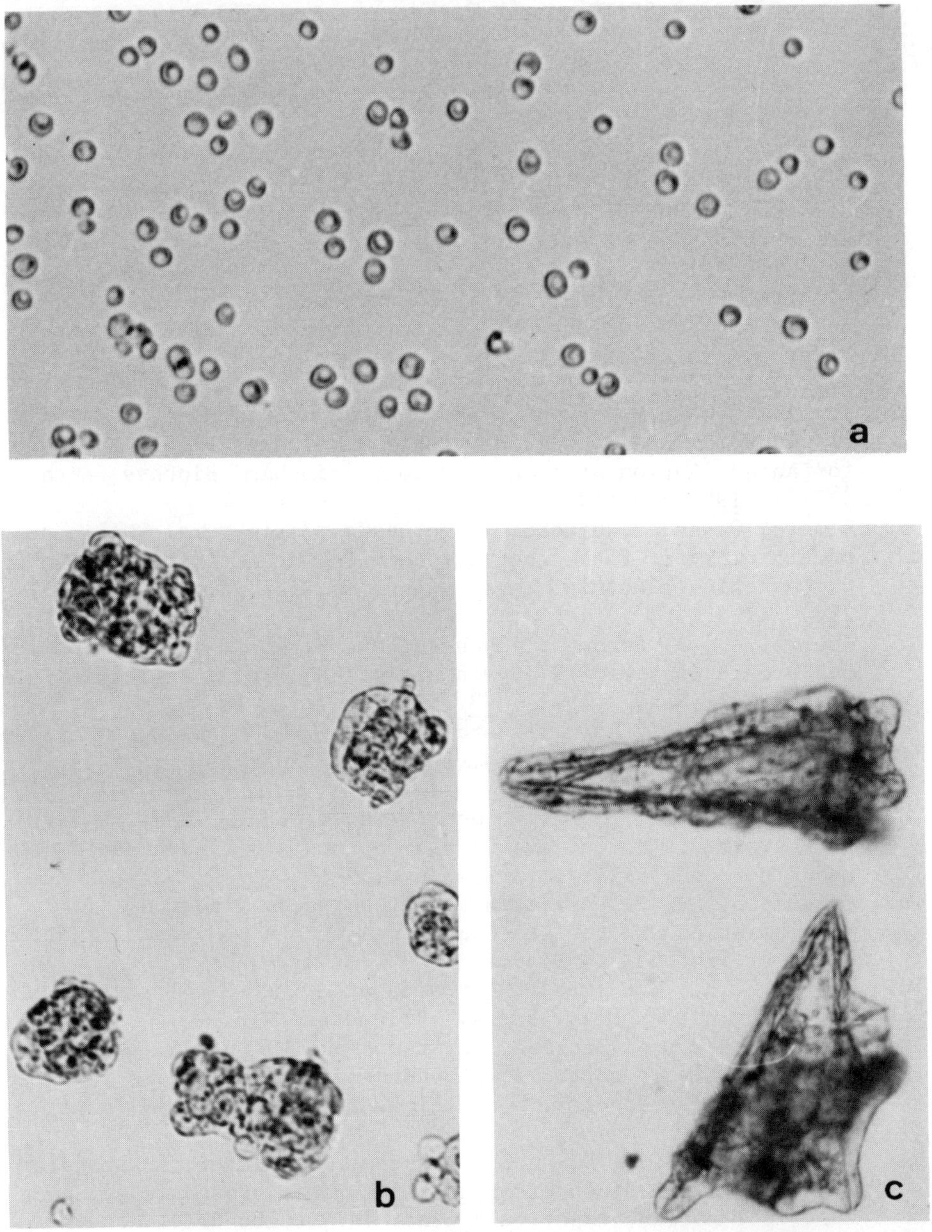

Fig. 3. a) Cells dissociated from sea urchin mesenchyme blastulae; (X 400).
b) The same after 6 hours of aggregation (X 400).
c) After 5 days of aggregation (X 250).

Part of the reported work has been supported by C.N.R. contract No. 77.00339.85 (Research Project on the Biology of Reproduction).

REFERENCES

1. Giudice, G. Developmental Biology of the sea urchin embryos. Academic Press Inc., New York, (1973) 1-469.
2. Giudice, G. and V. Mutolo. Synthesis of ribosomal RNA during sea urchin development. Biochim. Biophys. Acta (1967) 138 276-285.
3. Giudice, G. and V. Mutolo. Synthesis of ribosomal RNA during sea urchin development. II. Electrophoretic analysis of nuclear and cytoplasmic RNA's. Biochim. Biophys. Acta (1969) 179 341-347.
4. Sconzo, G., A.M. Pirrone, V. Mutolo and G. Giudice. Synthesis of ribosomal RNA during sea urchin development. III. Evidence for an activation of transcription. Biochim. Biophys. Acta (1970) 199 435-440.
5. Sconzo, G. and G. Giudice. Synthesis of ribosomal RNA in sea urchin embryos. V. Further evidence for an activation following the hatching blastula stage. Biochim. Biophys. Acta (1971) 254 447-451.
6. Sconzo, G., A. Bono, I. Albanese, and G. Giudice. Studies on sea urchin oocytes: II. Synthesis of RNA during oogenesis. Exptl. Cell Res. (1972) 72 95-100.
7. Cognetti, G., G. Spinelli and A. Vivoli. Synthesis of Histones during sea urchin oogenesis. Biochim. Biophys. Acta (1974) 349 447-455.
8. Cognetti, G., I. Di Liegro and F. Cavarretta. Studies of protein synthesis during sea urchin oogenesis. II. Synthesis of tubulin. Cell Differ. (1977) 6 159-165.
9. Matsumoto, L., H. Kasamatsu, L. Piko and J. Vinograd. Mitochondrial DNA replication in sea urchin oocytes. J. Cell Biol. (1974) 63 146-159.
10. Harvey, E.B. "The American Arbacia and Other Sea Urchins" (1966) Princeton Univ. Press., Princeton, N.J.
11. Rinaldi, A.M., A. Storace, A. Arzone and V. Mutolo. Cell nucleus negatively controls mitochondrial RNA synthesis in early sea urchin development. Cell Biol. Intern. Rep. (1977) 1 249-254.
12. Rinaldi, A.M., G. De Leo, A. Arzone, I. Salcher, A. Storace and V. Mutolo, Biochemical and electron microscopic evidence that cell nucleus negatively controls mitochondrial genomic activity in early sea urchin development. Submitted for publication.
13. Giudice, G. Restitution of whole larvae from disaggregated cells of sea urchin embryos. Dev. Biol (1962) 5 402-411.
14. Millonig, G. and G. Giudice. Electron microscopic study of the reaggregation of cells dissociated from sea urchin embryos. Dev. Biol. (1967) 15 91-101.

15. Giudice, G., V. Mutolo, G. Donatuti and M. Bosco. Reaggregation of mixtures of cells from different developmental stages of sea urchin embryos. Exptl. Cell Res. (1968) 54 279-281.
16. Giudice, G. and V. Mutolo. Reaggregation of dissociated cells of sea urchin embryos. Advances Morphogen. (1970) 8 115-158.
17. Giudice, G. Aggregation of cells isolated from vegetalized and animalized sea urchin embryos. Experientia (1963) 19 83-86.
18. Giudice, G. The mechanism of aggregation of embryonic sea urchin cells a biochemical approach. Dev. Biol. (1965) 12 233-247.
19. Vittorelli, M.L., G. Cannizzaro and G. Giudice. Trypsin treatment of cells dissociated from sea urchin embryos elicits DNA synthesis. Cell Differ. (1973) 2 279-284.
20. Matranga, V., C. Giarrusso, V. Vasile and M.L. Vittorelli. Trypsin treatment which elicits DNA synthesis, removes a high a high molecular weight glycoprotein from the plasma membrane of sea urchin embryonic cells. Cell Biol. Intern. Rep. (1978) 2 147-155.
21. Noll, H. and Vittorelli, M.L., manuscript in preparation.

NUCLEUS-ASSOCIATED POLYRIBOSOMES IN EARLY EMBRYOS OF LOACH

C.A. Kafiani, L.A. Strelkov, N.B. Chiaureli
and A.N. Davitashvili

Institute of Molecular Biology
Academy of Sciences of USSR, Moscow

The mechanism of nucleus-cytoplasmic exchange of macromolecules is one of the clues to the problem of control of genomic expression in eukaryotes. Early embryos of a fresh-water osseous fish, loach (Misgurnus fossilis), are among suitable models for studying this mechanism.

Extensive studies with loach embryos led to a detailed characterization of their development with regard to the cell cycle parameters, timing of synthesis of the main RNA classes as well as of nuclear proteins, and to a description of the schedule of RNA transport from nucleus to cytoplasm, and of proteins from cytoplasm to nucleus (reviewed in 1-3). One of these lines of research led to the discovery of informosomes, i.e. non-ribosomal RNP particles which contain RNA and protein in a 1:3 ratio (4). These studies have been facilitated by the absence of detectable new-formation of ribosomal RNA, in loach embryos, before the mid-gastrula stage, alleviating the necessity of using low actinomycin D doses usually employed in similar studies with many other cell systems.

During a systematic examination of RNP structures in early loach embryo cells, we found that a large fraction of polyribosomes is tightly associated with the cell nuclei. In this report, we present evidence that the nucleus-associated polyribosomes, abbreviated henceforth as "n-polysomes", represent a metabolically distinct sub-population of cell polyribosomes which translate newly-formed histone messenger RNAs.

On the basis of the present and published data, a model for the nuclear membrane-associated system of the chromatin biogenesis is proposed, where DNA replication and histone mRNA formation and translation are topologically coupled. The model involves "short-range" transport of transcription and translation products across the nuclear membrane, assuring efficient reproduction of chromatin in rapidly dividing cells, especially in embryonic cells that have large cytoplasm-to-nucleus volume ratios. Our findings suggest, furthermore, a reevaluation of earlier data on the timing of nucleus-cytoplasm transport of newly-formed RNA in early embryos.

RESULTS

A Large Fraction of Polyribosomes Is Associated with the Cell Nuclei

In this study, we used loach embryos at two stages: mid-blastula (8th hour of development at 20°C) and early gastrula (12th hour), referred to here as blastula and gastrula. Developing eggs were obtained as described (5) and the embryos (blastoderms) separated from the yolk by trypsin treatement (6). For labeling, the blastoderms were incubated, at room temperature, in a double-strength Holtfreter medium with labeled precursors, as defined in the figures.

In the experiments depicted in Figures 1-4, blastula stage blastoderms have been incubated 10 min with ^{14}C-lysine and ^{3}H-nucleosides, and homogenized in 1M sucrose based on a solution containing (in mM) NaCl, 100, KCl, 25, $MgCl_2$, 1.5, $CaCl_2$, 1.0, and Tris-HCl, 50, pH, 7.4 (all solutions used in this work have been sterilized with diethyl pyrocarbonate). The homogenate was layered on 1.8M sucrose underlayered with a cushion containing 50% Ficoll-400 (Pharmacia) and 2.4M sucrose, the two dense media being based on the salt mixture of the above composition. The nuclear pellet obtained upon spinning 90 min at 70,000 x g (+4°C), was resuspended and stirred for 10 min in a "low salt" medium containing (mM) NaCl, 100, KCl, 25, $MgCl_2$, 1.5, and Tris-HCl, 50, pH, 7.4. After collecting (800 x g, 10 min), the nuclei were re-extracted, and the combined extract was subjected to sedimentation analysis.

Fig. 1 demonstrates a striking amount of discretely sedimenting UV-absorbing and labeled material in the extract, the bulk of the material being concentrated in the 40S to 160S region. A proportion of the RNA and protein label was different in the 60S to 200S region and in the "light" region of the gradient: in sedimentation zones designated A and B in Fig. 1, the $^{3}H/^{14}C$ ratios were, respectively, 2.5 and 2.7, while in zone C the value was 3.8. This suggested a smaller proportion of newly-formed protein in the light (about 30S) particulate components of the extract as compared to heavier ones.

In an attempt to identify the particulate components released by low-salt extraction from the Ficoll-sucrose purified nuclei, we

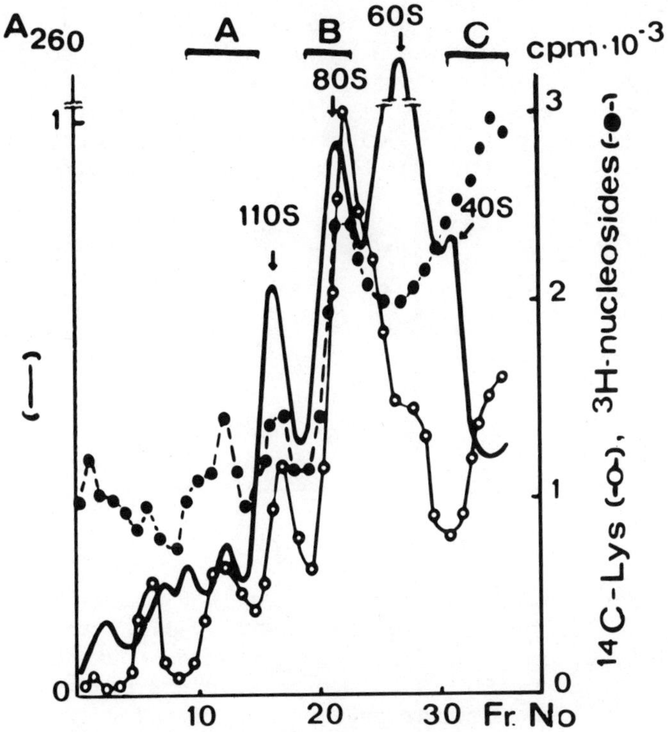

Fig. 1. Sedimentation of the "low-salt" extract of the Ficoll-sucrose-purified nuclei from blastula-stage embryos. Blastoderms have been incubated in the double-strength Holtfreter medium (composition, in mM: NaCl, 100, KCl, 10, $MgCl_2$, 3, $CaCl_2$, 3, $NaHCO_3$, 5, Tris-HCL, 50, pH 7.4) for 10 min with 7 μCi/ml ^{14}C-lysine (270 μCi/mmole) and 70 μCi/ml of a mixture of ^{3}H cytidine, uridine, and guanosine (sp. activities, respectively, 1.0, 5, and 10 Ci/mmole). The nuclei were isolated and extracted with the "low-salt" medium (composition, in mM: NaCl, 100, KCl, 25, $MgCl_2$, 1.5, Tris-HCl, 50, pH 7.4). The extract was layered on a 17-50% sucrose gradient based on the low-salt medium, and centrifuged in SW-27 rotor (Beckman) at 24,000 rpm for 16 h at +4°C. Fractions were collected through a UV-monitor (LKB) and aliquots counted directly in a dioxane-based scintillator in the Intertechnique SL-40 counter with a double-label program.

pooled separately fractions in zones A, B, and C, treated them with formalin, and subjected them to equilibrium centrifugation in CsCl (7). Results are shown in Fig. 2.

It can be seen that zones A and B each contain two double-labeled buoyant density components. Most of the label was recovered in a band peaking respectively at 1.51 and 1.53 g/cm^3, characteristic of monoribosomes and polyribosomes complexed with mRNA (4), which are referred to here, for brevity, as "engaged ribosomes".

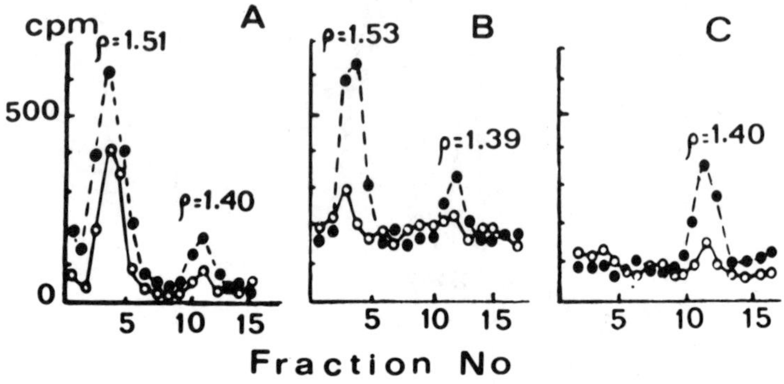

Fig. 2. Equilibrium centrifugation of RNP material of different sedimentation velocity from Fig. 1. Fractions in zones A, B, and C of Fig. 1 were pooled separately, treated with formalin (7), layered on preformed gradients of CsCl, and centrifuged in SW 50.1 rotor (Beckman) at 44,000 rpm and +20°C for 20 h. Fractions were collected through the bottom and precipitated with trichloroacetic acid. The material, collected on Millipore filters, was counted in a toluene-based scintillation mixture with SL-40 counter, with a double-label program. Symbols designating ^{14}C-lysine and ^{3}H-nucleoside labeled material are as in Fig. 1.

Both zones A and B contain in addition minor buoyant density components peaking respectively at 1.40 and 1.39 g/cm^3. This corresponds to an RNA/protein ratio of 1/3 which is characteristic of informosomes in general, and of loach embryo cytoplasmic informosomes in particular (4). The unique double-labeled buoyant density component of slowly sedimenting fractions (zone C) also peaked at 1.40 g/cm^3.

Comparisons of ^{3}H/^{14}C ratios in sedimentation and buoyant density fractions of Figs. 1 and 2 may be suggestive of the relations between the components involved. The main density components in zones A and B (1.51 and 1.53 g/cm^3), that have been tentatively identified as engaged ribosomes, displayed the same

$^3H/^{14}C$ ratio as that found in the 80S to 200S region in Fig. 1. In minor buoyant density components (1.40 and 1.39 g/cm^3) that have been revealed in the 80S to 200S region by equilibrium centrifugation, the $^3H/^{14}C$ ratio was 3.8, i.e. the same as was found in the "light" (C) zone particles, that are identifiable as informosomes.

The combined evidence of Fig. 1 and Fig. 2 indicates that the low-salt extract of loach embryo nuclei contains newly formed informosomes in both slowly and rapidly sedimenting fractions. The bulk of the rapidly (above 60S) sedimenting particles may be considered as engaged monoribosomes (80S) and polyribosomes that carry newly-formed RNA and protein (it has to be recalled that ribosome synthesis does not occur in blastula and gastrula stages, in loach, Refs. 1-3).

The polyribosomal nature of the bulk of the low-salt extracted material was confirmed by examining the effect of various treatments of the extracts. Fig. 3 shows that EDTA caused a quasi-total displacement of the UV absorbing and labeled material from the heavy to the light region of the gradient, and a similar shift was caused by RNAse treatment (results not shown). In contrast, treatment with sodium deoxycholate or DNAse did not cause such effects.

Hence, the bulk of the particulate material associated with, and extracted from, the Ficoll-sucrose purified nuclei of early loach embryos, represents polyribosomes (termed here, for brevity, "n-polysomes") and engaged monoribosomes. We wish to stress that very little such material could be found in low-salt extracts of the nuclei that had been subject to additional purification with 1% Triton X 100. This indicates that the polyribosomes and monoribosomes are directly or indirectly associated with nuclear membrane(s).

N-polysomes could be purified from low-salt extracts of the Ficoll-sucrose nuclear preparations by treating the extracts with sodium deoxycholate (DOC) and Tween-40 (0.5% each) in 0.3M KCl, followed by pelleting through 1.5M sucrose, as described in Fig. 4. High KCl brings about dissociation of non-engaged ribosomes, and the pellet contains engaged mono- and polyribosomes. Fig. 4 demonstrates the clear-cut pattern of distribution of UV-absorbing and accompanying double-labeled material of n-polysomes, which is similar to that reported for "light" free polyribosomes of early sea urchin embryos (8).

For comparison, cytoplasmic, membrane-bound and free polyribosomes were isolated and purified using the detergent-high KCl procedure (9) from gastrula-stage embryos (see Fig. 5). ^3H-uridine-pulsed blastoderms were homogenized in a "high-salt" medium (300 mM KCl, 5 mM $MgCl_2$, 0.32 M sucrose, 100 U/ml heparin, 50 mM Tris-HCl, pH 7.4). Nuclei and mitochondria were discarded, and cytoplasmic

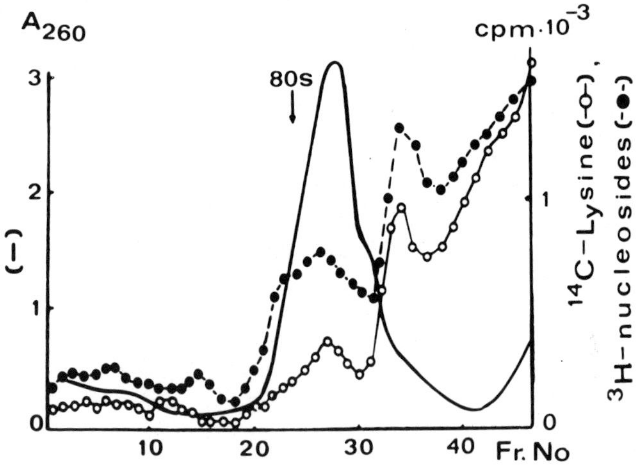

Fig. 3. Effect of EDTA (0.01M) on sedimentation of the low-salt extract of the blastula-stage, Ficoll-sucrose purified nuclei of Fig. 1. Conditions of centrifugation and counting as in Fig. 1.

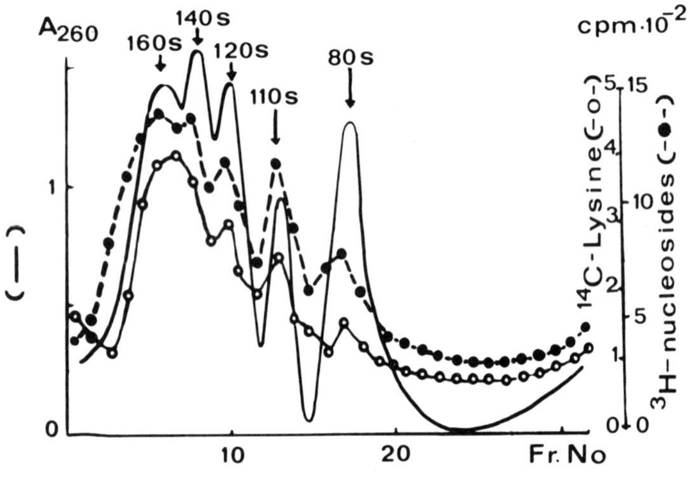

Fig. 4. Sedimentation of purified nucleus-associated polyribosomes to the blastula-stage embryos. The low-salt extract of Fig. 1 was treated with 0.5% Tween-40 and 0.5% DOC and centrifuged through 1.5M sucrose based on a "high-salt" buffer (0.3M KCl, 5 nM $MgCl_2$, 100 U/ml heparin, 25 mM Tris-HCl, pH 7.4) at 105,000 x g and +4°C for 4 h. The pellet was resuspended in the same buffer and layered on a 17-50% sucrose gradient based on the high-salt buffer.

polyribosomes were purified from the post-mitochondrial supernatant by 0.5% Tween-40 plus 0.5% DOC and high-salt medium treatment and sedimentation through 1.5 M sucrose essentially as described (9). Sedimentation of these polyribosomes, which will be referred to as "cytoplasmic" to distinguish them from the nucleus-associated ones, is shown in Fig. 5. Membrane-bound polyribosomes were, on average, heavier than free polysomes and n-polysomes, while free cytoplasmic polyribosomes were similar in size to n-polysomes but did not display comparable amplitudes of UV-absorbing and ^{3}H-uridine labeled peaks.

The amount of n-polysomes in early loach embryos is striking. As shown in Table 1, at the gastrula stage at least a half of the total cell complement of engaged ribosomes can be recovered in the nuclear fraction, the rest being divided unequally between free and membrane-bound polyribosomes. Table 1 also shows the gross distri-

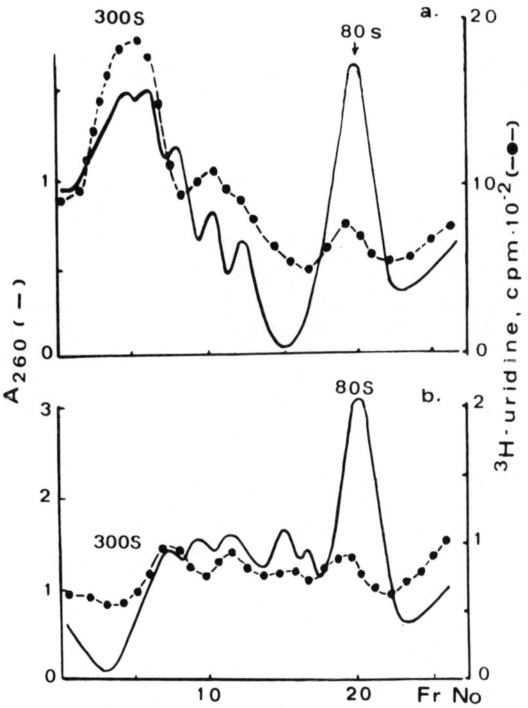

Fig. 5. Sedimentation of membrane-bound (a) and free (b) cytoplasmic polyribosomes from gastrula-stage embryos in 17-50% sucrose gradients. Blastoderms were incubated 60 min with 100 μCi/ml ^{3}H-uridine, and polyribosomes were purified from the post-mitochondrial supernatant as described in the text, and centrifuged as in Fig. 4.

Table 1. Distribution of engaged ribosomes and newly synthesized RNA and protein between particulate fractions of gastrula cells

	Content in the fractions, % of total			
	Free polysomes	Membrane bound polysomes	Nucleus-associated (n-) polysomes	Pure (n-polysome-free) nuclei
Engaged ribosomes	45	5	50	-
^{3}H-RNA	1	3	67	28
^{14}C-protein	13	4	27	55

Gastrula-stage blastoderms were labeled at 15 min with ^{3}H-uridine (50 μCi/ml) and ^{14}C-lysine (7 μCi/ml), after a 15 min pre-incubation with glucosamine (5 μM) to reduce the UTP pool (10), and divided in two parts. From one, cytoplasmic free and membrane-bound polyribosomes were purified as in Fig. 5. From the other part, Ficoll-sucrose nuclei were obtained and low-salt extracted as in Fig. 1 to give pure nuclei and nucleus-associated polyribosomes (n-polysomes). The latter were purified as in Fig. 4. Content of engaged mono- and polyribosomes was determined from UV absorption. For counting ^{3}H-RNA and ^{14}C-protein, aliquots of the particulate fractions were precipitated, respectively, with 10% or 20% cold trichloroacetic acid.

bution of 15-min labelled RNA and protein. Most of the total cell-labelled RNA was found in n-polysome fraction, and very little reached cytoplasmic polyribosomes. A large fraction of newly-made protein was likewise found in the n-polysomal compartment.

The data of this section indicate that cells of early loach embryos contain a large sub-population of light polyribosomes that are tightly associated with the periphery of the nucleus or otherwise strictly confined to the perinuclear compartment of the cell, and incorporate a large part of non-ribosomal RNA and of protein made in the cell.

Our next purpose was to examine some characteristics of the nucleus-associated polyribosomes that could be informative about the functional significance and metabolic properties of these polyribosomes as compared to polyribosomes of usual cytoplasmic localizations. The experiments of the next section are concerned with the nature of proteins and RNAs labeled in n-polysomes in early embryos.

Nucleus-Associated Polyribosomes Represent a Metabolically Distinct Sub-Population of Cell Polyribosomes

Examination of lysine and tryptophan incorporation in polyribosomes of early embryos revealed a high Lys/Trp incorporation ratio in n-polysomes. Fig. 6 shows the result obtained with blastula-stage embryos that have been shown by autoradiography (see Ref. 3) to incorporate lysine but no tryptophan into the nuclear proteins. However, a high Lys/Trp incorporation ratio was also found in n-polysomes of gastrula-stage embryos (Fig. 7) that reportedly incorporate the two amino acids into nuclear proteins (3). These results suggest that n-polysomes were synthesizing histone proteins.

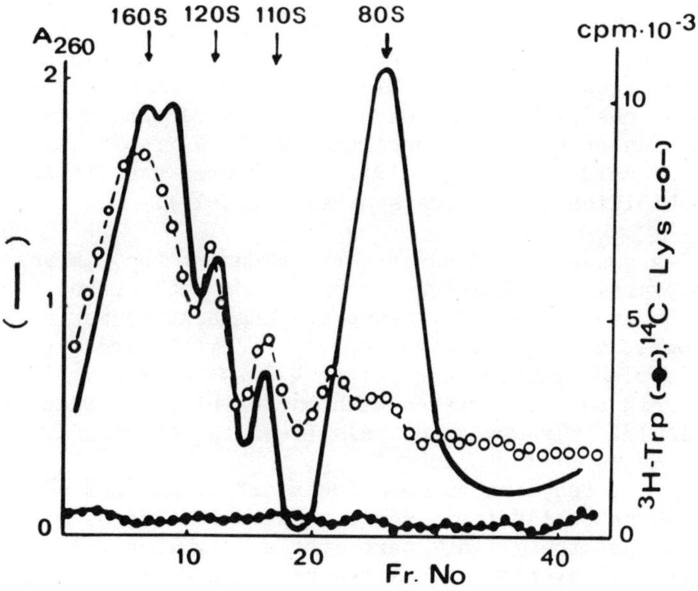

Fig. 6. Incorporation of ^{14}C-lysine and ^{3}H-tryptophan into nucleus-associated polysomes purified as in Fig. 4. Suspension of the blastula-stage embryo cells was incubated for 10 min with 50 μCi/ml ^{3}H-tryptophan (Amersham, 6.4 Ci/mmole) and ^{14}C-lysine (CSSR, 270 mCi/mmole). N-polysomes were isolated from the low-salt extract of nuclei, purified and centrifuged as in Fig. 4.

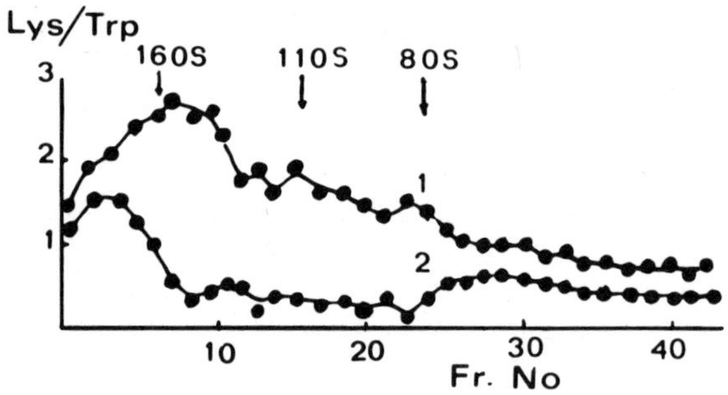

Fig. 7. Effect of hydroxyurea on the Lys/Trp incorporation ratio in sedimentation fractions of purified nucleus-associated polyribosomes of gastrula-stage embryos. Blastoderms have been incubated with ^{3}H-tryptophan and ^{14}C-lysine as in Fig. 6. N-polysomes were isolated, purified, and centrifuged in 17-50% sucrose as in Fig. 4. 1) Control; 2) After 90 min pre-incubation of the blastoderm with 5 mM hydroxyurea.

One way of testing such a proposal is to apply the hydroxyurea test. Hydroxyurea is known to block DNA replication through an inhibition of the ribonucleoside diphosphate reductase system (11). Due to coupling of histone translation with DNA replication in many kinds of animal cells (12-19), hydroxyurea usually leads to a specific inhibition of histone synthesis (13-19).

Fig. 7 shows that hydroxyurea treatment of the embryos caused a marked depression of Lys/Trp incorporation ratio throughout most sedimentation fractions. This supports the histone-making function of n-polysomes, which is further confirmed by the fact that 90% of lysine-labeled protein in n-polysomes was soluble in 5% perchloric acid and co-migrated with marker histones when electrophoresced in 15% polyacrylamide gels (results not shown).

The hydroxyurea test allowed us to detect another sub-population of embryo cell polyribosomes engaged in histone synthesis. Fig. 8 shows that the "light" part of the polyribosomal region is severely affected by hydroxyurea treatment of embryos: a large part of UV absorption was abolished (due to dissociation of polyribosomes that make histones, see Refs. 12,15-17,18,19,etc.), and the relative incorporation of lysine was suppressed. As the light region originates primarily from free polyribosomes (see Fig. 5), this result indicates that in loach embryos, as in sea urchin embryos (8,16) and in other animal cells (12-15,18,19,etc.), histones are synthesized in free polyribosomes. However, in loach embryo cells there is an additional, quantitatively comparable and synthetically more active (see Table 1) group of histone-making polyribosomes that are associated with the nuclei.

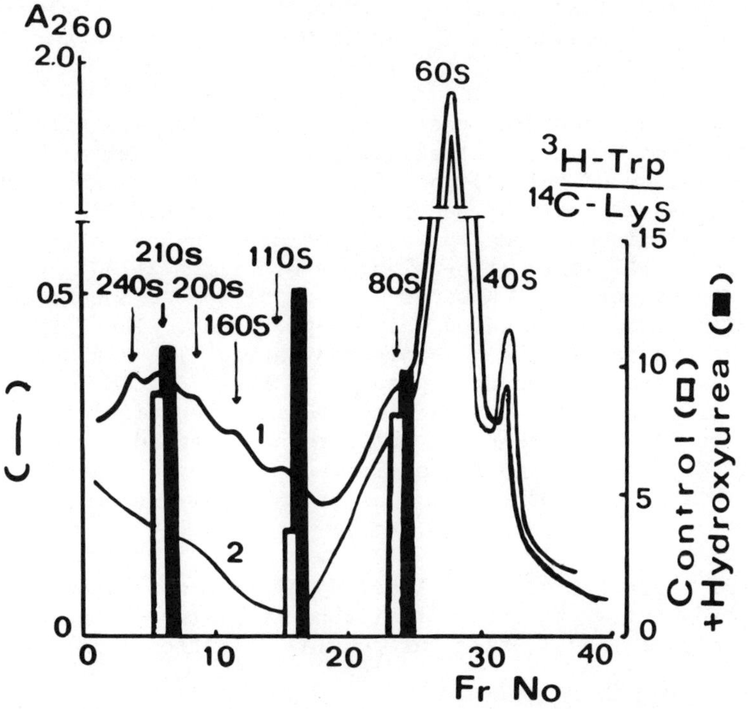

Fig. 8. Effect of hydroxyurea treatment of gastrula-stage embryos on sedimentation pattern and Trp/Lys incorporation ratio of total cytoplasmic polyribosomes. Blastoderms were incubated for 1 h with 50 μCi/ml ^{3}H-tryptophan and 7 μCi/ml ^{14}C-lysine and homogenized in the high-salt-sucrose medium (see the text and Fig. 5). The nuclei and mitochondria were discarded and the supernatant was treated with Tween-40+DOC (0.5% each), and centrifuged in 17-50% sucrose gradients based on the high-salt buffer of Fig. 4. Superimposed are: continuous UV tracing of total cytoplasmic polyribosomes from control embryos (1) and from embryos that have been preincubated with 5 mM hydroxyurea before adding labeled amino acids to the incubation (2). ^{3}H/^{14}C ratio was determined in all sedimentation fractions (cold 20% trichloroacetic acid insoluble material), but only the values representative of heavy and light polyribosome regions, and of engaged monoribosomes, are shown with the bars: white bars, control; black bars, hydroxyurea-treated embryo polyribosomes.

It was of obvious interest to compare mRNAs of the two groups of light polyribosomes, but free polyribosomes incorporated too little RNA label (see Table 1 and Figs. 5 and 10). Labeled RNA of n-polysomes was examined, and displayed features characteristic of histone mRNAs.

RNA was isolated from gastrula-stage n-polysomes with a modification of the Brawerman et al. (20) technique which allows a gross separation of poly(A)-poor from poly(A)-rich RNAs by phenol extraction respectively at neutral or alkaline pH. Table 2 shows that this technique, when applied to membrane-bound polyribosomes or to pure nuclei (feed of n-polysomes by Triton X-100), gave the expected fractionation. However with n-polysomes, both neutral and alkaline phenol extracted poly(A)-poor RNAs. Thus labeled RNA of n-polysomes is largely non-polyadenylated. The RNA extracted from n-polysomes with neutral phenol and freed of tRNA and 5S RNA by 1.5 M NaCl precipitation, sedimented in three distinct peaks, as shown in Fig. 9. Ribosomal RNAs are not labeled, as was known before (1,3), and labeled RNA sedimented in a unique band peaking at 9S. This is the pattern typical of RNA from free polysomes of early sea urchin embryos (8,17,etc.).

These characteristics of labeled RNA from n-polysomes of loach embryos are similar to those of histone templates that are reported to be devoid of long 3'-terminal poly(A) stretches (16,19,21-23,

Table 2. Proportion (%) of poly(U)-sepharose-retained ^{3}H-RNA from gastrula cell fractions

Phenol pH	Membrane-bound polysomes	Nucleus-associated polysomes	Pure nuclei
7.7	13	3	20
9.0	63	9	57

Blastoderms have been incubated 15 min with 100 μCi/ml ^{3}H-uridine. Cytoplasmic polyribosomes were isolated and fractionated as in Fig. 5. Nuclei were isolated with Ficoll-sucrose method, and washed with 1% Triton X-100 to yield n-polysomes and pure nuclei. RNA was extracted from the cell fractions with phenol at pH 7.7 and then again at pH 9.0 (20), reprecipitated with 1.5M NaCl and analysed with a poly(U)-sepharose column (21).

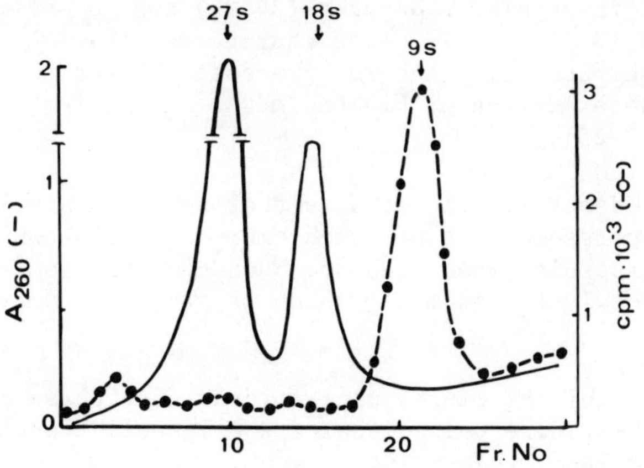

Fig. 9. Sedimentation of 10 min labeled RNA from nucleus-associated polyribosomes of gastrula-stage embryos. The pH 7.7 phenol-extracted n-polysomal RNA of Table 2 was centrifuged in 5-20% sucrose gradient based on 10 mM sodium acetate and 10 mM EDTA (pH 5.0).

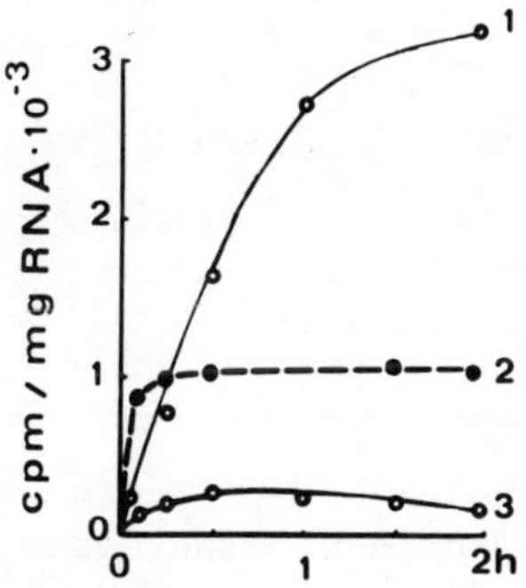

Fig. 10. Kinetics of RNA labeling in different polyribosome fractions of gastrula-stage embryos. Blastoderms have been incubated 15 min with 5 μM glucosamine, then ^{3}H-uridine was added to 50 μCi/ml, and incubation continued. At intervals samples of blastoderms were withdrawn and divided in two parts. From one part, n-polysomes were purified as in Fig. 4. From another part, cytoplasmic free and membrane-bound polyribosomes were purified as in Fig. 5. Total RNA was phenol-extracted from the polyribosomes, reprecipitated with 1.5 M NaCl and specific radioactivity assessed. 1) Membrane-bound polyribosomes; 2) Nucleus-associated polyribosomes; 3) Free polyribosomes.

etc.), except in amphibian oocytes (24,25) and to sediment as a 9S peak (8,12,17,24, etc.). Also characteristic of histone mRNAs are their high rate of synthesis, processing, transport, and decay, again with the exception of histone mRNAs in oocytes and early embryos (16,22-25).

We carried out kinetic studies that revealed similar properties in n-polysomal RNA of loach embryos. As shown in Fig. 10, at a stage (gastrula) when membrane-bound polyribosomes are accumulating newly-formed mRNA (which is about 80% polyadenylated)

Table 3. Effect of the glucosamine-cold uridine chase on the amount of ^{3}H-uridine pulse-labeled RNA in cell fractions of the gastrula-stage embryos

	Membrane-bound polysomes	Nucleus-associated polysomes	Pure nuclei
^{3}H-RNA after 10-min pulse with ^{3}H-uridine	600	27,000	13,000
^{3}H-RNA after 3 h chase with cold uridine	2,700	700	5,800
Percent ^{3}H-RNA left	450	3	44

Blastoderms have been pre-incubated 15 min with 5 mM glucosamine to reduce the UTP pool, the ^{3}H-uridine added to 100 μCi/ml. After 10 min, a sample of blastoderms was withdrawn, and the rest was chased by adding glucosamine to 200 mM and cold uridine to 2 mM and incubation for another 3 h. RNA was extracted from the cell fractions as in Table 2.

and free polyribosomes incorporate labeled RNA at a very low level, newly made RNA appears in n-polysomes at an extremely high rate and in a significant amount (see also Table 1). Specific radioactivity of n-polysomal RNA, however, levels off within a few minutes of incubation of the embryos with the labeled precursor, which suggests a rapid turnover of this RNA. This latter point has been confirmed by pulse-chase experiments. Table 3 shows that while

RNA of membrane-bound polyribosomes continued to accumulate label, thus exhibiting metabolic stability, and nuclear RNA conserved nearly half of original 10-min label, n-polysomes lost virtually all of their labeled RNA. Similar results were obtained in actinomycin D-chase experiments.

The combined evidence of this section indicates that nucleus-associated polyribosomes exist in cells of early loach embryos as a metabolically distinct sub-population of cell polyribosomes.

DISCUSSION

Ribosome or polyribosome material associated with the nuclei is routinely observed and is usually considered as cytoplasmic contaminant that can be removed by washing with detergents (27). In HeLa cells, polyanion-released polyribosomes have been reported (28,29) and shown to accept RNA from nuclei and to incorporate protein label (29). Polyribosomes have also been found in extracts of sea urchin blastula nuclei (30,31) and were originally reported to contain most newly-formed RNA of the nuclear fraction (30) but have not been ascribed any functional significance (31,32).

The data of the present paper appear to be the first demonstration of a functionally meaningful association of a metabolically distinct polyribosome fraction that has been characterized in its newly-formed RNA and protein components. Several lines of evidence show that this n-polysome fraction is engaged in histone synthesis. The size of these polysomes and characteristics of their RNA and newly-formed protein(s) are similar to those of classical light free polyribosomes that have been shown in many kinds of animal cells to be the sole cellular site of histone synthesis (12-19). The light free polyribosomes are also found in loach embryo cells in amounts comparable to those of light nucleus-associated polyribosomes. The two classes of light polyribosomes respond positively to hydroxyurea test but differ in that n-polysomes rapidly incorporate and turn over newly formed mRNA while free polyribosomes do not. This indicates that the former translate mainly newly transcribed templates, and the latter translate preformed templates, suggesting a distinct function for n-polysomes in view of possible developmental changes in subsets of synthesized histones.

There is an evident implication of these results on the problem of timing of nucleus-to-cytoplasm transport of genetic information in early embryogenesis. Mid-blastula was considered, in loach, as a stage where no newly synthesized nuclear RNA was transferred to the cytoplasm (33,34,etc.; reviewed in 2,3). However our present data reveal the existence, at this stage, of an efficient transfer of newly transcribed RNAs to a large population of ribosomes which reside in the immediate vicinity of the nucleus. This transfer, which we propose to term "short-range transport" to distinguish it from the ordinary transport to typical cytoplasmic locations (membrane-bound and free polyribosomes), apparently concerns mainly histone mRNAs, and seems to have gone undetected in earlier works in which the nuclear compartment has not been examined in sufficient detail. As to the transport to usual cytoplasmic locations, our data agree with previous ones (33,34): we did not observe any significant incorporation of labeled RNA into purified membrane-bound or free polyribosomes at the blastula stage and did observe the incorporation into membrane-bound polyribosomes at the gastrula stage (in addition to persistent short-range transport of histone mRNAs).

The kinetics of RNA labeling in nucleus-associated polyribosomes suggests that on their way to ribosomes, histone mRNAs are not subject to any significant delay similar to what has been observed with HeLa cells (15) and sea urchin embryos (17) in which, however, histones are reportedly synthesized in free polyribosomes. Newly made histones are also known to be transferred from the polyribosomes to chromatin with great speed and efficiency (35). These facts are difficult to reconcile with the notion of free, extranuclearly-located histone-making polyribosomes. With the extranuclear localization of these polyribosomes, it also seems difficult to solve the puzzle of the well-known functional coupling of histone mRNA translation and integrity with DNA replication in most animal cells (12-19,39,40) except oocytes and early cleavage blastomers (36,37). The attempts to find diffusible cytoplasmic or nuclear factors which could mediate this coupling have as yet been unsuccessful (38,39). Models for the coupling mechanism have been proposed (39-41) but the possibility of a spatial association of the biosynthetic systems involved has not been considered. We propose here a model for a nuclear envelope-associated chromatin-reproducing system which involves intimate interactions between its components and a non-diffusion transfer of transcription and translation products across the nuclear membranes.

Our hypothesis states that the replication-coupled histone synthesis occurs in polyribosomes which are more or less stably associated with the nuclear envelope. This association occurs due to the affinity of newly synthesized and/or nascent histone molecules to that one of the two newly formed daughter DNA strands which, according to the conservative mode of histone segregation during chromosome replication (42,43), is deficient in histones. The concentration of basic amino acid residues in N-terminal regions of histone molecules (except H1) (44) favors the possibility of the DNA-nascent histone interaction. The proposal about the interactions between the newly formed and/or nascent histone molecules and the newly replicated DNA regions is supported by the association of the DNA replication complex with the nuclear membrane(s) in cultured mammalian cells (45-47) and in the sea urchin embryos (48). The interactions may take place either at nuclear pores or directly through the nuclear membrane(s). The latter possibility does not seem unlikely in view of the properties of biological membranes.

The proposed DNA-histone interaction will occur at each replication site; hence the number of histone-making polyribosomes associated with the nucleus at a given moment will depend on the number of replicons acting in concert. Early embryo cells are known to have relatively short S-phase due to a shortened length and increased synchrony of replicons (49), and it is noteworthy that nucleus-associated polyribosomes are found in loach, and presumably in sea urchin (30) embryos. Demonstration of these polyribosomes in other objects will depend on the cell type and techniques used.

Our hypothesis states furthermore that newly transcribed histone messenger sequences are accepted by the perinuclearly located ribosomes immediately from their sites of transcription and processing. The kinetics of n-polysome RNA labeling suggests that the "short-range" transport might even operate with nascent or incompletely processed molecules of histone messenger precursor, so that the endonuclease cleavage of polycistronic pre-mRNA (50) sequence occurs when its 5'-terminal region has already formed the initiation complex with a ribosome. This coupled processing and short-range transfer of histone mRNA presumably occurs at those nuclear pores where histone gene clusters are situated. The possibility of existence of gene-specific nuclear pore-associated chromatin strands which provide for transcription, processing, and conveying the prospective messengers has been discussed by Lichtenstein and Shapot (51). Since there must be many more replicons acting at a time than histone gene-associated nuclear pores, we postulate that after having accepted the histone templates, the growing polyribosomes migrate along the nuclear periphery to those sites which oppose the sites where DNA is being replicated. This migration is directed by the affinity of the growing and/or newly made but still ribosome-attached histone molecules to histone-free DNA regions. Polyribosomes which migrate out from the pore make place for new ribosomes to accept new template molecules.

The proposed model accounts for observations which led to the view (39) that DNA replication plays a role in translation of histone mRNA. In addition to the normally observed coordination between histone and DNA synthesis during the cell cycle (12,19,etc.), an increase in the quantity of histone-deficient DNA elicited by a transient inhibition of protein synthesis leads to an increased histone synthesis (39,40). Instead, inhibition of DNA replication leads to a rapid blocking of histone synthesis, in fact much more rapid then after an inhibition of RNA transcription (12-19,39,47, etc.). Our hypothetical mechanism explains these observations by assuming that naked DNA serves as an immediate acceptor of newly made and/or nascent histone molecules, in addition to the earlier proposed product inhibition of histone synthesizing polyribosomes by accumulating histone molecules (39).

The proposed topology of the chromatin-reproducing system can also account for the kinetics of newly formed histone mRNA incorporation into the polyribosomes (15 and Fig. 10), which is presumed to occur in a non-diffusion manner as described. It can also account for the fact that blocking DNA replication brings about not only shut-off of histone synthesis but also dissociation of the histone-making polyribosomes and decay of histone mRNAs (12-19,47). We propose that histone mRNAs are stabilized when they are included in polyribosomes associated with the nucleus, but are rapidly destroyed when in the free polyribosome and free RNP state, except special cases of polyadenylated histone mRNAs (24,25). Blocking DNA replication causes detachment of polyribosomes that were attached to the nuclear envelope by histone-DNA interaction. This will also bring about a halt of histone polyribosome flow from the sites where they were engaged with histone mRNAs, which now will pass into the free informosome fraction where they are rapidly degraded. This view is consistent with the data (40) on accumulation of histone RNA in 40S RNPs of HeLa cells where initiation of protein synthesis had been inhibited. In loach embryo cells there are free RNP particles in the perinuclear compartment (Figs. 1 and 2), and their RNA was shown in preliminary experiments to be similar to n-polysomal RNA in its low proportion of polyadenylated molecules and in the pulse-chase behavior. Normally, the informosomal RNA makes a minor fraction of newly synthesized RNA of the nuclear wash. The model predicts that this fraction will increase when DNA synthesis is blocked under conditions of inhibited engagement and/or initiation of translation of histone messengers.

The model proposed in the present paper leads to a number of other predictions amenable to experimental testing that could eventually lead to a better understanding of the molecular and topological bases of the control of chromosome reproduction.

REFERENCES

1. Kafiani, C.A., Advan. Morphogenesis, 8, 209, 1970.
2. Neyfakh, A.A., Curr. Topics Devel. Biol., 6, 45, 1971.
3. Kafiani, A.A., and Kostomarova, A.A. Informational Macromolecules in Early Development of Animals, "Nauka", Moscow, 1978.
4. Spirin, A.S., Europ. J. Biochem., 10, 20, 1969.
5. Neyfakh, A.A., Zh. Obsch. Biol., 20, 203, 1959.
6. Ajtkhoghin, M.A., Belitsina, N.V. and Spirin, A.S., Biokhimia, 23, 169, 1964.

7. Spirin, A.S., Belitsina, N.V. and Lerman, M.I., J. Mol. Biol., 14, 611, 1965.
8. Nemer, M. and Lindsay, D.T., Biochem. Biophys. Res. Commun., 35, 156, 1969.
9. Uenoyama, K., Ono, T., Biochim. Biophys. Acta, 281, 125, 1972.
10. Scholtissek, C., Eur. J. Biochem., 24, 358, 1971.
11. Yeh, Y.-C., and Tessman, I., J. Biol. Chem., 253, 1323, 1978.
12. Robbins, E., and Borun, T.W., Proc. Nat. Acad. Sci. U.S.A., 57, 409, 1967.
13. Borun, T.W., Scharff, M.D., and Robbins, E., Proc. Nat. Acad. Sci. U.S.A., 58, 1977, 1967.
14. Pueyo, M.T., Bonaldo, M.F., and Lara, T.J.S., Cell. Differ., 4, 257, 1975.
15. Schoechteman, G., and Perry, R.P., J. Mol. Biol., 63, 591, 1972.
16. Zauderer, M., Liberti, P., and Baglioni, C., J. Mol. Biol., 79, 577, 1973.
17. Kedes, L.H., and Gross, P.R., Nature, 233, 1335, 1969.
18. Perry, R.P., and Kelley, D.E., J. Mol. Biol., 35, 37, 1968.
19. Gallwitz, D., and Mueller, G.C., J. Biol. Chem., 244, 5947, 1969.
20. Bramerman, G., Mendecki, J., and Lee, S.Y., Biochemistry, 11, 637, 1972.
21. Adesnik, M., Salditt, M., Thomas W., and Darnell, J.E., J. Mol. Biol., 71, 21, 1972.
22. Adesnik, M., and Darnell, J.E., J. Mol. Biol., 67, 397, 1972.
23. Wilson, M.C., and Melli, M.L., J. Mol. Biol., 110, 511, 1977.
24. Levenson, R.G., and Marku, K.B., Cell, 9, 311, 1976.
25. Ruderman, J.V., and Pardue, M.L., J. Cell Biol., 70, 89, 1976.
26. Gross, K.W., Jacobs-Lorena, M., Baglioni, C., and Gross, P.R., Proc. Nat. Acad. Sci. U.S.A., 70, 2614, 1973.
27. Penman, S., Smith, I., and Holtzman, E., Science, 154, 786, 1966.
28. Bach, M.K., and Johnson, H.G., Biochemistry, 6, 1916, 1967.
29. Goidl, J.A., Canaani, D., Boublik, M., Weissbach, H., and Dickerman, H., J. Biol. Chem., 250, 9198, 1975.
30. Aronson, A.I., and Witl, F.H., Proc. Nat. Acad. Sci. U.S.A., 62, 186, 1969.
31. Levner, M.H., Fromson, D., Ricklis, S., Graham, M., and Nemer, M., Arch. Biochem. Biophys., 169, 638, 1975.
32. Aronson, A.I., Wilt, F.H., and Wartiovaara, J., Exper. Cell Res., 72, 309, 1972.
33. Spirin, A.S., Belitsina, N.V., and Ajtkhoghin, M.A., Zh. Obshch. Biol., 25, 321, 1964.
34. Neyfakh, A.A., Kostomarova, A.A., and Burakova, T.A., Ontogenesis (USSR), 4, 331, 1973.

35. Jackson, V., Shires, A., Tanphaichitr, N., and Chalkley, R., J. Mol. Biol., 104, 471, 1976.
36. Arceci, R.J., and Gross, P.R., Proc. Nat. Acad. Sci. U.S.A., 74, 5016, 1977.
37. Adamson, E.O., and Woodland, H.R., Devel. Biol., 57, 136, 1977.
38. Pedersen, T., and Robbins, E., J. Cell Biol., 45, 509, 1970.
39. Butler, W.B., and Mueller, G.C., Biochim. Biophys. Acta, 294, 481, 1973.
40. Stahl, H., and Gallwitz, D., Eur. J. Biochem., 72, 385, 1977.
41. Weintraub, R., Cold Spring Harbor Symp. Quant. Biol., 38, 247, 1974.
42. Seale, R.L., Cell, 9, 423, 1976.
43. Weintraub, H., Cell, 9, 419, 1976.
44. DeLange, R.J., and Smith, E.L., Acc. Chem. Res., 5, 368, 1972.
45. Binkerd, P., and Toliver, A., Mol. Cell. Biochem., 5, 177, 1974.
46. Infante, A.A., Firstein, W., Hobart, P., and Murray, L., Biochemistry, 15, 4810, 1976.
47. Yoshida, S., and Cavalieri, L.F., Biochim. Biophys. Acta, 475, 42, 1977.
48. Hobard, P., Duncan, R., and Infante, A.A., Nature, 267, 542, 1977.
49. Callan, H.G., Proc. Roy. Soc. B 181, 19, 1972.
50. Melli, M.-L., Spinelli, G., Wyssling, H., and Arnold, E., Cell, 11, 651, 1977.
51. Lichtenstein, A.V., and Shapot, V.S., Biochem. J., 159, 783, 1976.

NUCLEIC ACIDS, HISTONES AND SPERMIOGENESIS:

THE POLY(ADENOSINE DIPHOSPHATE RIBOSE)POLYMERASE SYSTEM

Benedetta Farina, Maria Rosaria Faraone Mennella, and Enzo Leone

Istituto di Chimica Organica e Biologica
Facoltà di Scienze, Università di Napoli

The enzyme poly(adenosine diphosphate ribose)polymerase was discovered in chicken liver nuclei in 1966 (Chambon et al., 1) and thoroughly studied in a number of laboratories, where its main properties have been investigated; major contributions in this regard have been given by the groups of Hayaishi (2) and Sugimura (3). The enzyme has many peculiar properties.

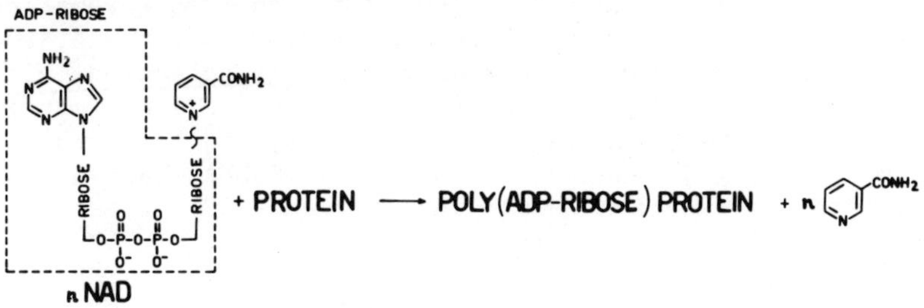

Poly (ADPR) polymerase reaction

It catalyzes the splitting of an NAD^+ molecule at the nicotinamide-ribose bond; the adenosine diphosphate ribose (ADPR) unit thus set free is then linked to a protein, most frequently a histone, and other ADPR residues are attached, end-to-end, to the first (protein-bound) residue so that a linear homopolymer is eventually formed which may comprise up to 30 ADPR residues (4). In the polymer, one ADPR unit is linked to the next by a 1'-2' glycosidic bond; the polymer structure has been confirmed by the use of specific enzymes, a poly(ADPR)glycohydrolase and a poly-

(ADPR)phosphodiesterase. The former enzyme carries out the hydrolysis of the ribose-ribose bond of poly(ADPR), and the reaction products are ADPR and oligo(ADPR) (5); the latter enzyme, which may act either by an exo- (6) or by an endo-nucleolytic mechanism (7), sets free phosphoribosyl-AMP residues, and one AMP residue per chain of polymer.

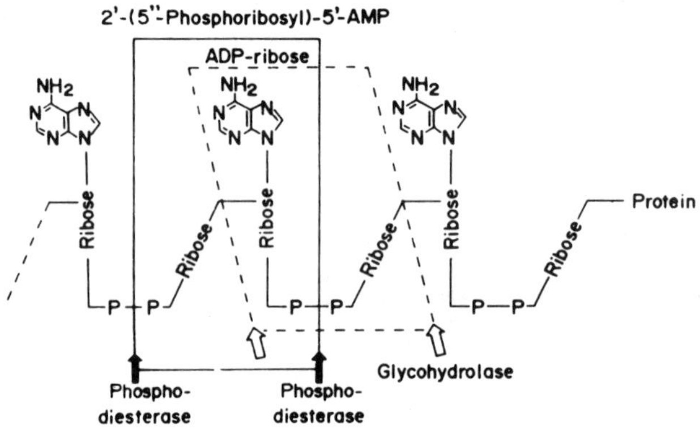

Enzymic cleavage of poly(ADPR)

Other characteristic properties of the enzyme are its endocellular localization and its dependence on DNA. In fact, poly (ADPR)polymerase activity is almost always associated with chromatin, and the enzyme can be isolated either as chromatin-bound (2), or after solubilization therefrom (8,9). As for DNA, there appears to be an absolute requirement for this polyanion, although no base sequence specificity seems to be required since single and double stranded DNAs from eukariotes, phages, and bacteria, as well as synthetic deoxyribonucleotide polymers and DNA-RNA hybrids, are effective in the reaction in varying degrees (8,9). The exact role of DNA still remains to be clarified.

Mg^{2+} ions are activators of the reaction. It has been shown by Stone & Shall (10) that the optimum Mg^{2+} concentration is a function of NAD^+ concentration, thus supporting the view that an NAD^+-Mg^{2+} complex, rather than NAD^+ itself, is the real substrate. It has also been proposed (8) that Mg^{2+} is an absolute requirement for enzyme activity.

Nicotinamide, which is a well known inhibitor of NAD^+-glycohydrolase (NADase, 11), is also an inhibitor for poly (ADPR) polymerase (2).

Histones appear to be the natural acceptor proteins for ADPR residues. There is now ample experimental evidence that histone

H1 is preferentially ADP-ribosylated. Smith & Stocken (12) have found that in rat liver nuclei, histone H1 is linked at a seryl residue with ADPR by an alkali-labile bond; Ueda et al. (13) have also reported histone H1 to be ADP-ribosylated in rat liver nuclei together with, to a lesser extent, histones H2 and H3. In HeLa cell nuclei also histone H1 is the natural acceptor for poly(ADPR) (14), and in these nuclei the formation of a complex has been observed which consists of an H1 dimer where two H1 molecules are linked by one chain of approximately 15 ADPR units (15). Wong et al. (16) have described attachment of poly(ADPR) in trout testis nuclei to histone H1 and also to histone H6 (histone T, a minor histone characteristic of trout testis, 17) and protamines.

It appears, therefore, that a novel role for NAD^+ has been revealed with the identification of the poly(ADPR)polymerase system. This possibility had already been implied, in fact, by some observations made in the past few years. Gholson (18) in 1966 had postulated an unknown function of NAD^+ in cellular metabolism during the main events of the cell cycle, and Rechsteiner et al. (19) have shown the nucleus in HeLa cells to be the site for rapid turnover of NAD^+, with this turnover about 20 times higher than that required to maintain the NAD^+ pool size during growth. The importance of poly(ADPR)polymerase in living cells on the other hand, has been confirmed by the occurrence of ADP-ribosylation of proteins *in vivo* (20,21). However, no unequivocal physiological function has been brought to light until now; some interesting examples of the most significant results are worth mentioning. Fragment A of diphtheria toxin, a protein with a molecular weight of 24,000 daltons, has been shown to be active in carrying out the ADP-ribosylation of Elongation Factor 2 (EF-2) protein, thereby producing a block at an elongation step of polypeptide assembly in the course of protein synthesis (22); the same mechanism has also been shown to be responsible for inhibition of protein synthesis in the case of *Pseudomonas aeruginosa* toxin (23). RNA polymerase in T-4 phage infected *E. coli* has been found to be modified by covalent attachment of ADPR to a specific arginine residue in the 40,000 dalton α-polypeptides (24). This last observation points to a function of poly(ADPR)polymerase in the control of gene expression. Results with the same implication, although from different experiments, have been obtained in studies on the differentiation mechanism of mesodermal cells of embryonic chick limbs; these cells can differentiate into either muscle or cartilage, and it was found that high NAD^+ levels are correlated with myogenic expression, whereas low NAD^+ levels are correlated with chondrogenic expression (25). The isolated chromatin from mesodermal cells was found to have poly(ADPR)polymerase activity, and the rate of net synthesis of poly(ADPR) was correlated with differentiation of chondrogenic cells from mesodermal cells *in vitro*. These and other observations have been interpreted as the poly(ADPR)polymerase system being a means of communicating, under the influence of fluctuations in cellular NAD^+ levels, with the genomic machinery, leading to differential phenotypic expression.

In still other investigations ADP-ribosylation of nuclear proteins appears to be involved in DNA synthesis (26,27), DNA repair (28), chromatin structure modification (29), and cell proliferation (30). Yoshihara et al. have described ADP-ribosylation of a (Ca^{2+}, Mg^{2+})-dependent nuclear endonuclease with resulting inhibition of the nuclease activity (31).

The process of spermiogenesis has been found to be closely connected with phosphorylation reactions. More precisely, it has been seen that in salmonids and related fish a complex mechanism of phosphorylation-dephosphorylation occurs whereby protamines, soon after their synthesis at the spermatid stage when they penetrate the nuclear membrane to substitute for the histones, are phosphorylated. In this phosphorylated state they bind to chromatin, an event which leads to the chromatin initial condensation at the middle spermatid stage. A sequential release of histone, probably accompanied by histone acetylation in the late spermatid stage, completes chromatin condensation; the final formation of sperm head nucleoprotamine, on the other hand, is closely linked to a controlled dephosphorylation (32-34).

The nature of protamine (and histone) phosphorylation such as involved in the phenomena just described has not been fully clarified, although the kinase-mediated mechanism of phosphorylation has received most attention and has been shown to be active in many systems. However, phosphorylation by the poly(ADPR)polymerase system may represent an alternative mechanism, and results are already available (as already quoted, 16) which show histones H1 and H6 and protamines in trout testis to be ADP-ribosylated; the obvious hypothesis has been advanced that such ADP-ribosylation is involved in the mechanism of chromosomal condensation occurring in spermiogenesis.

On the basis of this situation, we have directed our interest to a search for poly(ADPR)polymerase activity in testis cell nuclei of the bull and other mammals. Such an interest has been derived from a previous research on NAD^+-glycohydrolase (Leone and Bonaduce, 35) which had shown a high activity of that enzyme in bull seminal vesicles as well as in human seminal plasma (36); although there are differences between NAD^+-glycohydrolase and poly(ADPR)polymerase, the first attack on the NAD^+ molecule is common to both, and the possibility of the two enzymes having more than a casual similarity has in fact been taken into consideration (37). Furthermore, it has been found that epididymal spermatozoa, both from the bull and the rabbit, have a markedly higher content of NAD^+ than ejaculated spermatozoa (38), a very short incubation of epididymal spermatozoa being capable of reducing the NAD^+ content to the "ejaculated" level. Quite recently, a preliminary study on bull spermatozoa has shown them to possess poly(ADPR)polymerase activity; this is rather low, and seminal plasma has an inhibitory effect (39).

Poly(ADPR)polymerase Assay

This is based on the determination of TCA-insoluble radio-acivity after incubation of enzyme with ^{14}C-NAD^+. The assay contains, in a final volume of 125 μl, 90 mM Tris-HCl, pH 8.0, 5 mM NaF, 2 mM dithiothreitol, 8 mM Mg^{2+}, 39 picomoles of nicotinamide (U-^{14}C)adenine dinucleotide (from the Radiochemical Centre, Amersham, England; specific activity 266 mC/mmole) and enzyme, 50 to 100 μg protein. After 5 min at 25°C one adds to the mixture an equal volume of cold 20% TCA; the precipitate is taken to a Millipore filter (HAWP 00010) and washed with 20-30 volumes of 5% TCA. Radioactivity is counted with a Beckman LS-133 liquid scintillation counter.

One unit of enzyme corresponds to the incorporation of one μmole NAD^+ into acid-insoluble material in one minute at 25°C; for convenience, activity is expressed in milliunits.

Poly(ADPR)polymerase in Testis

We have found bull testis to have a high enzyme activity, the highest among the different species we have examined (Table I). This activity is confined to the nuclear fraction and has been found for every species to be higher than the activity of other tissues like liver or thymus, which are known for their high enzyme levels. When enzyme activity is expressed on the basis of DNA content rather than protein content, a more homogeneous pattern results (see columns 5 and 6 of Table I).

Poly(ADPR)polymerase Purification

Given the high activity of bull testis, we have worked out a purification procedure from this source which, although it doesn't yield a homogeneous enzyme preparation and is still liable to further modifications, permits one to reach a satisfactory specific activity, at a level seldom reported for any other purification procedure from other sources (Table II). The purification consists of a previous homogenization of testis tissue in a Waring Blendor with 3 volumes of 20 mM Tris-HCl, pH 7.6, containing 100 mM NaCl, 5 mM NaF and 10 mM β-mercaptoethanol; the homogenate is filtered through many layers of gauze, then centrifuged at 13,000 g for 30 min; the precipitate is washed with the same solution as above, centrifuged again, and the final pellet is extracted, by homogenizing with the Waring Blendor, with 4 volumes of the same solution as before, the NaCl concentration of which is raised to

Table I

Poly(ADPR)polymerase in Male Gonads from Various Species

	Protein (mg/ml)	DNA (mg/ml)	DNA/ protein	mU/ml	mU/mg protein	mU/mg DNA
BULL	25	5.5	0.22	30.0	1.20	5.4
HORSE	37	2.7	0.07	9.6	0.26	3.5
RABBIT	19	2.2	0.11	4.6	0.24	2.1
MOUSE	27	4.0	0.14	8.1	0.30	2.0
RAT	65	7.5	0.11	12.0	0.18	1.6
LACERTA SICULA	40	4.1	0.10	9.0	0.22	2.2
SPHAERECHINUS GRAN.	96	2.2	0.02	6.7	0.07	3.0

Table II

Purification of Poly(ADPR)polymerase from Bull Testis

	mg protein	mUnits	Specific activity (mU/mg)	Purification
HOMOGENATES*	535	650	1	1
NaCl EXTRACT**	153	650	4	4
AMMONIUM SULFATE				
75% SATUR. SUPERNATANT	25	480	19	19
CELLEX-P ELUATE	0.5	400	800	800

*5 g tissue in Tris-HCl 20 mM pH 7.4, NaCl 100 mM.
**Tris-HCl 20 mM pH 7.4, NaCl 1 M.

1 M. The extract, after centrifugation and dialysis, is brought to 75% ammonium sulfate saturation. After centrifugation, the supernatant is taken, after dialysis, to a column of Cellex-P, equilibrated with 20 mM K_2MPO_4, pH 8.4, containing 5 mM NaF and

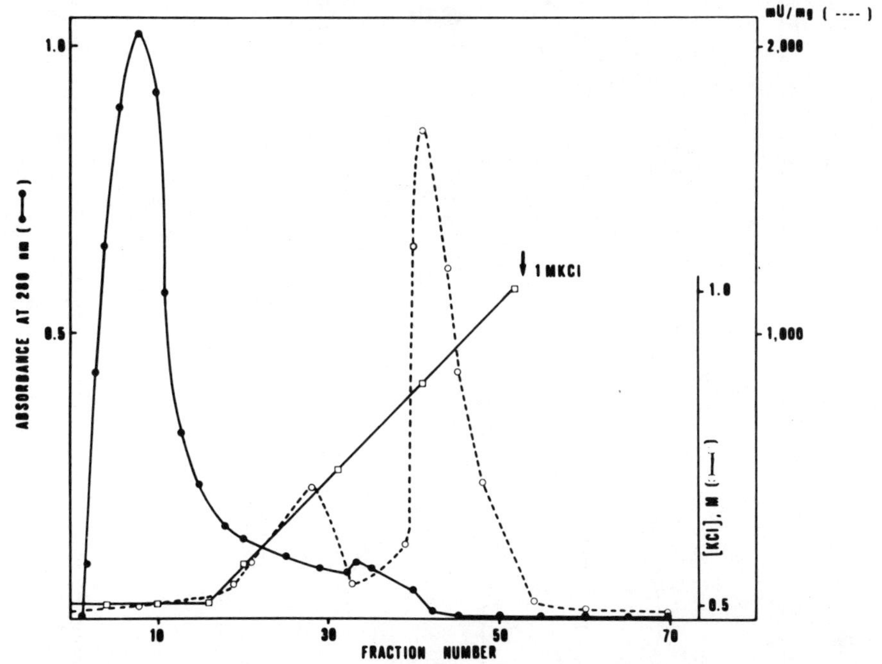

Fig. 1. Cellex-P column (cm 4x2.9) chromatography of 75% saturated ammonium sulfate supernatant (dialysed) from 1M NaCl extract of bull testes homogenate. Load, 25 mg protein (480 mU, specific activity 19). Column equilibrated with 20 mM K_2HPO_4, pH 8.4, containing 5 mM NaF and 10 mM β-mercaptoethanol. Elution by same solution, with a KCl gradient from 0.5 to 1.0 M. Flow rate, 72 ml/hr, 6 ml/fraction.

10 mM β-mercaptoethanol. Elution is performed by the same solution, with a KCl gradient from 0.5 to 1.0 M. The enzyme is eluted in a peak which contains very little protein (Fig. 1) thereby achieving an efficient purification.

An alternative procedure, which also can give satisfactory results, consists of BioGel A-5 chromatography of the NaCl extract, without ammonium sulfate precipitation. As it appears in Fig. 2, in this case also, enzyme is eluted with very little protein, thus effecting a useful purification step to which further steps can be added, including the Cellex-P column chromatography just described.

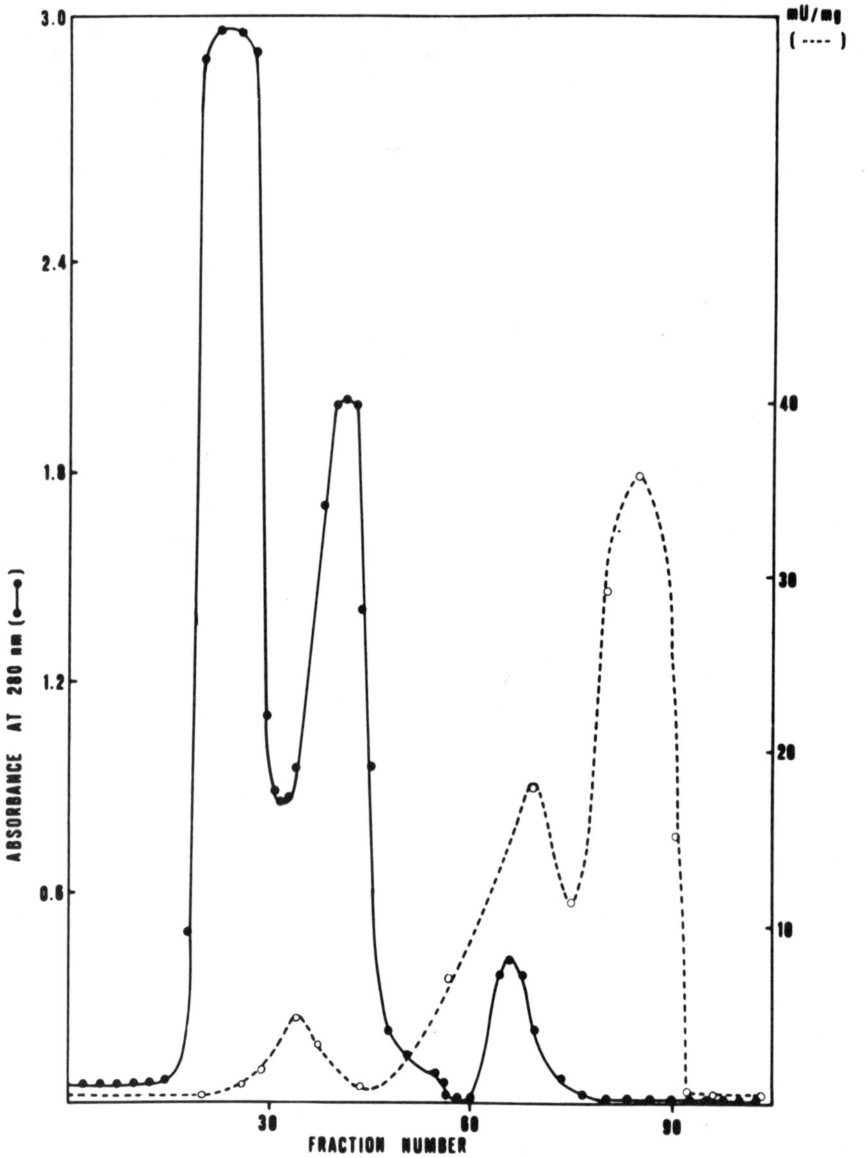

Fig. 2. BioGel A-5m column (cm 40 x 1.8) chromatography of 1 M NaCl extract from bull testes homogenate. Load, 20 mg protein (80 mU, specific activity 4). Column equilibrated with 20 mM Tris-HCl, pH 8.4, containing 500 mM KCl, 5 mM NaF, and 10 mM β-mercaptoethanol. Elution by same solution. Flow rate, 10 ml/hr, 1.1 ml/fraction.

TABLE III

Comparison Between Published Purification Methods for Poly(ADPR)polymerase and the Present Work

Reference	Source	Specific Source	Activity of Final Prep.	Yield, %	Purification -FOLD
Yamada et al. (8)	Rat Liver Nuclei	0.77	8.6	15	11
Ueda et al. (40)	Rat Liver Nuclei	0.3	1,650	20	5,500
Yoshihara et al. (41)	Calf Thymus	0.14	180	20	1,300
Mandel et al. (42)	Calf Thymus	0.12	363	6	3,111
Present Work	Bull Testes	1.0	800	61	800

Our purification procedure, as already stated, gives an enzyme preparation which represents an 800-fold purification over the starting bull testes homogenate, with a specific activity of 800. Both values compare well with the values reported for other purification procedures, as appears in Table III.

Kinetic Parameters

We have determined K_M values both for bull testis and rat testis enzyme; the values obtained are very close, i.e. 2.3×10^{-4}M for the bull testis and 3.5×10^{-4}M for the rat testis enzyme. These values are in agreement with the values known for enzymes from other tissues which range from 8.5×10^{-4}M (43) or 2.5×10^{-4}M (34) for rat liver nuclei, to 3.3×10^{-4}M for pigeon heart nuclei (44), 2.7×10^{-4}M for Ehrlich ascites-cell nuclei (45), 2.8×10^{-4}M for Physarum polycephalum nuclei (46), down to 0.8×10^{-4}M for the purified calf thymus enzyme (42). It is interesting to note that

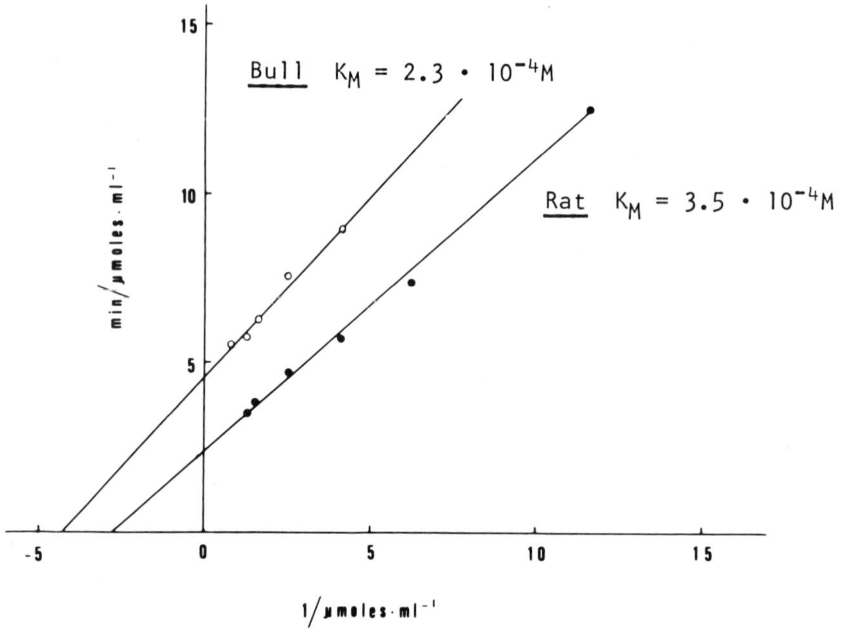

Fig. 3. Lineweaver-Burk plot of poly(ADPR)polymerase. Enzyme source was a 1 M NaCl extract of testis homogenate (bull enzyme, 80 µg protein, 0.12 mU; rat enzyme, 16 µg protein, 0.06 mU). NAD^+ was varied between 15 and 150 nanomoles per assay, to which 50 to 500 picomoles ^{14}C-NAD^+ had been added. Each assay contained 90 mM Tris-HCl, pH 8.0, 5 mM NaF, 15 mM Mg^{2+}, 2 mM dithiothreitol, in a final volume of 125 µl. Incubation at 25°C for 5 min.

most of these K_M values are in the physiological range of NAD^+ concentrations.

Activation by Divalent Cations

It has already been pointed out that Mg^{2+} ions are effectors of the poly (ADPR)polymerase reaction and that a possible complex between NAD^+ and Mg^{2+} has been suggested to be the substrate in that reaction (10). We have analyzed the effect of Mg^{2+} ions at various concentrations, and have found 15 mM Mg^{2+} to give maximum activation both for the rat testis and the bull testis enzyme (Fig. 4). Higher concentrations have less effect, and inhibition occurs at much higher concentrations (i.e. 70–80 mM).

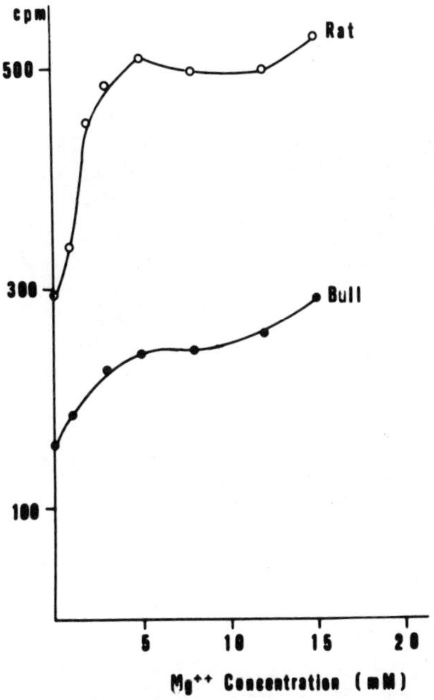

Fig. 4. Poly(ADPR)polymerase activation by Mg^{2+}. Enzyme used in these experiments was a 1.0 M NaCl extract of bull testis homogenate, 70 μg protein (0.126 mU). Assay was performed in the presence of 30 picomoles $^{14}C\text{-}NAD^+$ (25,000 cpm), 90 mM Tris-HCl, pH 8.0, 5 mM NaF and 2 mM dithiothreitol, plus varying amounts of a 0.4 M $MgCl_2$ solution, in a final volume of 125 μl. Incubation at 25°C for 5 min.

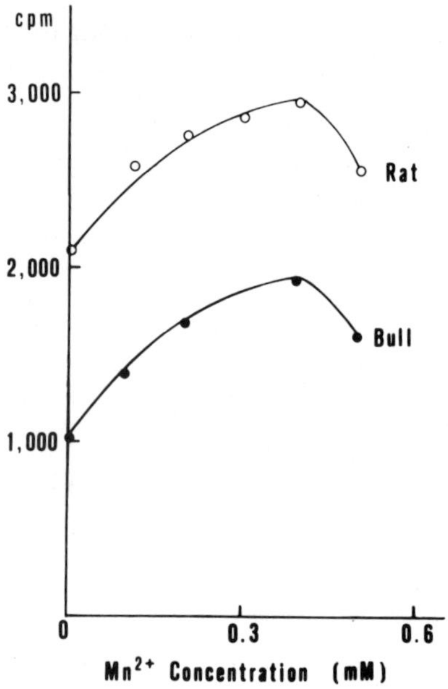

Fig. 5. Poly(ADPR)polymerase activation by Mn^{2+}. Same conditions as in Fig. 4; Mn^{2+} additions were made from a 0.4 M $MnCl_2$ solution.

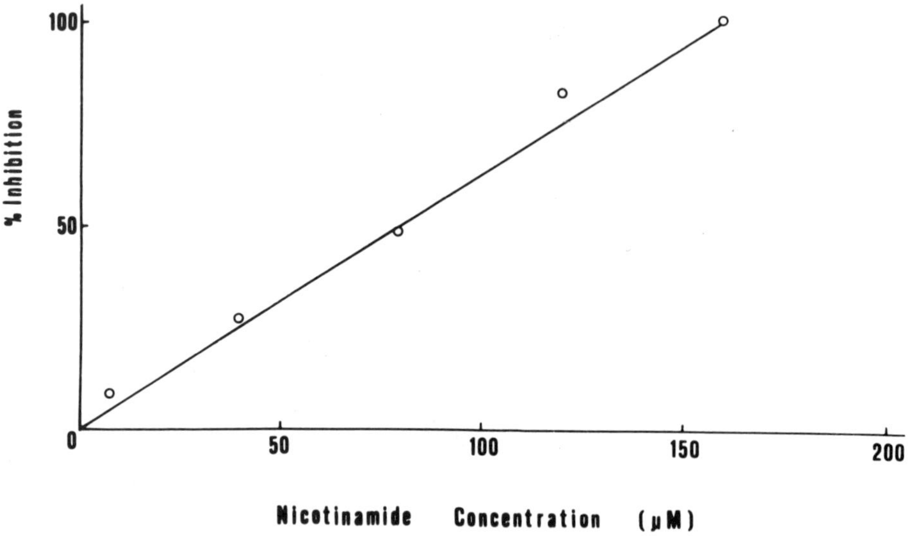

Fig. 6. Nicotinamide inhibition of poly(ADPR)polymerase. Same conditions as in Fig. 4, plus nicotinamide (from 0 to 160 μM).

A similar behavior is shown by Mn^{2+} ions; the activation produced by these ions is maximal at about 0.4 mM concentration, a value far lower than the optimum concentration of Mg^{2+}, but the magnitude of the activation produced is definitely lower too (Fig. 5).

Nicotinamide Inhibition

As it appears in Fig. 6, nicotinamide is a strong inhibitor of the enzyme. At 80 μM concentration a 50% inhibition is produced.

Reaction Rate as a Function of Time and of Salt Concentration

Increasing the incubation time leads, after a maximum value is obtained in about 30 min, to a progressive decline (Fig. 7). This behavior is to be connected with the presence of enzymes, like ADPR-glycohydrolase and ADPR-phosphodiesterase, in the crude preparations used for these assays, which on protracted incubation are allowed to hydrolyze the poly(ADPR) just formed.

When the enzyme is assayed in the presence of NaCl, the activity decreases as the salt concentration is increased beyond 0.1 M. At 0.5 M NaCl only 10% activity is left (Fig. 8). If the enzyme is kept in NaCl up to 2 M however, no detectable loss of activity occurs, provided, of course, the salt concentration is

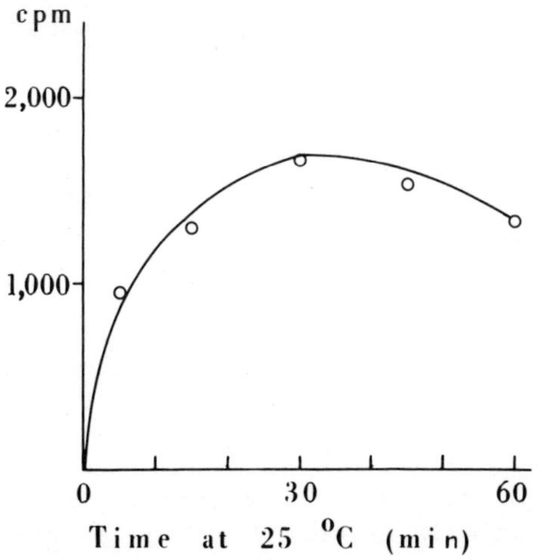

Fig. 7. Time-course of poly(ADPR)polymerase reaction. Same conditions as in Fig. 4, with incubation times as indicated.

adjusted when proceeding to the enzyme assay. The inhibitory effect may be explained on the basis of destabilization of the enzyme-NAD$^+$ complex (or enzyme-NAD$^+$-DNA complex).

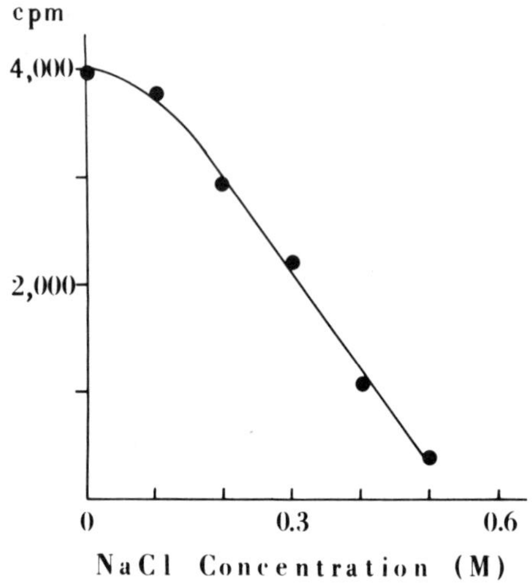

Fig. 8. Effect of NaCl concentration on the poly(ADPR)-polymerase assay. Same conditions as in Fig. 4, plus NaCl additions to give final concentrations in the assay mixtures as shown.

ADPR-Accepting Protein in the Bull Testis System

Incubation mixture of bull testis enzyme with ^{14}C-NAD$^+$ have been analyzed by polyacrylamide gel electrophoresis at pH 2.9 and 4.5 in the presence of 2.5 M urea. Radioactivity was located in correspondence with a band, which ran far ahead of the histone bands towards the cathode and which appears to correspond to the "testis specific protein" (TP) of Kistler et al. (47) as shown in Fig. 9. This new component of testis basic proteins is found in the testes of all mammals so far examined, while it is absent from sexually immature animals or when the tubular tissue has suffered damage (e.g. in experimental cryptorchidism). It is also absent from other organs like liver, spleen, etc. Recent research by Goldberg et al.(48) has assigned a more precise role to this protein, which appears to be specific to late spermatids formation. In mouse testis, these authors have observed the existence of three

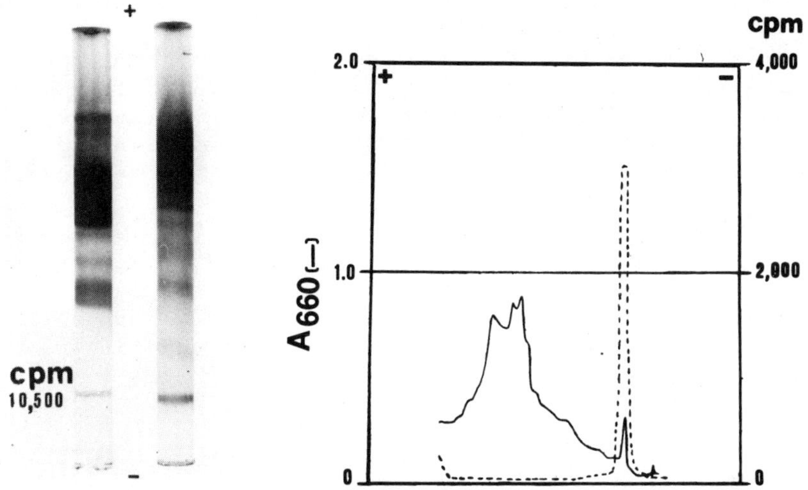

Fig. 9. ADP-ribosylation of basic protein from bull testis. From an incubation mixture of the same composition as described in Fig. 4 but on a larger scale (1 ml final volume) a 20% TCA fraction was prepared according to Kistler _et al_. (47) and analyzed by polyacrylamide (15% acrylamide) gel electrophoresis at pH 2.9 in the presence of 2.5 M urea and 0.9 M acetic acid, with a run for 180 min at 3 mA per tube. The left-hand gel shows, after Amido black staining, a band close to the cathode which corresponds to a high radioactivity level, which was measured in an identical unstained gel, cut into 1.5 mm thick sections, which were dissolved in 2% periodic acid prior to counting for radioactivity. The right-hand gel shows the result of a similar experiment, in which a 20% TCA fraction was analyzed, which was prepared from an enzyme preparation not incubated with ^{14}C-NAD^+. Alongside the gels is shown the densitometric profile corresponding to the left-hand gel, to which the radioactivity counts have been superimposed.

periods of histone synthesis during spermatogenesis, which is indicative of multiple roles for histones in differentiation; TP synthesis was found to be coincident with the 3rd wave of histone synthesis. While spermatid nuclei showed a relative reduction in histones H1 and H4 as compared to earlier spermatogenic events like primary spermatocytes, TP made its appearance with late spermatids; a mechanism therefore seems likely whereby TP replaces some components of histones H1 and H4 in spermiogenesis. Furthermore, although the initial condensation and shaping of the sperm head precedes TP synthesis, this protein may still have an important role in stabilizing the structure. For example, structural stabilization and induction of supercoiling in spermatids can be, at least partly, effects of histone replacement by TP.

Our finding of "acceptor" activity for ADPR residues in a basic protein fraction of testis closely resembling, and possibly identical with, the specific testis protein of Kistler *et al*. makes the poly(ADPR)polymerase system, with its high activity in testis tissue, a component which may play an important role in the complex process of spermiogenesis.

ACKNOWLEDGEMENT

This work has been done with the financial help of Consiglio Nazionale delle Ricerche, CNR, grant 77.00343.85 of Progetto Finalizzato della Biologia della Riproduzione.

REFERENCES

1. Chambon, P., Weill, J.D., Doly, J., Strosser, M.T., and Mandel, P., Biochem. Biophys. Res. Commun., 25, 638-643 (1966).
2. Nishizuka, Y., Ueda, K., Nakazawa, K., and Hayaishi, O., J. Biol. Chem., 242, 3164-3171 (1967).
3. Fujimura, S., Hasegawa, S., Shimizu, Y., and Sugimura, T., Biochim. Biophys. Acta, 145, 247-259 (1967).
4. Sugimura, T., Yoshimura, N., Miwa, M., Nagai, H., and Nagao, M., Arch. Biochem. Biophys., 147, 660-665 (1971).
5. Miwa, M., and Sugimura, T., J. Biol. Chem., 246, 6362-6364 (1971).
6. Matsubara, H., Hasegawa, S., Fujimura, S., Shima, T., Sugimura, T., and Futai, M., J. Biol. Chem., 245, 3606-3611 (1970).
7. Matsubara, H., Hasegawa, S., Fujimura, S. Shima, T., Sugimura, T., and Futai, M., J. Biol. Chem., 245, 4317-4320 (1970).
8. Yamada, M., Miwa, M., and Sugimura, T., Arch. Biochem. Biophys., 146, 579-586 (1971).
9. Yoshihara, K., Biochem. Biophys. Res. Commun., 47, 119-125 (1972).

10. Stone, P.R., and Shall, S., Eur.J. Biochem., 38, 146-152 (1973).
11. Mann, P.J.G., and Quastel, J.H., Biochem.J., 35, 502-517 (1941).
12. Smith, J.A., and Stocken, L.A., Biochem. J., 147, 523-529 (1975)
13. Ueda, K., Omachi, A., Kawaichi, M., and Hayaishi, O., Proc. Nat. Acad. Sci. USA, 72, 205-209 (1975)
14. Tanuma, S., Enomoto, T., and Yamada, M., Biochem. Biophys. Res. Commun., 74, 599-605 (1977)
15. Stone, P.R., Lorimer III, W.S., and Kidwell, W.R., Eur. J. Biochem., 81, 9-18 (1977).
16. Wong, N.C.W., Poirier, G.G., and Dixon, G.H., Eur. J. Biochem., 77, 11-21 (1977).
17. Wigle, D.T., and Dixon, G.H., J. Biol. Chem., 246, 5636-5644 (1971).
18. Gholson, R.K., Nature, 212, 933-935 (1966).
19. Rechsteiner, M., Hillyard, D., and Olivera, B.M., Nature, 259, 695-696 (1976).
20. Adamietz, P., Bredehorst, R., Oldekop, M., and Hilz, H., FEBS Lett., 43, 318-322 (1974)
21. Stone, P.R., Bredehorst, R., Kittler, M., Lengyel, H., and Milz, H., Hoppe-Seyler's Z. Physiol. Chem., 357, 51-56 (1976).
22. Collier, R.J., and Kandel, J., J. Biol. Chem., 246, 1496-1503 (1971).
23. Iglewski, B.H., and Kabat, D., Proc. Nat. Acad. Sci. USA, 72, 2284-2288 (1975).
24. Goff, C.G., J. Biol. Chem., 249, 6181-6190 (1974).
25. Caplan, A.I., and Rosenberg, M.J., Proc. Nat. Acad. Sci. USA, 72, 1852-1857 (1975)
26. Burzio, L., and Koide, S.S., Biochem. Biophys. Res. Commun., 40, 1013-1020 (1970).
27. Burzio, L., and Koide, S.S., Biochem. Biophys. Res. Commun., 53, 572-579 (1973).
28. Miller, E.G., Biochem. Biophys. Res. Commun., 66, 280-286 (1975).
29. Hilz, H., and Kittler, M., Hoppe-Seyler's Z. Physiol. Chem., 352, 1693-1704 (1971).
30. Kidwell, W.R., and Burdette, K.E., Biochem. Biophys. Res. Commun., 61, 766-773 (1974).
31. Yoshihara, K., Tanigawa, Y., Burzio, L., and Koide, S.S., Proc. Nat. Acad. Sci. USA, 72, 289-293 (1975).
32. Louie, A.J., Candido, E.P.M., and Dixon, G.H., Cold Spring Harb. Symp., 38, 803-819 (1974).
33. Louie, A.J., and Dixon, G.H., J. Biol. Chem., 247, 5498-5505 (1972)
34. Candido, E.P.M., and Dixon, G.H., J. Biol. Chem., 247, 5506-5510 (1972).

35. Leone, E., and Bonaduce, L., Biochim. Biophys. Acta, 31, 292-293 (1958).
36. Santoianni, P., and Leone, E., Bull. Soc. It. Biol. Sper., 34, 1943-1946 (1958).
37. Nakazawa, K., Ueda, K., Honjo, T., Yoshihara, K., Nishizuka, Y., and Hayaishi, O., Biochem. Biophys. Res. Commun., 32, 143-149 (1968).
38. Bistocchi, M., D'Alessio, G., and Leone, E., J. Reprod. Fert., 16, 223-231 (1968).
39. Farina, B., Faraone Mennella, M.R., and Leone, E., Rendic. Accad. Sci. Fis. Mat. Napoli, Sez. IV, 44, 257-263 (1977).
40. Ueda, K., Okayama, H., Fukushima, M., and Hayaishi, O., J. Biochem. (Tokyo), 77, 1p (1975).
41. Yoshihara, K., Hashida, T., Yoshihara, H., Tanaka, Y., and Ohgushi, H., Biochem. Biophys. Res. Commun., 78, 1281-1288 (1977).
42. Mandel, P., Okazaki, H., and Niedergang, C., FEBS Lett., 84, 331-336 (1977).
43. Clark, J.B., Ferris, G.M., and Pinder, S., Biochim. Biophys. Acta, 238, 82-85 (1971).
44. Ferro, A.M., and Kun, E., Biochem. Biophys. Res. Commun, 71, 150-154 (1976).
45. Romer, V., Lamprecht, J., Kittler, M., and Hilz, H., Hoppe-Seyler's Z. Physiol. Chem., 349, 109-115 (1968).
46. Brightwell, M.D., Leech, C.E., O'Farrel, M.K., Whish, W.J.D. and Shall, S., Biochem. J., 147, 119-129 (1975).
47. Kistler, W.S., Geroch, M.E., and Williams-Ashman, H.G., J. Biol. Chem., 248, 4532-4543 (1973).
48. Goldberg, R.B., Geremia, R., and Bruce, W.R., Differentiation, 7, 167-180 (1977).

ON THE PRINCIPLES OF NUCLEIC ACID - PROTEIN INTERACTION

W.A.Engelhardt

Institute of Molecular Biology, Academy of Sciences of the USSR
Moscow 117312, Vavilov str., 32, USSR

ABSTRACT

The difficulty of establishing recognition between partners with different molecular "languages", proteins and nucleic acids is by-passed by two mechanisms, in which the principle of coupling factors of opposite character, unity and diversity, integrated in one single molecular structure is operative. For free aminoacids it is the adaptor mechanism, the anticodon of tRNA playing a dual role. In the protein-DNA interaction hydrogen bonds are established with the components of the common, ever-repeating peptide bond. A kind of "recognition code" can be formulated.

The interaction between molecules of representatives of the two principal classes of biopolymers, the proteins and the nucleic acids, is one of the important problems of modern molecular biology. These interactions are characterized, as a rule, by their high degree of selectivity and specificity. This leads to the conclusion, that their directive force lies in phenomena of recognition, dependent on strictly fixed spatial localization and chemically determined properties which are responsible for the formation of the intermolecular forces producing close contact and bonds of various degree of stability between the interacting molecules.

The principle of complementarity, based on the concept of the Watson-Crick pairing, gives an adequate interpretation of the fundamental mechanism of interaction and recognition which finds place within the group of nucleotide substances on all levels, beginning with free, isolated nucleotides, proceeding to combinations between mononucleotides and macromolecular polymers, and ending with both partners of polymeric nature. In all these cases the recognition is based on the common molecular "language", the nucleotide language with its four-letter alphabet.

A new level of complexity appears when the recognition must occur between members of the two different classes of partners, on the one hand proteins or aminoacids, with their twenty-letter alphabet, and the four-letter nucleic language, on the other hand. This question will be discussed here, based on the results of research carried out in the Institute of Molecular Biology.

The central point of the process of biosynthesis of the protein molecule is the correct localization of each aminoacid on the mRNA template which carries the genetic information of their sequence and thus determines the primary structure of the protein to be formed. In this case the necessity of precise recognition between words of the aminoacid and the nucleotide languages appears in its sharpest form.

It can be said that Nature has surrendered to the difficulty of establishing a direct understanding between the two languages. But having given up the evidently unrealizable attempt to establish a direct recognition of the two incompatible languages Nature has found a highly ingenious way to solve this problem by an indirect mode of action, not to overcome the difficulty, but to circumvent, to py-pass it. This consists in the use of the well-known adaptor principle, based on the function of the transfer RNA. The structure of the tRNA's, as represented in its simpler form, by the clover-leaf pattern (Fig. 1) is characterized by the simultaneous presence of elements of opposite character - of unity and diversity, or singularity and multiplicity. This is achieved by the role of the two polar groups - the -C-C-A grouping of the acceptor end, and the anticodon on the other pole of the molecule.

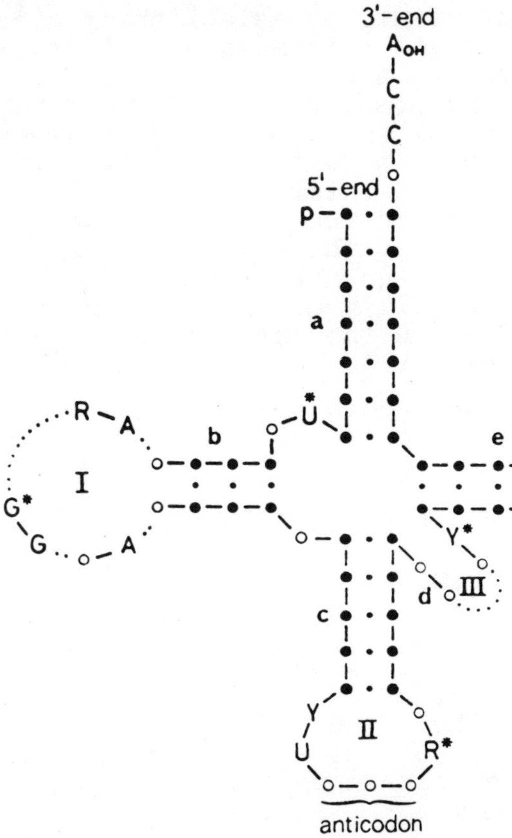

Fig.1. **Clover-leaf** pattern of tRNA

From the purely chemical point of view the acceptor end can react with any one of the twenty aminoacids, quite unspecifically, indiscriminately, in one and the same monotonous way. It represents the element of unity, similarity. In contrast, the anticodon contributes to the particular tRNA its unique character, its individuality among the sixty-odd members of the group. It strictly determines the recognition of the complementary codon of the mRNA, thus fixing the localization of the bound aminoacid in the correct point on the template.

It is the specific enzyme, the amino-acyl-tRNA-synthetase which transforms the indiscriminate, multivalent property of the acceptor grouping into a highly selective instrument, by establishing a connection between a definite aminoacid and its coding label, the anticodon. The shorter, more convenient name "codase" has been proposed for the enzyme, in accordance with its biological function and will be used (1) here. In the final count it is the codase that is responsible for producing the interrelation of the opposite properties inherent in the process of recognition. It is significant, that the philosophical statement, the unity of opposites, appears here as an illustration of its applicability as a leading principle in reasoning as well as in the interpretation of events of the material world.

The role of the anticodon of a tRNA molecule is not limited to its serving as coding label for recognizing the corresponding codon triplet on the mRNA. It appears that the anticodon plays a dual role, participating also in the recognition between the tRNA molecule and the corresponding codase. Codase being, as any enzyme, a protein, we have here to deal with a special case of nucleic acid-protein recognition with the advantage that at least in certain cases the full molecular structure, not only the primary, but the tri-dimensional as well, of one of the partners is known. Unfortunately the structure of the protein partner, the codase is not yet known in full detail, but nevertheless a certain amount of information is already available.

By special chemical means nucleotides can be selectively excised from the anticodon loop. The significant result was obtained (2): if the nucleotide was removed, which occupies the non-essential, so called "wobbling" position in the triplet, then the acceptor property was not lost. It disappeared if one of the essential nuclectides was removed. It needs hardly to be stressed, that in these experiments it was the principal function of the enzyme that was lost, the amino-acylation of the aminoacid and there are good reasons to regard this effect as the result of losing by the tRNA its property to recognize its protein partner due to changes of the anticodon, as the rest of the molecule remained unchanged.

These results lead to the conclusion that the anticodon of a tRNA molecule fulfills a dual role. First, it serves for the recognition of the necessary point on the mRNA, for the complementary binding to its respective codon. And on the other hand the anticodon plays a part in the effect of recognition between the nucleic acid, the tRNA, and the enzyme, codase. But it must be kept in mind, that the anticodon is not the only one point of interaction with the enzyme, even if perhaps the more important one. By systematic modification of several nucleotides, situated in different places of the clover leaf, other sites appear to be also involved in the interaction codase-tRNA (3). The results so far obtained are in excellent accord with conclusions reached by A.Rich in the course of his investigations on the tri-dimensional structure of tRNA. The view assigning a dual role to the anticodon grouping of the tRNA was at first met with considerable scepticism. But in the course of time this attitude has completely changed, and now this view has become almost generally accepted.

Greater, and quite new difficulties arize for studying the principles of recognition in the case when both partners are of the macromolecular kind. In a greater degree appears the duplicity of the factors involved in the recognition - the factors of chemistry and of geometry, or, in other words, of the composition, the chemical structure of the partners, on the one hand, and on the other - the conformation, the arrangement, localization in space of the different components between which the recognition takes place.

This problem has been attacked in the Institute of Molecular Biology by a group headed by B.Gottikh and consisting of chemists, experimental and theoretical physicists. The obvious condition for such a research is to possess a pair of interacting partners for which a complete knowledge of the structure of each of them would be available.

For this study was chosen a pair of substances which are known to possess a strong affinity to one another, thus being able to recognize each other. And for both these partners the complete primary structure is known, thus giving the possibility to draw conclusions concerning the principles, mechanism and conditions of the recognition.

One member of this pair is the stretch of DNA, which represents the so-called lac-operon, the part of the E.coli genome responsible for the operation of the gene or genes, involved in metabolism of lactose. The other partner is the specific protein, the lac-repressor, which has the property to recognize the DNA stretch, firmly bind to it, and inactivate it, thus to regulate its activity.

As would be expected, the relationships here are more complicated as compared with cases where we deal with substances of smaller molecular dimensions. In a greater degree appears the duplicity of the factors involved in the recognition - the factors of chemistry and of geometry, or, in other words, of the composition, the chemical structure of the partners, on the one hand, and on the other, the conformation, the arrangement, localization in space of the different components between which the recognition takes place.

The starting point was to use space-filling models, and find regions where the polynucleotide chain of the nucleic acid and the polypeptide chain of the protein could fit one another within the van-der-Waals distances. After time-consuming efforts this scrupulous task led to the desired results. Detailed mathematical analysis of all parameters helped by NMR studies, permitted to formulate the main principles which govern the recognition of the polymers, protein and nucleic acid.

Obviously, it is not possible to go here into any details of this study, and only the main conclusions that have been reached will be pointed out. Of great importance is the fact, that it appears that the specific side chains of the aminoacids, which contribute the individual character to each separate member of this numerous group do not directly participate in the act of recognition. Instead of this individual multiplicity the leading role belongs to the single structure, the peptide bond, which goes, as backbone, through the whole length of the protein molecule. The two components of the peptide bond, the C-O and NH groups serve to establish hydrogen bonds with definite members of the polynucleotide chain of the DNA. In short, the G-C pair reacts by forming a hydrogen bond with the carbonyl group and the A-T pair reacts by formation of a hydrogen bond with the amide group of the peptide bond (4). One might say that by thus transferring the role of recognition onto the peptide bond Nature has again by-passed the difficulty of using a language with a twenty-letter alphabet by replacing it with a two-letter language, the CO and NH letters. And here again we have the interplay of the two opposites, the multiplicity (side-chains) and singularity (peptide bond), integrated in one and the same molecular structure.

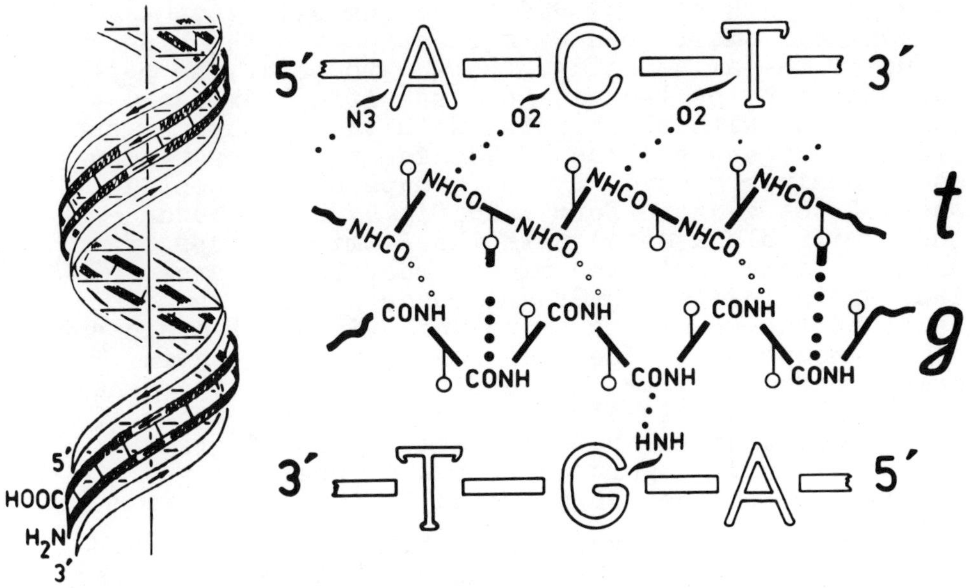

Fig.2. Scheme of interaction of protein and DNA

On Fig.2 are represented, in schematical form, the main features of the mechanism of recognition, as it now appears. In this scheme the DNA is in its normal two-stranded form. It is in its minor groove that the protein is situated. The polypeptide chain is not in its extended, alpha-helical form, as described by Linus Pauling, but in the pleated beta form, with two strands in antiparallel fashion. On the left is shown a hypothetical complex of DNA with the recognition site of a protein molecule localized in the minor groove in a form of the altered antiparallel β-structure.

On the right is demonstrated the system of hydrogen bonds. According to the model the stereospecific sites of the regulatory proteins form a helix isogeometric to the double helical DNA. Such a structure may occur as a consequence of disruptions of certain hydrogen bonds in the antiparallel β-structure. These perturbations give rise to two polypeptide chains having different spacial characteristics. They are marked with "t" and "g" (reacting with thymine or guanine). Small open circles stand for the side chains of the amino acids. The NH groups of the peptide bonds in the t-chain form specific hydrogen bonds (shown with black dots) with the O2 thymine and cytosine oxygens and the N3 adenine nitrogens facing into the minor groove. The CO groups of the peptide bonds of the g-chain form specific hydrogen bonds (shown with black dots) with the 2-amino groups of guanines facing into the minor groove. The large black spots indicate hydrogen bonds between the donor groups of the AT-coding amino acids in the t-chain and the CO-groups of the peptide bonds in the g-chain. The geometric factors, as they appear from Fig.1, are necessary for establishing the required close contact. It is clear that the role of geometry is of great, perhaps decisive importance for the possibility of establishing the set of hydrogen bonds.

But the role of geometry should not be over-emphasized. Obviously the chemical properties of the particular aminoacids forming a peptide bond, and depending on the character of their side-chains, exert a strong influence on the peptide bond. This influence is produced indirectly, determined by the geometric arrangement of the "atomic atmosphere" around the peptide bond, as well as by the arising chemical properties - the general reactivity, donor/acceptor properties, proton or electron displacements and so on. It is in this indirect way that the unspecific, uniform peptide bond acquires individual properties, which permit a selective recognition by the nucleotides of the DNA chain. Herewith the chemical factor finds its place in the mechanism of recognition alongside with the geometry.

Equally important is the fact shown in Table 1. The different aminoacids segregate in distinct groups in respect of their tendence to react with one or the other pair of nucleotides. At the proline residue, where a conformational kink occurs in the polypeptide chain, no interaction with the nucleotide chain is to be expected (5).

Table 1

THE CODE RULES

Base pair	Amino acid residues: Take part in the specific recognition	Take part in a recognition depending on a situation	Cannot be in the recognition site
A . . . T	SER, THR, ASN HIS, GLN, CYS	LYS, ARG	PRO
C . . . G	GLI, ALA, VAL, LEU, PHE, ILE MET, TYR, TRP	ASP, GLU	

The validity of the conclusions, obtained by the study of the particular pair of partners with fully known primary structure, the lac-operon and the lac-repressor, can be established only when more such pairs become known. But even now, when the structure of one pair of partners is known, certain predictions seem to be possible. So, for instance, for the pair lambda phage DNA, and the specific inhibitory protein, the structure of the nucleotide chain of the region of interaction is already known. The analysis of this structure permitted to Gottikh and his colleagues to make certain predictions concerning the probable structure of the corresponding length of the protein molecule (4,5).

Its actual primary structure has been studied by Ptashne and the recently published results are in very good accord with the predictions formulated by our workers (6). This brings strong support to the validity of the scheme, proposed for the regularities, which govern the interaction and recognition of proteins and nucleic acids. The data of Table 1 remind us, in a certain manner, of the classical genetic code. We must consider it as an "embrionic" prototype, the starting point of what could be regarded as a second kind of code. In addition to the genetic code we would have a "recognition code" or, more generally, another kind of "biological code".

REFERENCES

1. Kisselev, L.L. Priroda (Russ.) 24-33 (1974).
2. Chuguev, I.I., Axelrod, V.D., and Bayev, A.A. Biochem. Biophys. Res. Commun. 41, 108-114 (1970).
3. Kisselev, L.L., and Favorova, O.O. Adv. Enzymol. 40, 141-238 (1974).
4. Gursky, G.V., Tumanyan, V.G., Zasedatelev, A.S., Zhuze, A.L., Grokhovsky, S.L., and Gottikh, B.P. Molec. Biol. (Russ.) 9, 635-651 (1975).
5. Gursky, G.V., Tumanyan, V.G., Zasedatelev, A.S., Zhuze, A.L., Grokhovsky, S.L., and Gottikh, B.P. (1977) In "Nucleic Acid - Protein Recognition" (Vogel, H.J., Ed.), Acad. Press, N.Y., pp 189-217.
6. Sauer, R.T., and Anderegg, R. Biochemistry 17, 1092 (1978).

INTRODUCTORY TALK

Francesco Cedrangolo

I am glad, on behalf of the Italian Society of Biochemistry, to cordially welcome our Soviet friends who will participate in this Symposium. My personal greetings go first to Prof. Bayev, President of the I.U.B. A warm welcome to Prof. Engelhardt, whom I had the fortune to meet in Moscow in August, 1961, on the occasion of the 5th International Congress of Biochemistry. We owe him, among other findings, the discovery of the oxidative phosphorylation (see V.A. Engelhardt, Biochem. Zeit., 227, 116, 1930; ibidem, 251, 343, 1932) and of the identity of the myosin with the ATPase (see V.A. Engelhardt, Adenosyntriphosphatase Properties of Myosin, Adv. Enzymol., 6, 147, 1946).

Years of friendly relations and of common studies tie me to the Soviet biochemists. I remember, long ago, my first meeting in Naples with Prof. A.I. Oparin, member of the Academy of Sciences of the USSR, President from 1958 to 1961 of the International Union of Biochemistry. Oparin started the biochemical studies on the origin of life on earth, and his path was quickly followed by a great number of researchers. I also remember, and with pride, that I had the honour of being invited by Oparin himself to give a report at the First International Symposium on the origin of life, which was held in Moscow in the summer of 1957. The subject that I developed was the problem of the origin of proteins (F. Cedrangolo, "Reports at the International Symposium on the Origin of Life on the Earth", Moscow, 1957, Pergamon Press, London and New York, 1959).

Since 1949 I have had a long, lasting friendship with Prof. Arthur E. Braunstein, the discoverer of the transamination reactions. Actually, we have engaged in some verbal and written polemics, sometimes a bit lively, always friendly. The end of the story was at the 4th International Congress of Biochemistry, held in September, 1958, in Vienna, where Braunstein in the closing session of the Congress, agreed that we were both right! In fact, in the living cell, both direct deamination, as I have suggested since 1940 (F. Cedrangolo and G. Carandante, Arch. Sci. Biol., 28, 1, 1942), and indirect deamination, namely transdeamination, which Braunstein proposed on the basis of the model put forward in 1939 (see A.E. Braunstein, Adv. Enzymol., 19, 335, 1957), can take place at the same time.

And lastly let me mention the late Prof. Sergei Mardashev, whom I had the pleasure to meet in December, 1960, in Naples, and then in Moscow in August, 1961.

The choice of Villa Malaparte and Cerio Foundation as the location for this Symposium has been a brilliant idea. Capri has a long Soviet tradition. You can still feel alive in the air the memory of such great Soviet people as Lenin, Gorki, Lunaciarski, who, as Curzio Malaparte remembers in his posthumous book "Io in Russia e in Cina" (Ed. Vallecchi, Firenze), often met together in the "Pensione Weber" at the Marina Piccola. The Soviet tradition is so deeply felt in Capri that the presence of the great Soviets has been synthesized in a stele, with the image of Lenin, by a great Italian sculptor, Giacomo Manzù, located in the beautiful Augusto's Gardens.

Coming now to the object of this Symposium, I wish to recollect the ideas about molecular biology. In the report by H.A. Krebs and G.S.D. Haslewood, presented to the "Biochemical Society," July, 1969, the expression molecular biology is said to be an unfortunate one, since molecular biology, from the authors' point of view, cannot be separated from biochemistry. In other words biochemistry and molecular biology are the same thing, or better, molecular biology is a wider term, which includes biochemistry. Furthermore, as they stated, molecular biology is comprehensive of molecular genetics and cellular biology too: "We would go further and claim that all areas of biology will before long be approachable at the molecular level".

These considerations seem correct to me and I would like to add a few examples that might be appropriate. If one injects x mg of adrenalin into a dog and subsequently finds in its urine an increase of glucose, it is evident that one can look at this problem in molecular terms. This is also true when one shakes a certain weight of liver with x millimoles of L-alanine, and then finds that y millimoles of O_2 have been consumed and z millimoles of NH_3 have been produced. It is also evident that all research work on structure, synthesis, degradation, and biological function of macromolecules, should they be nucleic acids or not, have to be considered within the field of molecular biology.

If we look at other branches that aren't strictly biochemical, we will find the same thing.

At this state of the art an amusing phrase by Francis Crick (from G.S. Stent, Science, 160, 390, 1968) is symbolic: "I had to present myself as a molecular biologist because I was tired of explaining that I was at the same time a crystallographer, a biophysicist, a biochemist and a geneticist."

The expression molecular biology was coined by Astbury, who defined himself as a molecular biologist in 1939 (see Ann. Rev. Biochem., 8, 125, 1939). Therefore molecular biology starts, at least according to its paternity, with a physicist, who at the time was studying the shape of fibrous proteins with X-rays. From this prestigious name, other famous ones follow: Bernal, Perutz, Kendrew, Pauling, etc. (see P. Thuillier, La Recherche en Biologie Moleculaire, Societé d'Editions Scientifique, Paris).

According to other authors (see Mullins, Minerva, 10, 1, 1972), molecular biology originated with Delbrück around 1935 and developed mainly through the work of the so called "phage group". In this group we find the names of scientists from different countries and working in different laboratories: Luria, Anderson, Hershey, Lwoff, Monod, etc. Among the important results obtained by this group after 1954 is the central dogma of molecular biology. Therefore, we have the birth of molecular biology from genetics, and its development and strengthening through the work of geneticists, as well as biochemists.

Finally, according to H.V. Wyatt (Nature, 86, 235, 1972), the paper by Avery, Macleod and MacCarty (J. Exp. Med., 79, 137, 1944) was instrumental in the development of molecular biology. In fact, for the first time it was demonstrated that DNA holds the genetic patrimony of the cell, being transmissible from one cell to another. I may therefore point to bacteriology as yet another origin of molecular biology.

In conclusion, without going further, it seems to me that the alternatives might be the following. First, we may keep the literal meaning of molecular biology, so that everything is molecular biology as it is stated in the above-mentioned report to the Biochemical Society; in this case molecular biology cannot be considered as a science in itself, because it is split into biochemistry, molecular genetics, cellular biology, etc. Second, to molecular biology must be assigned tighter borders; the object of molecular biology should be limited to those studies concerning genetics at the molecular level and biochemical studies of DNA and RNA. Therefore, all the studies concerning structure and function of small molecules, as well as of macromolecules different from DNA and RNA, would be outside the interest of molecular biology.

On the other hand, an exception should be made for proteins, keeping in mind that it was during the study of their structure that the term molecular biology was coined.

It seems to me that these confusions have also been experienced by the organizers of this symposium. In fact, while silently admitting that this is a molecular biology symposium, they did not specify it, but defined the meeting as a symposium on "Macromolecules in the Functioning Cell".

Finally, I hope that a solid friendship among Soviet and Italian colleagues will arise, since, as I have already said on other occasions, even for the scientist the heart may often be more valuable than the brain!

CLOSING TALK

Antonio Cajano

It is a pleasure for me, as an oncologist devoted to the care of patients, to be a moderator among so many scientists of basic sciences.

I would like to express my appreciation to the "Consiglio Nazionale delle Ricerche," to the Italian Society of Biochemistry, and to Prof. Pietro Volpe, for making possible the realization of this Symposium, which originated in the small interdisciplinary nucleus of the "Capri Center for Cell Biology and Natural Sciences."

I believe it is necessary to maintain an open dialogue between basic and applied research: physicians should be more "biologists" than they are, and biologists should aim a little more to the problems of medicine. As a matter of fact, medical care of patients is chiefly applied research, whose general principles derive from basic research. This difference, obviously, is merely formal and theoretical, on the basis of operational and teaching grounds, and I would like to quote here the words of Emil Freireich, now in Houston, a dear friend with whom I worked in Boston. While commemorating Dr. Karnofoki, the well-known chemotherapist, Freireich said: "Who took the word clinical out of the phrase clinical research?" His intent was to emphasize research in the field of diseases, which are tied to the changes and evolution of man and environment.

In this context I cannot help considering that the papers presented at this symposium are all connected with problems of human diseases.

Prof. Bayev and his coworkers open the way to the study of DNA sequences in human normal and cancer cells. Dr. Cecilia Saccone and colleagues remind me, in their paper, of some of my old experiments with rabbit antisera to rat liver mitochondria (Bollett. Fondaz. Pascale, 29: 13, 1955 and 32: 294, 1958; Riv. Emoter. Immunoematol. 6: 215, 1959; 7th International Cancer Congress, London, 1958), in which selective disfunction of mitochondrial system was investigated in relation to its influence on cell morphology. Prof. Zbarsky and coworkers clarify some of the mechanisms which could gradually lead to cancerization of human cells. Sottocasa's paper is concerned with calcium transport, which could be critical in antibody formation, and everyone knows the relationship between immunology and cancer. The problem of cell response to many inducers, discussed by Salganik, is similar to what immunologists discuss about "one cell - one or more antibodies?" As far as the "differential gene expression during the cell life cycle" is concerned, in the paper by Eremenko and colleagues the concept that doubling the time of cancer cells is increased puts forth the question that the aging of cancer cells is dependent also on the impairment of macrophage functions. Finally, Farina and colleagues' paper emphasizes the role of some peculiar enzyme system in spermiogenesis. This topic is concerned with the relationship between mesenchymal and parenchymal cells. Some studies of mine with Scyllium canicula at the "Stazione Zoologica" in Naples (Atti Soc. Peloritana Scienze fis. matem. e natur. 19: fasc. I, II, 1972-73), show that each Sertoli cell takes care of a group of spermatozoa by enveloping them with its mantle of hyaline and dense cytoplasm. Gradually, the morphology of the phenomenon goes through nine divisions both of Sertoli cells and spermatogonia. Whereas Sertoli cells divide no more, spermatogonia divide four times more and then start their maturation to spermatocytes, spermatids, and spermatozoa (Stanley, M.P., Z. Zell, 75: 453, 1966). Similar investigations could be undertaken in humans in order to ascertain whether, in cancer, mesenchymal cells are initially involved.

I believe it is worthwhile to quote these contributions, in order to show the need for more intimate relationships between biologists and physicians. Moreover, I firmly believe that in all the institutions and hospitals in which patients are taken in care, the medical staff should be aided by responsive and responsible biologists: this collaboration would allow a better understanding of the experiments mother nature makes on humans. Medicine is not the more or less reliable application of laboratory tests, but rather is chiefly a continuing biological research work. Progress in medical science and in medical art is realized through proper routine work coupled with investigative work which represents the routine of tomorrow.

CLOSING TALK

Ernesto Quagliariello

At the end of this Symposium I wish to try to summarize some of the highlights of the lectures and the discussions held at the meeting. First of all, however, I would like to express my appreciation for the opportunity to attend such an outstanding symposium, which brought together leading authorities in the field of biochemistry and molecular biology from both countries, the USSR and Italy. I wish also to congratulate both speakers and attendants, who made possible such a lively interaction of scientific thoughts which I am sure will produce active catalysis of ideas and experimentation in the areas which have been discussed at the meeting.

I think the basic idea on the minds of the organizers of the symposium was a valid one, since the science of molecular biology in the two countries has reached a high standard level due to the activity of some outstanding research groups, which have actively contributed to the development of several aspects of this science in the recent years.

Because of the structure of the symposium a rather wide range of topics has been covered: this may at first sight appear to be a pitfall, since such different subjects had to be treated. However, the aim of the symposium was to give a restricted audience the chance to obtain knowledge about the main lines of research being carried out in Italy and USSR: this aim has been successful and the book containing the proceedings of the symposium, which will be published by Plenum Publishing Co. in a few months will make this knowledge available to a world-wide audience of scientists.

Now, switching to the content of the various contributions, I wish to make a few observations on some of the concepts which I consider to be among the most relevant ones emerging from the essays. As a matter of fact, each contribution has been presented in the form of a short review with some emphasis given to previous work in the field and it appears most likely it will increase the number of the symposium-book readers.

The structure of yeast ribosomal genes has been described clearly by Bayev, and some hints as to the possible processing mechanism of 35 S rRNA precursor have been discussed. Timofeeva, from the same group as Professor Bayev, has also given a thorough, well presented paper on the organization of ribosomal gene cluster in the loach, showing, among other facts, that amplified rDNA lacks the heterogeneity of non-transcribed spacer. Zbarsky and his group from the Institute of Developmental Biology have characterized nuclear matrix proteins in normal rat liver and in hepatoma tissue, showing some typical differences which do not appear in regenerating rat liver. Mirzabekov has discussed nucleosome structure by presenting a symmetrical model deduced from the studies on the histone components situated in the central part of nucleosomal DNA. He has also presented Professor Georgiev's contribution which was concerned with multiple dispersed genes in _Drosophila_ and their expression. An accurate review of mitochondrial organization of the genome in rat liver has been presented by Saccone. Tocchini-Valentini discussed the formation of complex DNA in cell-free systems from _Xenopus_ oocytes, also by describing the insertion of superhelical turns into relaxed DNA. The dynamic relationships in macromolecular synthesis during the life cycle of a eukariote cell have been discussed extensively by Volpe, who has presented a generalized model for the metabolism of DNA, RNA and protein macromolecules during the cellular cycle.

Another set of essays deals mainly with macromolecule structure and function. Ovchinnikov has described the several proteins functioning as translation factors in eukariotes, which he proposes are RNA-binding proteins. A clear comprehensive review on tRNA maturation with special emphasis on the process of post-transcriptional methylation has been presented by Cimino, who has discussed the purification and the properties of several tRNA methyltransferases prepared from a bacterial source. Sverdlov, of the Bioorganic Chemistry Institute of the USSR Academy of Sciences, has stressed the roles that different subunits of RNA-polymerase display in the initiation of RNA synthesis: σ subunits seem to be directly involved at the start of this process. Whitehead has reviewed extensively the mechanism and the role of ATP-dependent DNAses and DNA-dependent ATPases: their functions now appear to be relevant to several molecular genetics processes. Cerletti has presented studies on the biosynthesis of iron-sulfur structures in proteins: a possible enzymic mechanism for posttranscriptional insertion of iron-sulfur structures has been discussed. Panfili, from Sottocasa's group, has illustrated the properties of some proteins involved in active transport: particularly, a glycoprotein involved in active transport of calcium across the mitochondrial membrane and a protein involved in bilirubin transport across the plasma membrane have been described and their role and properties discussed. Salganik reviewed the role of enzyme induction and deinduction by several effectors, and the role that this mechanism exerts to provoke changes in enzymatic programs which take place in cell associations.

Molecular embryology is a rapidly expanding field; however, much has yet to be discovered. It is certainly an area where the scientific community expects a burst of new developments in the next decade. Some important contributions have been presented in this field at this symposium. Scarano discussed his hypothesis of DNA methylation as a critical process in cell determination in higher metazoa on the basis of experiments performed in early embryonic development of the sea urchin. In the same organism the regulation of macromolecular synthesis, particularly the uncoupling of nuclear and mitochondrial DNA synthesis, has been discussed by Giudice. Kafiani has presented results on the nucleus-associated polyribosomes, which seem to be engaged in histone synthesis and are different from other polyribosomes present in the cytoplasm of the loach embryo cells and also involved in histone biosynthesis. Leone discussed the role of the poly-(adenosine diphosphate ribose) polymerase system in spermiogenesis, aiming to identify the target protein for the action of this enzyme system.

Finally, there is no sufficient word to express admiration for the elegant and stimulating conclusive presentation by Professor Engelhardt on nucleic acid-protein interactions: the flowing time seemed to the whole audience only to add, if possible, brilliancy and depth to his scientific thoughts and activity.

On behalf of the Italian Society of Biochemistry and of the National Research Council of Italy, I wish to thank deeply the organizers and the Scientific Committee for having brought here, to this magnificent "magic-blue" spot of Capri, such an outstanding group of scientists, both as speakers and as discussants. I am sure this meeting will result in a catalytic process for the development of new achievements in molecular biology.

CONTRIBUTORS

ANANIEV, E.V.	Institute of Molecular Genetics USSR Academy of Sciences, Moscow
ARZONE, A.	Institute of Comparative Anatomy, University of Palermo
BAKKER, H.	Laboratory of Physiological Chemistry, University of Groningen
BAYEV, A.A.	Institute of Molecular Biology USSR Academy of Sciences, Moscow
BEABEALASHVILLI, R.	Institute of Molecular Biology USSR Academy of Sciences, Moscow
BELYAVSKY, A.	Institute of Molecular Biology USSR Academy of Sciences, Moscow
BONOMI, F.	Department of General Biochemistry, University of Milan
BULDYAEVA, T.V.	N.K. Koltzov Institute of Developmental Biology, USSR Academy of Sciences, Moscow
CERLETTI, P.	Department of General Biochemistry, University of Milan
CERIO-VENTURA, G.	Institute of Biological Chemistry, University of Rome

CHIAURELI, N.B.	Institute of Molecular Biology, USSR Academy of Sciences, Moscow
CIMINO, F.	Institute of Biological Chemistry, 2nd Medical School, University of Naples
DAVITASHVILI, A.N.	Institute of Molecular Biology USSR Academy of Sciences, Moscow
DE GIORGI, C.	Institute of Biological Chemistry, University of Bari
DOMOGATSKY, S.P.	Institute of Protein Research USSR Academy of Sciences, Poustchino
EISNER, G.I.	Institute of Molecular Biology USSR Academy of Sciences, Moscow
ENGELHARDT, W.A.	Institute of Molecular Biology USSR Academy of Sciences Moscow
EREMENKO, T.	International Institute of Genetics and Biophysics, CNR, Naples
FARAONE-MENNELLA, M.R.	Institute of Organic and Biological Chemistry, University of Naples
FARINA, B.	Institute of Organic and Biological Chemistry, University of Naples
FASELLA, P.M.	Institute of Biological Chemistry, University of Rome
GEORGIEV, G.P.	Institute of Molecular Biology USSR Acacemy of Sciences, Moscow
GIUDICE, G.	Institute of Comparative Anatomy, University of Palermo

GRACHEV, M.A.	Institute of Organic Chemistry Siberian Division of the Academy of Sciences, Novosibirsk
GRECO, M.	Institute of Biological Chemistry, University of Bari
GVODZEV, V.A.	Institute of Molecular Genetics USSR Academy of Sciences, Moscow
ILYIN, Y.V.	Institute of Molecular Biology USSR Academy of Sciences, Moscow
IZZO, P.	Institute of Biological Chemistry, 2nd Medical School, University of Naples
KAFIANI, C.A.	Institute of Molecular Biology USSR Academy of Sciences, Moscow
KOLCHINSKY, A.	Institute of Molecular Biology USSR Academy of Sciences, Moscow
KRAYEV, A.S.	Institute of Molecular Biology USSR Academy of Sciences, Moscow
KROON, A.M.	Laboratory of Physiological Chemistry, University of Groningen
KUPRIYANOVA, N.S.	Institute of Molecular Biology USSR Academy of Sciences, Moscow
KUZMINA, S.N.	N.K. Koltzov Institute of Developmental Biology, USSR Academy of Sciences, Moscow
LEONE, E.	Institute of Organic and Biological Chemistry, University of Naples

LEVITAN, T.L.	Shemyakin Institute of Bioorganic Chemistry, USSR Academy of Sciences, Moscow
LIPKIN, V.M.	Shemyakin Institute of Bioorganic Chemistry, USSR Academy of Sciences, Moscow
LIUT, G.F.	Institute of Biological Chemistry, University of Trieste
LUCIANI, M.	Institute of Biological Chemistry, University of Trieste
LUNAZZI, G.C.	Institute of Biological Chemistry, University of Trieste
MATRANGA, V.	Institute of Comparative Anatomy, University of Palermo
MELNIKOVA, A.	Institute of Molecular Biology USSR Academy of Sciences, Moscow
MENNA, T.	International Institute of Genetics and Biophysics, CNR, Naples
MIRZABEKOV, A.	Institute of Molecular Biology USSR Academy of Sciences, Moscow
MODYANOV, N.N.	Shemyakin Institute of Bioorganic Chemistry, USSR Academy of Sciences, Moscow
MUTOLO, V.	Institute of Comparative Anatomy, University of Palermo
NOLL, H.	Institute of Comparative Anatomy, University of Palermo
OVCHINNIKOV, L.P.	Institute of Protein Research USSR Academy of Sciences, Poustchino
OVCHINNIKOV, Yu. A.	Shemyakin Institute of Bioorganic Chemistry, USSR Academy of Sciences, Moscow

PAGANI, S.	Department of General Biochemistry, University of Milan
PALITTI, F.	Institute of Biological Chemistry, University of Rome
PANFILI, E.	Institute of Biological Chemistry, University of Trieste
PEPE, G.	Institute of Biological Chemistry, University of Bari
PLETNEV, A.G.	Institute of Organic Chemistry Siberian Division of Academy of Sciences, Novosibirsk
RINALDI, A.M.	Institute of Comparative Anatomy, University of Palermo
RUBTSOV, P.M.	Institute of Molecular Biology USSR Academy of Sciences, Moscow
SACCONE, C.	Institute of Biological Chemistry, University of Bari
SALCHER, I.	Institute of Comparative Anatomy, University of Palermo
SALGANIK, R.I.	Institute of Cytology and Genetics, Siberian Branch of the USSR Academy of Sciences, Novosibirsk
SALVATORE, F.	Institute of Biological Chemistry, 2nd Medical School, University of Naples
SANDRI, G.	Institute of Biological Chemistry, University of Trieste
SAYCHIKOV, E.F.	Institute of Organic Chemistry Siberian Division of the Academy of Sciences, Novosibirsk
SCARANO, E.	Laboratory of Molecular Embryology, CNR, Naples

SERYAKOVA, T.A.	Institute of Protein Research USSR Academy of Sciences, Poustchino
SHICK, V.	Institute of Molecular Biology USSR Academy of Sciences, Moscow
SKRYABIN, K.G.	Institute of Molecular Biology USSR Academy of Sciences, Moscow
SOTTOCASA, G.L.	Institute of Biological Chemistry, University of Trieste
SPIRIN, A.S.	Institute of Protein Research USSR Academy of Sciences, Poustchino
STRELKOV, L.A.	Institute of Molecular Biology USSR Academy of Sciences, Moscow
SVERDLOV, E.D.	Shemyakin Institute of Bioorganic Chemistry, USSR Academy of Sciences, Moscow
TCHURIKOV, N.A.	Institute of Molecular Biology USSR Academy of Sciences, Moscow
TIMOFEEVA, M. Ya.	Institute of Molecular Biology USSR Academy of Sciences, Moscow
TIRIBELLI, C.	Institute of Medical Pathology University of Trieste
TOSI, L.	"Stazione Zoologica," Naples
TRABONI, C.	Institute of Biological Chemistry, 2nd Medical School, University of Naples
TSAREV, S.A.	Shemyakin Institute of Bioorganic Chemistry, USSR Academy of Sciences, Moscow

VELLANTE, A.	Institute of Biological Chemistry, University of Rome
VITTORELLI, M.L.	Institute of Comparative Anatomy, University of Palermo
VLASIK, T.N.	Institute of Protein Research USSR Academy of Sciences, Poustchino
VOLPE, P.	International Institute of Genetics and Biophysics, CNR, Naples
WHITEHEAD, E.P.	Biology Department Commission of the European Communities
ZAHARYEV, V.M.	Institute of Molecular Biology USSR Academy of Sciences, Moscow
ZBARSKY, I.B.	N.K. Koltzov Institute of Developmental Biology, USSR Academy of Sciences, Moscow

PARTICIPANTS

BANERJI, Shobhona	Tata Institute of Fundamental Research, Molecular Biology Unit, Bombay, India
BAYEV, A.A.	Institute of Molecular Biology, USSR Academy od Sciences, Moscow, USSR
BONOMI, F.	Istituto di Biochimica Generale, Università di Milano, Milano, Italy
BUONO, C.	Istituto Internazionale di Genetica e Biofisica del CNR, Napoli, Italy
BUONOCORE, V.	Istituto di Chimica Organica e Biologica, Università di Napoli, Napoli, Italy
CACACE, M.G.	Laboratorio di Embriologia Molecolare del CNR, Arco Felice - Napoli, Italy
CAJANO, A.	Divisione di Medicina, Ematologia Oncologica e Immunologia Applicata, Istituto per la Cura dei Tumori, Fondazione Senatore Pascale, Napoli, Italy
CARSANA, Antonella	Istituto di Chimica Organica e Biologica, Università di Napoli, Napoli, Italy

CEDRANGOLO, F.	Istituto di Chimica Biologica, I Facoltà di Medicina e Chirurgia, Università di Napoli, Napoli, Italy
CERLETTI, P.	Istituto di Biochimica Generale, Università di Milano, Milano, Italy
CIMINO, F.	Istituto di Chimica Biologica, II Facoltà di Medicina e Chirurgia, Università di Napoli, Napoli, Italy
D'ALESSIO G.	Istituto di Chimica Organica e Biologica, Università di Napoli, Napoli, Italy
DE GIORGI, Carla	Centro di Studio sui Mitocondri e Metabolismo Energetico, Istituto di Chimica Biologica, Università di Bari, Bari, Italy
DELFINI, C.	Istituto Superiore di Sanità, Roma, Italy
DE PETROCELLIS, Benita	Laboratorio di Embriologia Molecolare del CNR, Arco Felice - Napoli, Italy
DE ROSA, M.	Laboratorio di Chimica delle Molecole di Interesse Biologico del CNR, Arco Felice - Napoli, Italy
DE VINCENTIIS, M.	Istituto di Biologia Generale e Genetica, Università di Napoli, Napoli, Italy
DI DONATO, A.	Istituto di Chimica Organica e Biologica, Università di Napoli, Napoli, Italy
EREMENKO, Tamilla	Istituto Internazionale di Genetica e Biofisica del CNR, Napoli, Italy

ENGELHARDT, V.A.	Institute of Molecular Biology, USSR Academy of Sciences, Moscow, USSR
FARAONE MENNELLA, M.Rosaria	Istituto di Chimica Organica e Biologica, Università di Napoli, Napoli, Italy
FARINA, Benedetta	Istituto di Chimica Organica e Biologica, Università di Napoli, Napoli, Italy
FONTANA, A.	Istituto di Chimica Organica, Università di Padova, Padova, Italy
FURIA, Adriana	Istituto di Chimica Organica e Biologica, Università di Napoli, Napoli, Italy
GAMBACORTA, Agata	Laboratorio per la Chimica di Molecole di Interesse Biologico del CNR, Arco Felice - Napoli, Italy
GERACI, G.	Laboratorio di Embriologia Molecolare del CNR, Arco Felice - Napoli, Italy
GIUDICE, G.	Istituto di Anatomia Comparata, Università di Palermo, Palermo, Italy
GLISIN, V.	Center for Multidisciplinary Studies, University of Belgrade, Belgrade, Yugoslavia
GRANIERI, A.	Istituto Internazionale di Genetica e Biofisica del CNR, Napoli, Italy
GRECO, Margherita	Centro di Studi sui Mitocondri e Metabolismo Energetico, Istituto di Chimica Biologica, Università di Bari, Bari, Italy

IZZO, Paola	Istituto di Chimica Biologica, II Facoltà di Medicina e Chirurgia, Università di Napoli, Napoli, Italy
KAFIANI, K.A.	Institute of Molecular Biology, USSR Academy of Sciences, Moscow, USSR
LEONE, E.	Istituto di Chimica Organica e Biologica, Università di Napoli, Napoli, Italy
MARINO, G.	Istituto di Chimica Organica e Biologica, Università di Napoli, Napoli, Italy
MENNA, T.	Istituto Internazionale di Genetica e Biofisica del CNR, Napoli, Italy
MIRZABEKOV, A.D.	Institute of Molecular Biology, USSR Academy of Sciences, Moscow, USSR
MONROY, A.	Stazione Zoologica, Napoli, Italy
NUCCI, R.	Istituto Internazionale di Genetica e Biofisica del CNR, Napoli, Italy
ORUNESU, M.	Istituto di Fisiologia Generale, Università di Genova, Genova, Italy
OVCHINNIKOV, L.	Institute of Protein Research, USSR Academy of Sciences, Poustchino-Moscow, USSR
PAGANI, Silvia	Istituto di Biochimica Generale, Università di Milano, Milano, Italy
PANFILI, E.	Istituto di Chimica Biologica, Università di Trieste, Trieste, Italy

PICCOLI, Renata	Istituto di Chimica Organica e Biologica, Università di Napoli, Napoli, Italy
QUAGLIARIELLO, E.	Istituto di Chimica Biologica Università di Bari, Bari, Italy
ROMANO, Marta	Istituto per la Cura dei Tumori, Fondazione Senatore Pascale, Napoli, Italy
ROSSI, M.	Istituto Internazionale di Genetica e Biofisica del CNR, Napoli, Italy
RUFFO, A.	Istituto di Bioorganica, Facoltà di Farmacia, Università di Napoli, Napoli, Italy
SACCONE, Cecilia	Istituto di Chimica Biologica, Università di Bari, Bari, Italy
SALGANIK, R.I.	Institute of Cytology and Genetics, Siberian Branch of the USSR Academy of Sciences, Novosibirsk, USSR
SALVATORE, F.	Istituto di Chimica Biologica II Facoltà di Medicina e Chirurgia, Università di Napoli, Napoli, Italy
SCANDURRA, R.	Istituto di Chimica Biologica, Facoltà di Medicina e Chirurgia, Università di Roma, Roma, Italy
SCARANO, E.	Laboratorio di Embriologia Molecolare del CNR, Arco Felice - Napoli, Italy
SEPE, S.	Istituto Internazionale di Genetica e Biofisica del CNR, Napoli, Italy

SERRA, Virginia	Dipartimento di Biologia Cellulare, Università della Calabria, Arcavacata di Rende, Cosenza, Italy
SUZUKI, H.	Istituto di Chimica Organica e Biologica, Università di Napoli, Napoli, Italy
SVERDLOV, E.D.	M.M. Shemyakin Institute of Bioorganic Chemistry, USSR Academy of Sciences, Moscow, USSR
TIMOFEEVA, Marguerita Ya.	Institute of Molecular Biology USSR Academy of Sciences, Moscow, USSR
TOCCHINI-VALENTINI, G.	Laboratorio di Biologia Cellulare del CNR, Roma, Italy
TRABONI, Cinzia	Istituto di Chimica Biologica II Facoltà di Medicina e Chirurgia, Università di Napoli, Napoli, Italy
VOLPE, P.	Istituto Internazionale di Genetica e Biofisica del CNR, Napoli, Italy
WHITEHEAD, E.	Biology Department Commission of the European Communities, c/o Istituto di Chimica Biologica, Università di Roma, Roma, Italy
ZIKOV, I.N.	USSR Academy of Sciences, Moscow, Italy
ZBARSKY, I.B.	N.K. Koltzov Institute of Developmental Biology, USSR Academy of Sciences, Moscow, USSR

INDEX

	DATE DUE		